AF388974

# Plasma for Energy and Catalytic Nanomaterials

# Plasma for Energy and Catalytic Nanomaterials

Special Issue Editors

**Feng Yu**
**Lanbo Di**

MDPI • Basel • Beijing • Wuhan • Barcelona • Belgrade • Manchester • Tokyo • Cluj • Tianjin

*Special Issue Editors*

Feng Yu
Shihezi University
China

Lanbo Di
Dalian University
China

*Editorial Office*
MDPI
St. Alban-Anlage 66
4052 Basel, Switzerland

This is a reprint of articles from the Special Issue published online in the open access journal *Nanomaterials* (ISSN 2079-4991) (available at: https://www.mdpi.com/journal/nanomaterials/special_issues/plasma_energy_nano).

For citation purposes, cite each article independently as indicated on the article page online and as indicated below:

LastName, A.A.; LastName, B.B.; LastName, C.C. Article Title. *Journal Name* **Year**, *Article Number*, Page Range.

ISBN 978-3-03928-654-6 (Hbk)
ISBN 978-3-03928-655-3 (PDF)

# Contents

# About the Special Issue Editors

**Feng Yu** received his B.S. degree in Applied Chemistry from the University of Jinan in 2003, and obtained his Ph.D. in physical chemistry from Technical Institute of Physics and Chemistry, Chinese Academy of Sciences (TIPC, CAS) in 2010. He then joined Nanyang Technological University (NTU) as a research fellow and Institute of Chemical and Engineering Sciences, Agency for Science, Technology and Research (ICES, A*STAR) as Scientist I. Then, he joined Shihezi University and worked as one scientist of the Recruitment Program of Global Experts (1000 Talent Plan). Now, Prof. Yu continues to undertake research in advanced functional materials for electrochemical catalysis and heterogeneous catalysis.

**Lanbo Di** received his Ph.D. in Plasma Physics from Dalian University of Technology in 2012. Since then, he has worked at the College of Physical Science and Technology at Dalian University. He has been a professor at Dalian University since 2019. He is a research professor in the Department of Chemistry and Chemical Engineering in Inha University, co-working with Professor Dong-Wha Park during 2016 to 2017, and a doctoral supervisor in Inha University since 2017. He won the first Dalian Youth Science and Technology award in 2014. He has been selected as being in the 1000-level of Liaoning BaiQianWan Talents and the candidate of the Eni Award in 2018. His research interests are focused on AP cold plasma for synthesizing supported metal catalysts and their energy and environmental applications, gas-liquid discharge, as well as plasma enhanced chemical vapor deposition (PECVD).

# Preface to "Plasma for Energy and Catalytic Nanomaterials"

Plasma for energy and catalytic nanomaterials is a hotspot in the interdisciplinary research between chemistry, materials, engineering, environment, mathematics, and physics. Compared with conventional preparation methods, the plasma method has been proven to be a fast, facile, and environmentally friendly method for synthesizing highly efficient nanomaterials. Plasma-synthesized nanomaterials generally show enhanced metal—support interactions, small-sized metal nanoparticles, specific metal structures, and abundant oxygen vacancies. Therefore, they exhibit high catalytic activity and stability in energy and catalytic applications. Now, the plasma-assisted preparation of nanomaterials is receiving increasing interest for energy and catalytic applications, such as methane reforming, Fischer–Tropsch synthesis, oxygen reduction reaction (ORR), hydrogen evolution reaction (HER), the removal of volatile organic compounds (VOCs), and CO preferential oxidation (PROX), and many others. Despite the growing interest in plasma for energy and catalytic nanomaterials, the synthesis mechanisms of nanomaterials using plasma still remain obscure due to the complicated physical and chemical reactions that occur during plasma preparation. Considerable amount of research are needed to better understand the controllable preparation mechanisms of the plasma method and to widen its application scope in synthesizing energy and catalytic nanomaterials.

In a conference, Prof. Lanbo Di, who works on plasma technology, and Prof. Feng Yu, who works on energy materials and catalysts, stated that something wonderful could happen between plasma and materials. Thus, they accepted the invitation to serve as Guest Editors for the Special Issue "Plasma for Energy and Catalytic Nanomaterials". Since the Special Issue was launched, it has attracted much attention around the world. Many researchers submitted their original work to this Special Issue. Finally, this Special Issue published 10 research papers and 2 review papers as high-quality studies that offer sufficient novelty and impact to appeal to our readership.

This Special Issue reviews original methods including plasma-assisted synthesized nanomaterials, a plasma-modified interface of nanomaterials, plasma-assisted catalysis, and the mechanism of action of plasma. Firstly, the plasma method allows thermodynamically and dynamically difficult reactions to proceed at low temperatures due to the activation of energetic electrons. Advanced nanomaterials, with superior particle size and good dispersion, could be synthesized by plasma-assisted preparation methods, including plasma-enhanced atomic layer deposition technology, coaxial pulse arc plasma deposition, plasma sputtering, solution plasma sputtering, etc. Gas plasma is also employed to provide high energy state gas with free radicals, ions, and electrons, which endow nanomaterials with surface active groups, heteroatom doping, surface etching, chemical oxidation/reduction, and high dispersed components. Plasma could be directly used in catalytic reactions either with or without catalysts. However, understanding the mechanism through which plasma acts and precisely controlling the plasma process remain sizable challenges.

We and the MDPI staff are pleased to offer this Special Issue to all interested readers including research scientists, postdoctoral researchers, and graduate and Ph.D. students. The Special Issue can serve as a useful reference for libraries. We hope that this new contribution will lead to further developments and advancements in the synthesis and applications of energy and catalytic nanomaterials with plasma. We think that the plasma method provides an additional strategy to easily address energy and catalytic nanomaterials, showing potential for developments in scientific and applicative field.

**Feng Yu, Lanbo Di**
*Special Issue Editors*

 *nanomaterials*

# Plasma for Energy and Catalytic Nanomaterials


**Feng Yu [1,2,*] and Lanbo Di [3,*]**

[1]  Key Laboratory for Green Processing of Chemical Engineering of Xinjiang Bingtuan, School of Chemistry and Chemical Engineering, Shihezi University, Shihezi 832003, China

[2]  Bingtuan Industrial Technology Research Institute, Shihezi University, Shihezi 832003, China

[3]  College of Physical Science and Technology, Dalian University, Dalian 116622, China

*  Correspondence: yufeng05@mail.ipc.ac.cn (F.Y.); dilanbo@163.com (L.D.); Tel.: +86-0993-205-8775 (F.Y.)


Received: 30 January 2020; Accepted: 13 February 2020; Published: 15 February 2020

This Special Issue "Plasma for Energy and Catalytic Nanomaterials" of Nanomaterials is focused on advancements in synthesis and applications of energy and catalytic nanomaterials by plasma. The preparation of nanomaterials is gaining increasing interest for energy and catalytic applications, such as methane reforming, Fischer-Tropsch synthesis, an oxygen reduction reaction (ORR), a hydrogen evolution reaction (HER), the removal of volatile organic compounds (VOCs), and CO preferential oxidation (PROX), etc. The plasma method allows thermodynamically and dynamically difficult reactions to proceed at low temperatures due to the activation of energetic electrons. Compared to conventional preparation methods, it has been proven to be a fast, facile, and environmentally friendly method for synthesizing highly efficient nanomaterials. The synthesized nanomaterials generally show enhanced metal–support interactions, small sizes of metal nanoparticles, specific metal structures, abundant oxygen vacancies, etc. Therefore, they exhibit high catalytic activity and stability in energy and catalytic applications. In spite of the growing interest in plasma for energy and catalytic nanomaterials, the synthesis mechanisms of nanomaterials using plasma still remain obscure due to the complicated physical and chemical reactions that occur during plasma preparation. A great deal of research is needed to better understand the controllable preparation mechanisms of the plasma method and to widen its application scope in synthesizing energy and catalytic nanomaterials.

Generally, solution plasma sputtering was used to prepared metal nanoparticles supported on carbon materials. Li et al. [1] used various organic quinolone (Q), aniline (A), and quinoline-aniline 1:1: mixed solution (QA) as nitrogen and carbon resources, synthesized different cobalt nanoparticles/nitrogen-doped carbon (Co/N-C) by solution plasma sputtering, and used as ORR electrocatalysts for zinc–air (Zn–Air) batteries. For ORR catalysts, N-doped species are crucial to electrocatalytic active sites. It is found that the as-obtained QA-Co/N-C sample with dominant quaternary-N and amino-N gave an onset potential of 0.87 V (vs. RHE) and a limit current density of 6.39 mA/cm$^2$. When used in a primary aqueous Zn–Air battery, the QA-Co/N-C exhibited an open-current voltage of 1.43 V and the peak power density of 87 mW/cm$^2$, which is comparable to those of the commercial 20 wt % Pt/C electrocatalyst. Moreover, QA-Co/N-C performed a stable galvanostatic discharge for 30,000 s at 20 mA/cm$^2$ and showed a great potential to be a Pt-free ORR electrocatalyst.

With the help of gas plasma, active components and catalyst supports can be optimized for achieving great catalytic activity. Zhang et al. [2] employed gas plasma to activate Au/P25-As prepared by a modified impregnation method with different working gases (H$_2$, Ar, O$_2$, Air). The Au/P25-O$_2$P catalyst activated by oxygen plasma showed excellent CO oxidation activity mainly due to the small size of gold nanoparticles and the high concentration of [O]$_s$ species. The Au/P25-O$_2$P exhibited a CO conversion of 100% at 40 °C, which is 30 °C lower than that of Au/P25-As catalyst. Liu et al. [3] etched nitrogen-doped carbon anchored by FeCo alloys (FeCo@NC) using dielectric barrier discharge (DBD) plasma in Ar atmosphere. Compared with FeCo@NC, the as-obtained DBD-FeCo@NC exposed more active sites, such as Fe/Co-N-C sites and enriched defect sites. The DBD−FeCo@NC performed an onset potential of 0.95 V as ORR electrocatalyst and an initial potential of 1.49 V as OER electrocatalyst, both of which are much better than those measured for FeCo@NC without Ar-plasma etching.

Gas plasma has also been proved to be a fast, facile, and green method for introducing groups to nanomaterials. Mi et al. [4] modified boron nitride nanosheets (BNNSs) by atmospheric pressure Ar+$H_2O$ low-temperature plasma initiated by bipolar nanosecond pulse DBD. The as-obtained plasma-modified BNNSs (P-BNNSs) contained nearly twice the content of surface hydroxyl than BNNSs. Moreover, the coating amount of silane coupling agent (SCA) on the surface of P-BNNSs increased by 45% more than the BNNSs, which enhanced the dehydration condensation reaction of P-BNNSs with SCA. Due to the P-BNNSs, the BN/epoxy resin (EP) insulating nanocomposites performed high thermal conductivity and high breakdown strength. Furthermore, some small molecule gases can be produced to exfoliate bulk particles and synthesize two-dimensional (2D) nanosheets via a plasma process. Zhang et al. [5] have successfully obtained 2D $MoS_2$ nanosheets and 2D g-$C_3N_4$ nanosheets. Using $H_2$/Ar plasma, the corresponding $NH_3$ and $H_2S$ generated and expanded to the layers of bulk $(NH_4)_2MoS_4$, while the bulk g-$C_3N_4$ was oxidized into corresponding $CO_x$ and $NO_x$ and generated 2D g-$C_3N_4$ nanosheets via air plasma. The prepared $MoS_2$ and g-$C_3N_4$ nanosheets showed the thickness of 2–3 and 1.2 nm, respectively. They exhibited excellent photocatalytic activity due to the nanosheet structure, larger surface area, more flexible photophysical properties, and longer charge carrier average lifetime. It can be predicted that plasma as the environmentally benign approach provides a general platform for fabricating ultrathin nanosheet materials, which will greatly help the practical application and scientific research of 2D catalytic materials.

Unlike conventional preparation methods, which generally need excess toxic reducing chemical agents, plasma can be used to generate redox species with a green engineering. Fan et al. [6] prepared Ag nanoparticles supported on cotton fabric (Ag/Cotton) with high antibacterial activity against both the Gram-negative bacterium *E. coli* and the Gram-positive bacterium *B. subtilis* by a surface plasma at atmospheric pressure for the first time. Ag/Cotton exhibited remarkable unusual physical and chemical properties, and excellent antibacterial performance against a wide scope of pathogens. Xie et al. [7] synthesized aqueous gold nanoparticles (AuNPs) using a $HAuCl_4$/sodium citrate solution via alternating the current plasma jet (A-Jet) and the pulse power driven plasma jet (P-Jet), respectively. Due to the high concentration of $Cl^-$ and $H_2O_2$ in the A-Jet, the AuNP growth rate is more than 40 times faster than that in the P-Jet. Moreover, there is a broad size control range and a narrow AuNP size distribution in the A-Jet.

In DBD plasma, the electrode is crucial to the experimental apparatus as well as the packing materials, reactor, discharge power, etc. Li et al. [8] studied $CO_2$ decomposition using DBD plasma with different metal foam electrodes, including Al foam, Fe foam, and Ti foam. For example, the Fe foam electrode exhibited more discharge area compared with the Fe rod electrode. The $CO_2$ conversion using Fe foam electrode reached 44.84% (with a corresponding energy efficiency of 6.86%), which is much better than the 21.15% $CO_2$ conversion reached using the Fe rod electrode (with a corresponding energy efficiency of 3.92%). Taheraslani et al. [9] investigated the deposits formed by $CH_4$+Ar plasma processing in a packed bed reactor with packing materials including γ-alumina, Pd/γ-alumina, $BaTiO_3$, silica-SBA-15, $MgO/Al_2O_3$, and α-alumina. Usually, the deposits mainly consist of carbon content (91 at. %) with the H/C molar radio around 1.7. Different from other packing materials, Pd/γ-alumina could restrain carbon-rich agglomerates due to the fast hydrogenation of deposit-precursors. Zhang et al. [10] measured vibrational energy distribution and electron energy distribution by high resolution temporal-spatial spectra emitted from the plasma, studied the characteristic evolution and discharge regimes transition of nanosecond pulsed DBD plasma, and distinguished the three main stages in the discharge, namely the streamer breakdown, the transition from streamer to diffuse regime, and the propagation of surface discharge on the plate electrode surface. It is important to develop plasma sources in material synthesis applications.

In summary, the papers published in this Special Issue include: plasma-assisted synthesized nanomaterials, a plasma-modified interface of nanomaterials, plasma-assisted catalysis, and the mechanism of plasma. Yu and co-workers [11,12] gave the brief overview of the advanced progress of plasma for energy and catalytic nanomaterials. The advanced nanomaterials with superior particle size and good dispersion could be synthesized by plasma-assisted preparation methods, including plasma enhanced atomic layer deposition technology, coaxial pulse arc plasma deposition, plasma sputtering, solution plasma sputtering, etc. Furthermore, gas plasma is employed to provide high energy state gas with free radicals, ions, and electrons, which could endow nanomaterials with surface active groups, heteroatom doping, surface etching, chemical oxidation/reduction, and high dispersed components. In addition, plasma could be directly used in catalytic reactions either with or without catalysts. Up until now, understanding the mechanism of plasma and precisely controlling the process of plasma has been a challenge.

**Funding:** This research was funded by the National Natural Science Foundation of China (Nos.21663022, 21773020) and Science and Technology Innovation Talents Program of Bingtuan (No.2019CB025).

**Acknowledgments:** F.Y. and L.D. would like to thank all the authors for their contributions to this Special Issue and the reviewers assisted in evaluating and improving quality of submitted manuscripts. We also very much appreciate the help of the editorial staff at Nanomaterials.

**Conflicts of Interest:** The authors declare no conflict of interest.

## References


1. Kim, S.; Park, H.; Li, O.L. Cobalt Nanoparticles on Plasma-Controlled Nitrogen-Doped Carbon as High-Performance ORR Electrocatalyst for Primary Zn-Air Battery. *Nanomaterials* **2020**, *10*, 223. [CrossRef] [PubMed]

2. Zhang, J.; Di, L.; Yu, F.; Duan, D.; Zhang, X. Atmospheric-Pressure Cold Plasma Activating Au/P25 for CO Oxidation: Effect of Working Gas. *Nanomaterials* **2018**, *8*, 742. [CrossRef] [PubMed]

3. Liu, M.; Yu, F.; Ma, C.; Xue, X.; Fu, H.; Yuan, H.; Yang, S.; Wang, G.; Guo, X.; Zhang, L. Effective Oxygen Reduction Reaction Performance of FeCo Alloys In Situ Anchored on Nitrogen-Doped Carbon by the Microwave-Assistant Carbon Bath Method and Subsequent Plasma Etching. *Nanomaterials* **2019**, *9*, 1284. [CrossRef] [PubMed]

4. Mi, Y.; Gou, J.; Liu, L.; Ge, X.; Wan, H.; Liu, Q. Enhanced Breakdown Strength and Thermal Conductivity of BN/EP Nanocomposites with Bipolar Nanosecond Pulse DBD Plasma Modified BNNSs. *Nanomaterials* **2019**, *9*, 1396. [CrossRef] [PubMed]

5. Zhang, B.; Wang, Z.; Peng, X.; Wang, Z.; Zhou, L.; Yin, Q. A Novel Route to Manufacture 2D Layer $MoS_2$ and g-$C_3N_4$ by Atmospheric Plasma with Enhanced Visible-Light-Driven Photocatalysis. *Nanomaterials* **2019**, *9*, 1139. [CrossRef] [PubMed]

6. Fan, Z.; Di, L.; Zhang, X.; Wang, H. A Surface Dielectric Barrier Discharge Plasma for Preparing Cotton-Fabric-Supported Silver Nanoparticles. *Nanomaterials* **2019**, *9*, 961. [CrossRef] [PubMed]

7. Xie, P.; Qi, Y.; Wang, R.; Wu, J.; Li, X. Aqueous Gold Nanoparticles Generated by AC and Pulse-Power-Driven Plasma Jet. *Nanomaterials* **2019**, *9*, 1488. [CrossRef] [PubMed]

8. Li, J.; Zhai, X.; Ma, C.; Zhu, S.; Yu, F.; Dai, B.; Ge, G.; Yang, D. DBD Plasma Combined with Different Foam Metal Electrodes for $CO_2$ Decomposition: Experimental Results and DFT Validations. *Nanomaterials* **2019**, *9*, 1595. [CrossRef] [PubMed]

9. Taheraslani, M.; Gardeniers, H. High-Resolution SEM and EDX Characterization of Deposits Formed by $CH_4$+Ar DBD Plasma Processing in a Packed Bed Reactor. *Nanomaterials* **2019**, *9*, 589. [CrossRef] [PubMed]

10. Zhang, L.; Yang, D.; Wang, S.; Jia, Z.; Yuan, H.; Zhao, Z.; Wang, W. Discharge Regimes Transition and Characteristics Evolution of Nanosecond Pulsed Dielectric Barrier Discharge. *Nanomaterials* **2019**, *9*, 1381. [CrossRef] [PubMed]

11. Yu, F.; Liu, M.; Ma, C.; Di, L.; Dai, B.; Zhang, L. A Review on the Promising Plasma-Assisted Preparation of Electrocatalysts. *Nanomaterials* **2019**, *9*, 1436. [CrossRef] [PubMed]
12. Li, J.; Ma, C.; Zhu, S.; Yu, F.; Dai, B.; Yang, D. A Review of Recent Advances of Dielectric Barrier Discharge Plasma in Catalysis. *Nanomaterials* **2019**, *9*, 1428. [CrossRef] [PubMed]

*Article*

# Cobalt Nanoparticles on Plasma-Controlled Nitrogen-Doped Carbon as High-Performance ORR Electrocatalyst for Primary Zn-Air Battery

**Seonghee Kim [1], Hyun Park [2] and Oi Lun Li [1,*]**

[1]  School of Materials Science and Engineering, Pusan National University, Busan 46241, Korea; ksh09290@pusan.ac.kr

[2]  Department of Naval Architecture and Ocean Engineering, Pusan National University, Busan 46241, Korea; hyunpark@pusan.ac.kr

*  Correspondence: helenali@pusan.ac.kr

Received: 24 December 2019; Accepted: 26 January 2020; Published: 28 January 2020

**Abstract:** Metal–air batteries and fuel cells have attracted much attention as powerful candidates for a renewable energy conversion system for the last few decades. However, the high cost and low durability of platinum-based catalysts used to enhance sluggish oxygen reduction reaction (ORR) at air electrodes prevents its wide application to industry. In this work, we applied a plasma process to synthesize cobalt nanoparticles catalysts on nitrogen-doped carbon support with controllable quaternary-N and amino-N structure. In the electrochemical test, the quaternary-N and amino-N-doped carbon (Q-A)/Co catalyst with dominant quaternary-N and amino-N showed the best onset potential (0.87 V vs. RHE) and highest limiting current density ($-6.39$ mA/cm$^2$). Moreover, Q-A/Co was employed as the air catalyst of a primary zinc–air battery with comparable peak power density to a commercial 20 wt.% Pt/C catalyst with the same loading, as well as a stable galvanostatic discharge at $-20$ mA/cm$^2$ for over 30,000 s. With this result, we proposed the synergetic effect of transitional metal nanoparticles with controllable nitrogen-bonding can improve the catalytic activity of the catalyst, which provides a new strategy to develop a Pt-free ORR electrocatalyst.

**Keywords:** cobalt nanoparticles; nitrogen-doped carbon; highly durable electrocatalysts; Zn-air battery

---

## 1. Introduction

Eco-friendly energy devices such as metal-to-air batteries and fuel cells receive much attention, while the importance of air electrodes for oxygen reduction reactions is increasingly prominent. Mainly, air electrodes often require platinum catalysts to improve their slow oxygen reduction reaction (ORR). However, many studies are attempting to reduce the Pt content or entirely replace Pt due to its low durability, high price, and rarity. Among them, transition metal-based electrocatalysts have exhibited promising ORR performance in basic electrolytes and active research is being done in using these as Pt catalyst replacements [1–5]. Among the various transition metals, cobalt is the most widely used as an alternative catalyst material due to its decent oxygen reduction reaction activity, higher four-electron selectivity, high durability, and low price [6–9]. On the other hand, cobalt nanoparticles as single active sites supported on a pristine carbon matrix often exhibit low activity compared with platinum catalysts [10,11]. Thus, there are many attempting to improve the intrinsic activity of the catalysts. One of the most typical methods is to synthesize a catalyst using a heteroatom-doped carbon matrix [12,13]. Among them, nitrogen-doped (N-doped) carbon represents a far better performance compared with pristine carbon support [14–18]. The catalyst often shows relatively high catalytic performance in alkaline electrolytes. In the case of N-doped carbon, nitrogen reduces the charge density of the nearby carbon atoms due to the differences in electronegativity of carbon and nitrogen [19]. Most recently, a few studies reported enhanced catalytic activity when combining cobalt nanoparticles with N-doped carbon support, and some of their ORR activity might even be comparable to that of platinum-based catalysts [20–24].

To date, there are different arguments about how various types of nitrogen functional groups affect the activity of catalysts [25–34]. Chatterjee et al. reported nitrogen-carbon nano-ions with pyridinic-N and pyrrolic-N as dominant catalysts. The catalysts contained a nitrogen content of up to 7.5%, and the author suggested that pyridinic-N worked as an active site for oxygen reduction with notably high activity [31]. On the other hand, Wang et al. found a correlation between the potential cycle and the diminishing concentration of quaternary-N in the catalyst. Based on the density functional theory (DFT) calculation and experimental results, quaternary-N displayed lower Gibbs free energy on the rate-liming step in the ORR reaction compared with pyridinic-N or pyrrolic-N. The author claimed that quaternary-N was responsible for the ORR activity in N-doped carbon [35]. Not only for single controlled C-N bonding, Yan also reported the synergetic effect of quaternary-N- and pyridinic-N-doping on the oxygen reduction reaction by using theoretical calculation [17]. After that, Ning et al. reported the synergetic effect of pyridinic-N and quaternary-N by measuring the transferable electrons of N-doped carbon. They suggested that the kinetic current density of the ORR in alkaline media is depended on the ratio of pyridinic-N and quaternary-N [36]. Additionally, Li et al. reported another synergic effect of amino-N and quaternary-N on ORR, where the experimental results showed that dominant amino-N-doped carbon indicated a higher onset potential, and the incorporation of quaternary-N into amino-N improved the 4-electron reaction selectivity and limiting current density [28].

Recently, a great deal of research has been reported on the development of a heterogeneous atomic-doped carbon catalyst through bottom-up synthesis using heterocyclic compounds through liquid plasma engineering. The approach has many advantages, such as being conducted at room temperature and being able to synthesize a heterogeneous atomic-doped carbon catalyst simply by selecting different precursors [37–44]. On the other hand, due to the low thermal stability of amino-N, it is hard to retain amino-N on the carbon matrix by the conventional synthesis route [27]. Thus, we applied a low-temperature facile plasma synthesizing method to fabricate amino-N and quaternary-N selectively on carbon support by careful selection of the precursors. Although a few studies have reported on metal-free tunable N-doped carbon electrocatalysts and/or metal nanoparticles doped on N-doped carbon, the synergic effect of transitional metal nanoparticles with precisely tunable amino-N and quaternary-N has rarely been reported.

In this study, we fabricated cobalt nanoparticles on a nitrogen-doped carbon catalyst through two heterocyclic compounds, quinoline and aniline, via plasma engineering. A pair of cobalt electrodes were applied as the precursor of metal nanoparticles, while quinoline and aniline were the sources of quaternary-N and amino-N, respectively, in the N-doped carbon support. Through XPS results, we can confirm that quaternary-N and amino-N are successfully retained within the carbon matrix from their corresponding precursors. Further, in the electrochemical performance test, cobalt nanoparticles supported on the mixture of quaternary-N and amino-N-doped carbon ((Q-A)/Co) had higher ORR onset potential and limiting current density than those of a single amino-N-doped carbon (A/Co) or quaternary-N-doped carbon (Q/Co) catalyst.

## 2. Materials and Methods

### 2.1. Synthesis of Co-N/C Catalyst by a Plasma Process

Plasma synthesis was performed between a pair of high purity transition cobalt electrodes (99.999%, Nilaco Co., Ltd., Tokyo, Japan, diameter of 1 mm) inside an organic solution of aniline (A-Co), quinolone (Q-Co), and aniline-quinoline 1:1: mixed solution (Quinoline, Aniline, >99%, Junsei Chemical Co., Ltd., Tokyo, Japan) using a bipolar pulse power supply (MPP-HV02, KURITA, Kyoto, Japan). The plasma was discharged at a voltage of ~4 kV, a frequency of 50 kHz, and a pulse width of 1.0 µs. Stable plasma was discharged for 30 min to obtain cobalt nanoparticles/N-doped carbon (Co-N/C). The solution was filtered using a φ 55 mm polytetrafluoroethylene filter, and the resulting filtered carbon powder samples were dried in an oven for 10 h at 80 °C, then heated at 700 °C for 1 h with 1.0 cc/min nitrogen atmosphere to improve their electrical conductivity. The schematic of the plasma synthesis is illustrated in Figure 1.

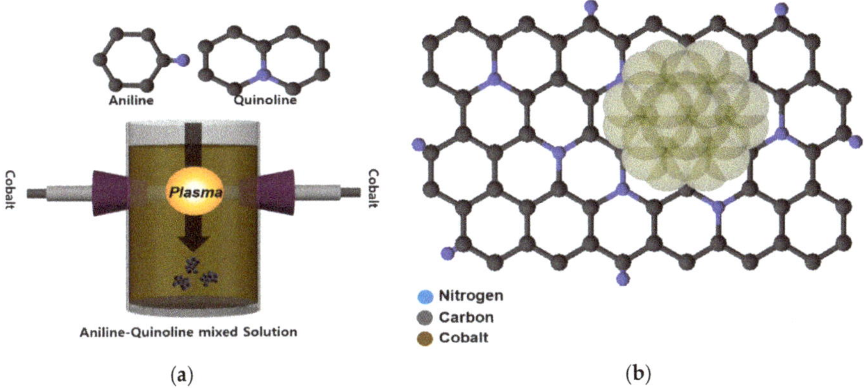


**Figure 1.** (**a**) Schematic illustration of the solution plasma process and synthesis of the cobalt nanoparticles/N-doped carbon (Co-N/C) catalyst. (**b**) Structure of the N-doped carbon catalyst.

### 2.2. Structure and Chemical Composition Analysis

The nitrogen absorption-desorption method (BET, Brunauer Emmett Teller; Shimadzu, TriStar-II 3020, Tokyo, Japan) was used for analyzing the surface area, pore volume, and pore diameter of the various Co-N/C catalysts. For morphology and chemical composition analysis, the synthesized carbon samples were characterized using scanning electron microscopy (SEM; JEOL, JSM-7100F, Tokyo, Japan), X-ray diffraction (XRD; Rigaku, Ultima IV, Tokyo, Japan), and X-ray photoelectron spectroscopy (XPS; JEOL, JPS-9010MC, Tokyo, Japan).

### 2.3. Electrochemical Measurements

An electrochemical analyzer (Biologic, VSP, Grenoble, France) was used to analyze the ORR electrochemical properties of the synthesized cobalt-nitrogen-doped carbon. The catalyst ink for the electrochemical analysis was made by adding 4 mg of well-ground Co-N/C-doped carbon catalysts into a mixture composed of 480 µL of distilled water, 480 µL of ethanol, and 40 µL of Nafion®117 Solution that was then ultrasonicated for 30 min. A total of 800 µg/cm$^2$ of well-dispersed catalyst was applied on a well-polished glass carbon (GC) disk (diameter: 4 mm) electrode (working electrode), where a platinum coil and Hg/HgO (1 M NaOH) were used as the counter and reference electrodes, respectively. After the three-electrode cell was prepared, ORR activity was measured by linear sweep voltammetry (LSV) in O$_2$ saturated 0.1 M KOH with a scan rate of 5 mV/s and rotating speed of 1600 rpm, between a potential range of 0.2 to 1.2 V vs. RHE. Chronoamperometry (CA) was conducted at 0.6 V vs. RHE for 30,000 s. In order to investigate the cycle durability of the synthesized catalysts, cyclic voltammetry (CV) was conducted at 50 mV/s for 3000 cycles between a potential range of 0.4 and 1.0 V vs. RHE. The kinetic current of the prepared electrocatalyst was calculated by a rotation rind disk electrode (RRDE). Where $i_d$ is disk current, $i_r$ is ring current, and $C_e$ is collection efficiency (0.42) [45]:

$$\text{Electron transfer number(n)} = 4 * (id / (id + (\frac{ir}{Ce}))) \tag{1}$$

$$H_2O_2 \text{ yielding}(\%) \; = \; 2*\left(\frac{\frac{ir}{Ce}}{Id+\left(\frac{ir}{Ce}\right)}\right)*100 \tag{2}$$

### 2.4. Primary Zn–Air Battery Measurement

The home-made Zn–air battery was assembled with the as-prepared catalysts and loaded on a gas diffusion layer electrode, with a Zn foil as the metal electrode, and 6 M KOH + 0.2 M $ZnCl_2$ as the electrolyte. For comparison, a benchmark catalyst (20 wt.% Pt/C) was also measured as the oxygen electrocatalyst. The catalyst ink (same as electrochemical ORR measurement) were well-suspended and dropped onto one face of carbon paper, with a mass loading of 1 mg/cm$^{-2}$. The power density was measured at 10 mV/s from OCV to 0.4 and discharge stability of the sample was measured by discharging at −20 mA/cm$^2$ for 30,000 s.

## 3. Results

### 3.1. Properties of Co Nanoparticles/N-Doped Carbon Catalyst

The SEM images of Q-A/Co, A/Co, and Q/Co are shown in Figure 2a–c. The three electrocatalysts exhibited very similar morphology, and carbon spherical nanoparticles were heavily agglomerated. There were no obvious structural differences from applying various types of nitrogen-carbon precursors during the plasma process. Figure 3 demonstrates the low resolution SEM images with EDS elemental mapping for Q-A/Co, Table S1 summarizes the atomic percentage of each element. It was found that nitrogen from the heterocyclic compound precursors (aniline and quinoline) was successfully retained with a high concentration of around 5% within the carbon matrix. Although the percentage of Co (at.%) was quite small (0.01–0.03%), cobalt nanoparticles formed by electrode sputtering during the plasma process were uniformly deposited on the synthesized carbon matrix (Figure 3c).

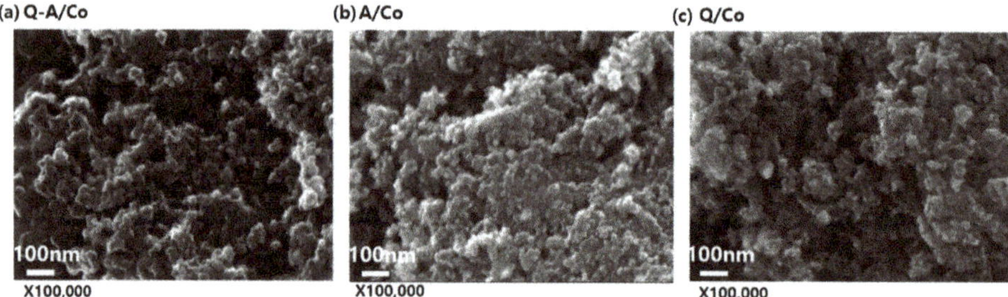


**Figure 2.** SEM images of the (**a**) quaternary-N and amino-N-doped carbon (Q-A/Co), (**b**) amino-N-doped carbon (A/Co), and (**c**) quaternary-N-doped carbon (Q/Co) catalysts.

Figure 4 and Table 1 summarize the porous structure of the synthesized N-doped carbon obtained from the $N_2$ adsorption-desorption method. The BET isotherm linear plot in Figure 4a confirms that all synthesized N-doped carbon had a meso-macro porous structure from the hysteresis graph. This is coming from the interparticle voids between the primary carbon particles in the usual solution plasma method synthesized carbon [46]. The BET surface areas of Q-A/Co, A/Co, and Q/Co were 211.1, 210.2, and 206.2 m$^2$/g, respectively. This result implies that the porous and structural properties of synthesized Co-N-doped carbon are not affected by the original precursor. In Figure 4b, Q-A/Co exhibits a slightly higher pore volume of 0.67 cm$^3$/g and has mainly mesopores with an average diameter of 10.1 nm. To confirm the crystalline structure of the deposited cobalt particle, XRD analysis in Figure 5 clearly indicates pure cobalt metal peaks (Co$^0$) at 44°([110], 52°([200]), and 76°([220]). Regardless of the liquid precursor, the composition of the cobalt nanoparticles is identical with similar crystallinity. Combined with the EDS picture and XRD profile, it is clear that the cobalt nanoparticles in their pristine form are successfully fabricated through the plasma process and incorporated into N-doped carbon support.

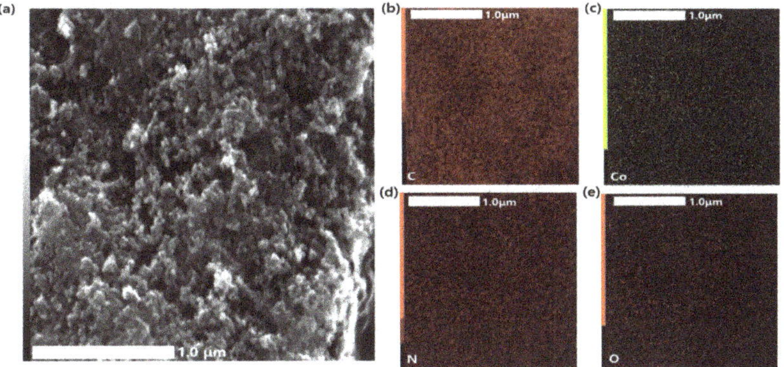

**Figure 3.** (**a**) SEM image of the Q-A/Co catalyst. (**b–e**) EDS mapping of the synthesized Q-A/Co catalyst: (**b**) carbon, (**c**) cobalt, (**d**) nitrogen, (**e**) oxygen.

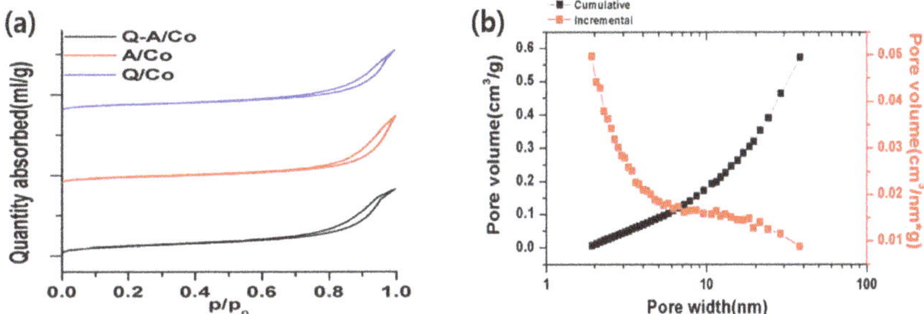

**Figure 4.** (**a**) BET isotherm linear plot of the three types of synthesized N-doped carbon. (**b**) Barrett-Joyner-Halenda (BJH) adsorption pore distribution of Q-A/Co.

**Table 1.** Textural parameters of Q/A-Co derived from the N2 adsorption-desorption isotherms.

| | BET Surface Area | BJH Adsorption Pore Volume | BJH Adsorption Average Pore Width |
|---|---|---|---|
| Q-A/Co | 211.1 m$^2$/g | 0.6711 cm$^3$/g | 10.1 nm |
| A/Co | 210.2 m$^2$/g | 0.6708 cm$^3$/g | 13.08 nm |
| Q/Co | 206.2 m$^2$/g | 0.5871 cm$^3$/g | 15.3 nm |

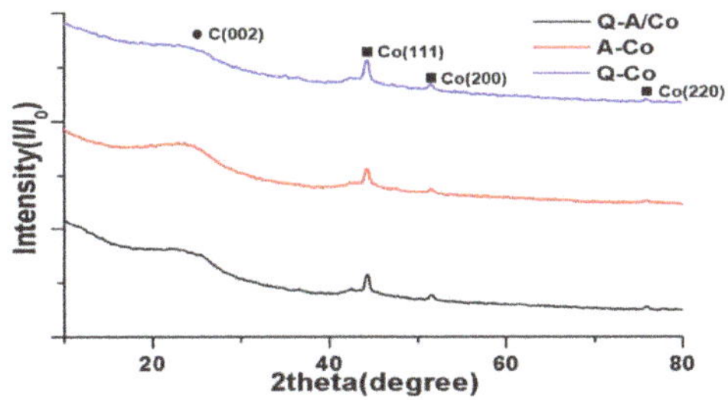

**Figure 5.** XRD patterns obtained for metal cobalt-N doped carbon.

### 3.2. Chemical Bonding States of Co Nanoparticles/N-Doped Carbon Catalyst

In order to analyze and understand the effect of nitrogen functional groups on ORR catalytic activity, we applied XPS to figure out the surface nitrogen-carbon bonding states in detail. First, we can confirm that nitrogen is successfully doped into the carbon matrix by the presence of C–N bonding in all synthesized samples from C1s XPS spectra in Figure 6a–c. In order to identify the types of C–N bonding, N1s narrow spectra of Q-A/Co, A/Co, and Q/Co are demonstrated in the corresponding Figure 6d–f. As a reference, the binding energy of amino-N is around 399.4–399.6 eV and quaternary-N is around 401.1–401.5 eV [47]. A/Co, which was synthesized from the aniline precursor, shows a highest amino-N peak of 35%. Additionally, Q/Co fabricated from the quinoline precursor shows a dominant quaternary-N peak of 32%. For Q-A/Co, which was synthesized from a mixed aniline and quinoline solution, two major peaks for quaternary-N (30%) and amino-N (31%) are present equally. From the above XPS results, it is clear that C–N bonding is retained from the heterocyclic compound precursor, and plasma engineering is a facile route to control various types of nitrogen-carbon bonding by choosing the corresponding single or multiple C–N precursors. The relative percentages of nitrogen bonding states of three Co-N-doped catalysts are summarized in Table 2.

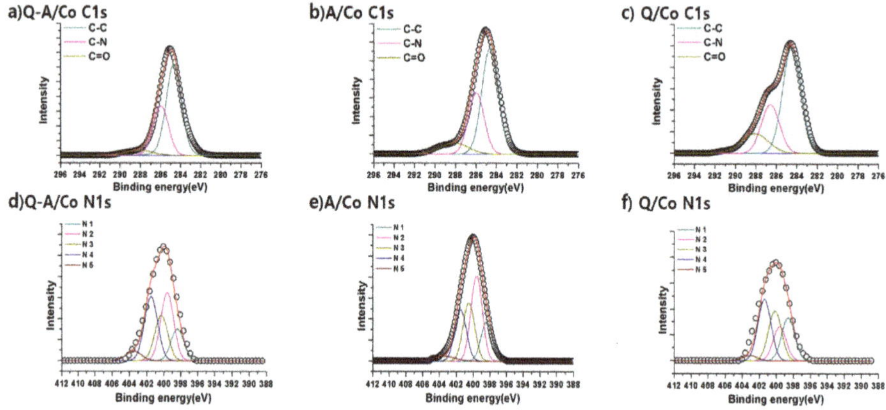

**Figure 6.** (**a–c**) High-resolution C 1s XPS spectra with peak deconvolution of Co-N/C. (**d–f**) N 1s XPS spectra with peak deconvolution of Co-N/C.

**Table 2.** Nitrogen bonding states of three Co-N-doped catalysts from the deconvolution of N 1s spectra.

| Bonding | Binding Energy | Q-A/Co | A-Co | Q-Co |
|---|---|---|---|---|
| | | Relative Percentage (%) | | |
| N1(Pyridinic-N) | 398.4–398.6 eV | 14 | 16 | 22 |
| N2(Amino-N) | 399.4–399.6 eV | 31 | 35 | 18 |
| N3(Pyrrolic-N) | 400.1–400.3 eV | 20 | 24 | 25 |
| N4(Quaternary-N) | 401.1–401.5 eV | 30 | 21 | 32 |
| N5(Oxide-N) | 403.3–403.7 eV | 5 | 4 | 3 |

### 3.3. Electrochemical Properties of Co Nanoparticles/N-Doped Carbon Catalyst

Figure 7a–d shows the electrochemical ORR catalytic activity of synthesized Co-N-doped carbon in 0.1 M KOH. In the polarization curve, A/Co with dominant amino-N demonstrates a slightly higher onset potential (0.85 V vs. RHE) than that of the Q/Co (0.84 V vs. RHE). Figure 7b demonstrates a lower Tafel-slope of Q/Co (67 mV/dec) and Q-A/Co (78 mV/dec) compared with that of A/Co (89 mV/dec), which indicates that quaternary-N doping can improve the ORR reaction kinetics. Figure S1a–c further evaluated the CV curves of each electrocatalyst with a scan rate of 10–80 mV/s, and the double layer capacitance ($C_{dl}$) values are derived from each CV based on the equation $C_{dl} = \frac{d(\Delta j)}{2d(Vb)}$, where $Vb$ is the scan rate (Figure S1). The electrochemically active surface areas (ECSA) are calculated by ECSA=$C_{dl}$/$Cs$, of which $C_s$ is the specific capacitance of a flat surface with 1 cm$^2$ of surface area, and the rugosity can be estimated from ECSA/geometric surface area (0.1256 cm$^2$) [48,49]. The calculated $C_{dl}$ values of A/Co, Q/Co, and Q-A/Co were 14 mF/cm$^2$, 14 mF/cm$^2$, and 11 mF/cm$^2$, respectively. Although ESCA, of all synthesized catalysts, exhibited a similar value of 11–14 mF/cm$^2$, the limiting current density of the Q/Co at 0.6 V vs. RHE (−6.41 mA/cm$^2$) was much higher than that of the A/Co (−5.17 mA/cm$^2$). This result clearly implies that amino-N has a positive effect on the onset potential, whereas quaternary-N improves the reaction kinetics, both of which are similar to previously reported findings [28,50]. Interestingly, Q-A/Co with both dominant amino-N and quaternary-N shows certain differences in ORR activity compared with single amino-N-doped carbon or quaternary-N-doped carbon. The Q-A/Co exhibits better onset potential (0.87 V vs. RHE) and limiting current density (−6.27 mA/cm$^2$), which is quite comparable to a commercial 20 wt.% Pt/C catalyst. (0.95 V vs. RHE and −5.43 mA/cm$^2$). In reaction selectivity results based on the electron transfer number calculation in Figure S2, Q-A/Co shows the higher 4 e− selectivity of 3.91 compared with those of Q/Co and A/Co (~3.8). The above findings prove the synergetic effect of amino-N and quaternary-N on ORR activities. In the durability test as displayed in Figure 7c,d, the onset ORR potential of Q-A/Co and 20 wt.% Pt/C shift negatively by 10 mV and 20 mV, respectively, after 3000 cycles. Moreover, the relative current density of Q-A/Co reduces by merely 9% while 20 wt.% Pt/C decreases by 16% after 30,000 s at 0.6 V vs. RHE. Both experimental results confirmed that the amino-N and quaternary-N are incorporated into the carbon matrix with high stability. The detailed electrochemical catalytic activity of Co-N-doped carbon and 20wt.% Pt/C are summarized in Table 3.

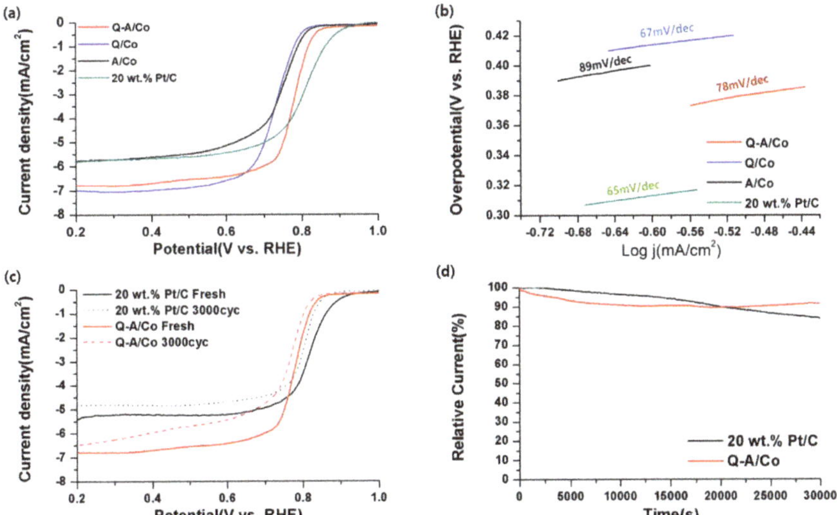


**Figure 7.** (**a**) Linear sweep voltammetry (LSV) curve of Co-N-doped carbon and 20 wt.% Pt/C at 5 mV/s in O$_2$ saturated 0.1 M KOH in rotation speed 1600 rpm. (**b**) Correlated Tafel slope from each LSV curve. (**c**) Cycle durability of Q-A/Co and 20 wt.% Pt/C after 3000 cycles. (**d**) Chronoamperometry (CA) of Q-A/Co and 20 wt.% Pt/C at 0.6 V vs. RHE.

**Table 3.** Summary of the electrochemical catalytic activity of Co-N-doped carbon and 20 wt.% Pt/C.

|  | Q-A/Co | Q/Co | A/Co | 20 wt.% Pt/C |
|---|---|---|---|---|
| Onset Potential | 0.87 V vs. RHE | 0.84 V vs. RHE | 0.85 V vs. RHE | 0.95 V vs. RHE |
| Potential at $-3$ mA/cm$^2$ | 0.78 V vs. RHE | 0.74 V vs. RHE | 0.74 V vs. RHE | 0.81 V vs. RHE |
| Current density At 0.6 V vs. RHE | $-6.27$ mA/cm$^2$ | $-6.41$ mA/cm$^2$ | $-5.17$ mA/cm$^2$ | $-5.43$ mA/cm$^2$ |
| Potential at $-3$ mA/cm$^2$ After 3000 cycles Figure 7c | 0.77 V vs. RHE | - | - | 0.79 V vs. RHE |
| Current density After 3000 cycles Figure 7d | 9% decrease | - | - | 16% decrease |

### 3.4. Primary Zn–Air Battery Test of Co Nanoparticles/N-Doped Carbon Catalyst (Q-A/Co)

Since Q-A/Co outperformed other as-synthesized electrocatalysts, it was further applied as the air cathode in a home-made primary aqueous Zn–Air battery cell, in order to examine its performance in a practical application. In open circuit voltage, the Q-A/Co catalyst exhibited an open-current voltage (OCV) of 1.43 V, which is only slightly lower than that of 20 wt.% Pt/C (1.49 V). On the other hand, the peak power density of Q-A/Co (87 mW/cm$^2$) is almost similar to the Pt/C (89 mW/cm$^2$) due to the higher discharge current density (Figure 8b). Although the step-voltage discharge of Q-A/Co (Figure 8c) shows lower voltage in the low current density range between 1 and 20 mA/cm$^2$, Q-A/Co works comparably to the commercial 20 wt.% Pt/C catalyst in a higher current density region above 50 mA/cm$^2$. During the galvanostatic discharge at $-20$ mA/cm$^2$, the Q-A/Co catalyst displays a stable discharge for 30,000 s (Figure 8d). Based on the Zn–air battery test, Q-A/Co is a potential noble-free ORR catalyst to replace expensive noble Pt/C in primary Zn–air batteries with the high peak power density and durable discharge performance.

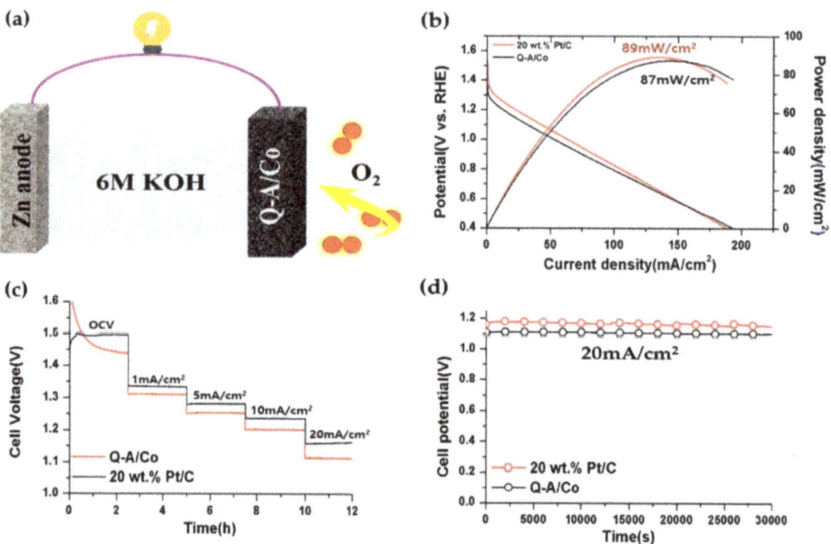


**Figure 8.** (**a**) Graph for the home-made Zn-air battery. (**b**) Discharge power density of Q-A/Co at 10 mV/s. (**c**) OCV and step voltage of Q-A/Co. (**d**) Discharge stability of Q-A/Co at 20 mA/cm$^2$.

## 4. Conclusions

Consequently, in this study, cobalt with dominant amino-N- and quaternary-N-doped carbon was synthesized through a plasma process. Both EDS and XRD results indicated that Co particles were deposited on the carbon matrix with nitrogen. From the XPS surface analysis, Q-A/Co exhibited both major amino-N and quaternary-N peaks evenly. This result clearly indicated that aniline and quinoline precursors can be applicable as selective N-C bonding precursors. From the electrocatalytic activity measurement, we confirmed that amino-N had a positive effect on ORR onset potential, whereas quaternary-N improved the limit current density, as reported. In particular, the combination of amino-N and quaternary-N provided a synergetic effect to better the onset potential and limit current density. Q-A/Co displayed a relatively high catalytic activity in terms of onset potential (0.87 V vs. RHE) and limit current density ($-6.39$ mA/cm$^2$), which is only slightly inferior to that of commercial 20 wt.% Pt/C. In a primary Zn–air electrode test, Q-A/Co further exhibited high potential from its high peak power density (87 mW/cm$^2$) and durable discharge performance. From the above results, we suggested that plasma engineering is a facile route to fabricate transition metal nanoparticles on selective amino-N and quaternary-N. When combining the synergic effect of transitional metal-based nanoparticles with the controlled N-doped carbon matrix, the noble metal-free catalysts might be further developed as promising candidates to replace the state-of-the-art Pt/C in fuel cell and metal-air batteries.

**Supplementary Materials:** The following are available online at http://www.mdpi.com/2079-4991/10/2/223/s1. Table S1: Atomic percent of A/Co, Q/Co, Q-A/Co from EDS, Figure S1. (a–c) CV curves of synthesized catalysts with scan rate 10–0 mV/s in 0.1M KOH, (d) Current density differences plot with scan rate. ΔJ is current differences between Ja (anodic current) and Jc (cathodic current) at potential 0.5 V vs. RHE which is non faradaic region; Figure S2. (a) Electron transfer number and, (b) peroxide yielding of Q-A/Co, A/Co, Q/Co.

**Author Contributions:** S.K. performed carbon characterization and electrochemical data; H.P. and O.L.L. helped in the data analysis of electrochemistry, discussion of the results and wrote the manuscript. All authors discussed the results and contributed to manuscript preparation. All authors have read and agreed to the published version of the manuscript.

**Funding:** This work was supported by Basic Science Research Program through the National Research Foundation of Korea (NRF) grant funded by the Korea government (MSIP) through GCRC-SOP (No. 2011-0030013).

**Conflicts of Interest:** The authors declare no conflict of interest.

## References

1. Fan, W.; Li, Z.; You, C.; Zong, X.; Tian, X.; Miao, S.; Shu, T.; Li, C.; Liao, S. Binary Fe, Cu-doped bamboo-like carbon nanotubes as efficient catalyst for the oxygen reduction reaction. *Nano Energy* **2017**, *37*, 187–194. [CrossRef]
2. Du, J.; Cheng, F.; Zhang, T.; Chen, J. M(Salen)-derived Nitrogen-doped M/C(M = Fe, Co, Ni) Porous Nanocomposites for electrocatalytic Oxygen Reduction. *Sci. Rep.* **2014**, *4*, 4386. [CrossRef] [PubMed]
3. Peng, H.; Liu, F.; Lut, X.; Liao, S.; You, C.; Tian, X.; Nan, H.; Luo, F.; Song, H.; Fu, Z.; et al. Effect of Transition Metals on the structure and Performance of the Doped Carbon Catalysts Derived from Polyaniline and Melamine for ORR Application. *ACS Catal.* **2014**, *5*, 3797–3805. [CrossRef]
4. Ratso, S.; Kruusenberg, I.; Kaarik, M.; Kook, M.; Puust, L.; Saar, R.; Leis, G.; Tammeveski, K. Highly efficient transition metal and nitrogen co-doped carbide-derived carbon electrocatalysts for anion exvhange membrane fuel cells. *J. Power Sources* **2018**, *375*, 233–243. [CrossRef]
5. Martinez, U.; Babu, S.K.; Holby, E.F.; Chung, H.T.; Yin, X.; Zelenay, P. Progress in the Development of Fe-Based PGM-free Electrocatalysts for the Oxygen Reduction Reaction. *Adv. Mater.* **2019**, *31*, 1806545. [CrossRef]
6. Cai, S.; Meng, Z.; Tang, H.; Wang, Y.; Tsiakaras, P. 3D Co-N-doped hollow carbon sphere as excellent bifunctional electrocatalysts for oxygen reduction reaction and oxygen evolution reaction. *Appl. Catal. B Environ.* **2017**, *217*, 477–484. [CrossRef]
7. Fu, G.; Liu, Y.; Chen, Y.; Tang, Y.; Goodenough, J.B.; Lee, J.M. Robust N-doped carbon aerogels strongly coupled with iron-cobalt particles as efficient bifunctional catalysts for rechargeable Zn-air batteries. *Nanoscale* **2018**, *10*, 19937. [CrossRef]

8.    Ghanbarlou, H.; Rowshanzair, S.; Kazeminasab, B.; Parnian, M.J. Non-precious metal nanoparticles supported on nitrogen0doped graphene as a promising catalyst for oxygen reduction reaction: Synthesis, characterization and electrocatalytic performance. *J. Power Sources* **2015**, *273*, 981–989. [CrossRef]

9.    Sharma, M.; Jang, J.H.; Shin, D.Y.; Kwon, J.A.; Lim, D.H.; Choi, D.; Sung, H.; Jang, J.H.; Lee, S.Y.; Lee, K.Y.; et al. Work function-tailored graphene via transition metal encapsulation as a highly active and durable catalyst for the oxygen reduction reaction. *Energy Environ. Sci.* **2019**, *12*, 2200–2211. [CrossRef]

10.    Guo, H.; Feng, Q.; Zhu, J.; Xu, J.; Li, Q.; Liu, S.; Xu, K.; Zhang, C.; Liu, T. Cobalt nanoparticle-embedded nitrogen-doped carbon/carbon nanotube frameworks derived from a metal-organic framework for tri-functional ORR, OER and HER electrocatalysis. *J. Mater. Chem. A* **2019**, *7*, 3664. [CrossRef]

11.    Zhang, C.L.; Lu, B.R.; Cao, F.H.; Wu, Z.Y.; Zhang, W.; Cong, H.P.; Yu, S.H. Electrospun metal-organic framework nanoparticle fibers and their derived electrocatalysts for oxygen reduction reaction. *Nano Energy* **2019**, *55*, 226–233. [CrossRef]

12.    Guo, B.; Ma, R.; Li, Z.; Guo, S.; Luo, J.; Yang, M.; Liu, Q.; Thomas, T.; Wang, J. Hierarchical N-Doped Porous Carbons for Zn-Air Batteries and Supercapacitors. *Nano-Micro Lett.* **2020**, *12*, 20. [CrossRef]

13.    Khan, Z.; Park, S.O.; Yang, J.; Park, S.; Shanker, R.; Song, H.K.; Kim, Y.; Kwak, S.K.; Ko, H. Binary N,S-doped carbon nanospheres from bio-inspired artificial melanosomes: A route to efficient air electrodes for seawater batteries. *J. Mater. Chem. A* **2018**, *6*, 24459. [CrossRef]

14.    Zhao, S.; Wang, D.W.; Amal, R.; Dai, L. Carbon-Based Metal-Free Catalyst for Key Reactions Involved in Energy Conversion and Storage. *Adv. Mater.* **2019**, *31*, 1801526. [CrossRef] [PubMed]

15.    Guo, D.; Shibuya, R.; Akiba, C.; Saji, S.; Kondo, T.; Nakamura, J. Active sites of nitrogen-doped carbon materials for oxygen reduction reaction clarified using model catalysts. *Science* **2016**, *351*, 6271. [CrossRef] [PubMed]

16.    Yan, P.; Liu, J.; Yuan, S.; Liu, Y.; Cen, W.; Chen, Y. The promotion effects of graphitic and pyridinic N combinational doping on graphene for ORR. *Appl. Surf. Sci.* **2018**, *445*, 398–403. [CrossRef]

17.    Wu, G.; Santandreu, A.; Kellogg, W.; Gupta, S.; Ogoke, O.; Zhang, H.; Wang, H.L.; Dai, L. Carbon nanocomposite catalysts for oxygen reduction and evolution reactions: From nitrogen doping to transition-metal addition. *Nano Energy* **2016**, *29*, 83–110. [CrossRef]

18.    Dumont, J.H.; Martinez, U.; Artyushkove, K.; Purdy, G.M.; Dattelbaum, A.M.; Zelenay, P.; Mohite, A.; Atanassov, P.; Gupta, G. Nitrogen-Doped Graphene Oxide Electrocatalysts for the Oxygen Reduction Reaction. *ACS Appl. Nano Mater.* **2019**, *2*, 1675–16785. [CrossRef]

19.    Vilaplana, A.F.; Herrero, E. Understanding the chemisorption-based activation mechanism of the oxygen reduction reaction on nitrogen-doped graphitic materials. *Electrochim. Acta* **2016**, *204*, 245–254. [CrossRef]

20.    Hu, F.; Yang, H.; Wang, C.; Zhang, Y.; Lu, H.; Wang, Q. Co-N-Doped Mesoporous Carbon Hollow Spheres as Highly Efficient Electrocatalysts for Oxygen Reduction Reaction. *Small* **2017**, *13*, 1602507. [CrossRef]

21.    Ma, X.; Zhao, X.; Huang, J.; Sun, L.; Li, Q.; Yang, X. Fine Co Nanoparticles Encapsulated in a N-Doped Porous Carbon Matrix with Superfical N-doped Porous Carbon nanofibers for Efficient Oxygen Reduction. *ACS Appl. Mater. Interfaces* **2017**, *9*, 21747–21755. [CrossRef] [PubMed]

22.    Liu, S.; Wang, Z.; Zhou, S.; Yu, F.; Yu, M.; Chiang, C.Y.; Zhou, W.; Zhao, J.; Qiu, J. Metal-Organic-Framework-Derived Hybrid Carbon Nanocages as a Bifunctional Electrocatalyst for Oxygen Reduction and Evolution. *Adv. Mater.* **2017**, *29*, 1700874. [CrossRef] [PubMed]

23.    Li, R.; Wang, X.; Dong, Y.; Pan, X.; Liu, X.; Zhao, Z.; Qiu, J. Nitrogen-doped carbon nanotubes decorated with cobalt nanoparticles derived from zeolitic imidazolate framework-67 for highly efficient oxygen reduction reaction electrocatalyst. *Carbon* **2018**, *132*, 580–588. [CrossRef]

24.    Shu, J.; Niu, Q.; Wang, N.; Nie, J.; Ma, G. Alginate derived Co/N doped hierarchical porous carbon microspheres for efficient oxygen reduction reaction. *Appl. Surf. Sci.* **2019**, *485*, 520–528. [CrossRef]

25.    Sun, M.; Wu, X.; Deng, X.; Zhang, W.; Xie, Z.; Huang, Q.; Huang, B. Synthesis of pyridinic-N doped carbon nanofibers and its electro-catalytic activity for oxygen reduction reaction. *Mater. Lett.* **2018**, *220*, 313–316. [CrossRef]

26.    Muthuswamy, N.; Buan, M.E.M.; Walmsley, J.C.; Ronning, M. Evaluation of ORR active sites in nitrogen-doped carbon nanofibers by KOH post treatment. *Catalysis Today* **2018**, *301*, 11–16. [CrossRef]

27.    Zhang, C.; Hao, R.; Liao, H.; Hou, Y. Synthesis of amino-functionalized graphene as metal-free catalyst and exploration of the roles of various nitrogen stated in oxygen reduction reaction. *Nano Energy* **2013**, *2*, 88–97. [CrossRef]

28. Li, O.L.; China, S.; Wada, Y.; Panomsuwan, G.; Ishizaki, T. synthesis of graphitic -N and amino-N in nitrogen-doped carbon via a solution plasma process and exploration of their synergic effect for advanced oxygen reduction reaction. *J. Mater. Chem. A* **2017**, *5*, 2073. [CrossRef]

29. Mamtani, K.; Jain, D.; Dogu, D.; Gustin, V.; Gunduz, S.; Co, A.C.; Ozkan, U.S. Insights into oxygen reduction reaction (ORR) and oxygen evolution reaction (OER) active sites for nitrogen-doped carbon nanostructures (CNx) in acidic media. *Appl. Catal. B Environ.* **2018**, *220*, 88–97. [CrossRef]

30. Singh, S.K.; Takeyasu, K.; Nakamura, J. Active sites and Mechanism of Oxygen Reduction Reaction Electrocatalysis on Nitrogen-Doped Carbon Materials. *Adv. Mater.* **2019**, *31*, 1804297. [CrossRef]

31. Chatterjee, K.; Ashokkumar, M.; gullapalli, H.; Gong, Y.; Vajtal, R.; Thanikaivelan, P.; Ajayan, P.M. Nitrogen-rich carbon nano-onions for oxygen reduction reaction. *Carbon* **2018**, *130*, 645–651. [CrossRef]

32. Li, Z.; Gao, Q.; Qian, W.; Tian, W.; Zhang, H.; Zhang, Q.; Liu, Z. Ultrahigh Oxygen Reduction Reaction Electrocatalytic Activity and Stability over Hierarchical Nanoporous N-doped Carbon. *Sci. Rep.* **2018**, *8*, 2863. [CrossRef] [PubMed]

33. Lv, Q.; Si, W.; He, J.; Sun, L.; Zhang, C.; Wang, N.; Yang, Z.; Li, X.; Wang, X.; Deng, W.; et al. Selectively nitrogen-doped carbon materials as superior Metal-free catalysts for oxygen reduction. *Nat. Commun.* **2018**, *9*, 3376. [CrossRef] [PubMed]

34. Yang, H.B.; Miao, J.; Hung, S.F.; Chen, J.; Tao, H.B.; Wang, X.; Zhang, L.; Chen, R.; Gao, J.; Chen, H.M.; et al. Identification of catalytic sites for oxygen reduction and oxygen evolution in N-doped graphene materials: Development of highly efficient metal-free bifunctional electrocatalyst. *Sci. Adv.* **2016**, *2*, 1501122. [CrossRef] [PubMed]

35. Wang, N.; Lu, B.; Li, L.; Niu, W.; Tang, Z.; Kang, X.; Chen, S. Graphitic Nitrogen is responsible for oxygen electroreduction on Nitrogen-Doped Carbons in Alkaline Electrolytes: Insights from Activity Attenuation Studies and Theoretical Calculations. *ACS Catal.* **2018**, *8*, 6827–6836. [CrossRef]

36. Ning, X.; Li, Y.; Ming, J.; Wang, Q.; Wang, H.; Cao, Y.; Peng, F.; Yang, Y.; Yu, H. Electronic synergism of pyridinic- and graphitic- nitrogen on N-doped carbons for the oxygen reduction reaction. *Chem. Sci.* **2019**, *10*, 1589. [CrossRef] [PubMed]

37. Kim, D.W.; Li, O.L.; Pootawang, P.; Saito, N. Solution plasma synthesis process of tungsten carbide on N-doped carbon nanocomposite with enhanced catalytic ORR activity and durability. *RSC Adv.* **2014**, *4*, 16813. [CrossRef]

38. Panomsuwan, G.; Saito, N.; Ishizaki, T. Nitrogen-doped carbon nanoparticles derived from acrylonitrile plasma for electrochemical oxygen reduction. *Phys. Chem. Chem. Phys.* **2015**, *17*, 6227. [CrossRef]

39. Panomsuwan, G.; China, S.; Kaneko, Y.; Saito, N.; Ishizaki, T. In situ solution plasma synthesis of nitrogen-doped carbon nanoparticles as metal-free electrocatalysts for the oxygen reduction reaction. *J. Mater. Chem. A* **2014**, *2*, 18677. [CrossRef]

40. Panomsuwan, G.; Saito, N.; Ishizaki, T. Simple one-step synthesis of fluorine-doped carbon nanoparticles as potential alternative metal-free electrocatalysts for oxygen reduction reaction. *J. Mater. Chem. A* **2015**, *3*, 9972. [CrossRef]

41. Panomsuwan, G.; Saito, N.; Ishizaki, T. Electrocatalytic oxygen reduction activity of boron-doped carbon nanoparticles synthesized via solution plasma process. *Electrochem. Commun.* **2015**, *59*, 81–85. [CrossRef]

42. Li, O.L.; Wada, Y.; Kaneko, A.; Lee, H.; Ishizaki, T. Oxygen Reduction Reaction Activity of Thermally Tailored Nitrogen-Doped Carbon Electrocatalysts Prepared through Plasma Synthesis. *ChemElectroChem* **2018**, *5*, 1995–2001. [CrossRef]

43. Li, O.L.; prabaka, K.; Kaneko, A.; Park, H.; Ishizaki, T. Exploration of Lewis basicity and oxygen reduction reaction activity activity in plasma-tailored nitrogen-doped carbon electrocatalysts. *Catal. Today* **2019**, *337*, 102–109. [CrossRef]

44. Li, O.L.; Shi, Z.; Lee, H.; Ishizaki, T. Enhanced Electrocatalytic Stability of Platinum Nanoparticles Supported on Sulfur-Doped Carbon using in-situ solution plasma. *Sci. Rep.* **2019**, *9*, 12704. [CrossRef]

45. Zhou, R.; Zheng, Y.; Jaroniec, M.; Qiao, S.Z. Determination of the Electron Transfer Number for the Oxygen Reduction Reaction: From Theory to Experiment. *ACS Catal.* **2016**, *6*, 4720–4728. [CrossRef]

46. Kang, J.; Li, O.L.; Saito, N. Synthesis of structure-controlled carbon nano spheres by solution plasma process. *Carbon* **2013**, *60*, 292–298. [CrossRef]

47. Ferrero, G.A.; Fuertes, A.B.; Sevilla, M.; Titrici, M.M. Efficient metal-free N-doped mesoporous carbon catalyst for ORR by a template-free approach. *Carbon* **2016**, *106*, 179–187. [CrossRef]

48. An, L.; Jiang, N.; Li, B.; Hua, S.; Fu, Y.; Liu, J.; Hao, W.; Xia, D.; Sun, Z. A highly active and durable iron/cobalt alloy catalyst encapsulated in N-doped graphitic carbon nanotubes for oxygen reduction reaction by a nanofibrous dicyandiamide template. *J. Mater. Chem. A* **2018**, *6*, 5962. [CrossRef]

49. Guan, B.Y.; Yu, L.; Lou, X.W. Formation of Single-Holed Cobalt/N-Doped Carbon Hollow Particles with Enhanced Electrocatalytic Activity toward Oxygen Reduction Reaction in Alkaline Media. *Adv. Sci.* **2017**, *4*, 1700247. [CrossRef]

50. Lai, L.; Potts, J.R.; Zhan, D.; Wang, L.; Poh, C.K.; Tang, C.; Gong, H.; Shen, Z.; Lin, J.; Ruoff, R.S. Exploration of the active center structure of nitrogen-doped graphene-based catalysts for oxygen reduction reaction. *Energy Environ. Sci.* **2012**, *5*, 7936. [CrossRef]

 *nanomaterials*

*Article*

# Atmospheric-Pressure Cold Plasma Activating Au/P25 for CO Oxidation: Effect of Working Gas

**Jingsen Zhang [1], Lanbo Di [1,*], Feng Yu [2], Dongzhi Duan [1] and Xiuling Zhang [1,*]**

[1]   College of Physical Science and Technology, Dalian University, Dalian 116622, China; zhangjingsen0708@163.com (J.Z.); duandongzhi0529@163.com (D.D.)

[2]   School of Chemistry and Chemical Engineering, Shihezi University, Shihezi 832003, China; yufeng05@mail.ipc.ac.cn

*   Correspondence: dilanbo@163.com (L.D.); xiulz@sina.com (X.Z.); Tel.: +86-411-8740-2712 (L.D. & X.Z.)

Received: 6 August 2018; Accepted: 17 September 2018; Published: 19 September 2018

**Abstract:** Commercial $TiO_2$ (P25) supported gold (Au/P25) attracts increasing attention. In this work, atmospheric-pressure (AP) cold plasma was employed to activate the Au/P25-As catalyst prepared by a modified impregnation method. The influence of cold plasma working gas (oxygen, argon, hydrogen, and air) on the structure and performance of the obtained Au/P25 catalysts was investigated. X-ray diffraction (XRD), UV-Vis diffuse reflectance spectroscopy (DRS), transmission electron microscopy (TEM), and X-ray spectroscopy (XPS) were adopted to characterize the Au/P25 catalysts. CO oxidation was used as model reaction probe to test the Au/P25 catalyst. XRD results reveal that supporting gold and AP cold plasma activation have little effect on the P25 support. CO oxidation activity over the Au/P25 catalysts follows the order:  $Au/P25-O_2P > Au/P25-As > Au/P25-ArP \approx Au/P25-H_2P > Au/P25-AirP$. Au/P25-AirP presents the poorest CO oxidation catalytic activity among the Au/P25 catalysts, which may be ascribed to the larger size of gold nanoparticles, low concentration of active $[O]_s$, as well as the poisoning $[NO_x]_s$. The poor catalytic performance of Au/P25-ArP and $Au/P25-H_2P$ is ascribed to the lower concentration of $[O]_s$ species. 100% CO conversion temperatures for $Au/P25-O_2P$ is 40 °C, which is 30 °C lower than that over the as-prepared Au/P25-As catalyst. The excellent CO oxidation activity over $Au/P25-O_2P$ is mainly attributed to the efficient decomposition of gold precursor species, small size of gold nanoparticles, and the high concentration of $[O]_s$ species.

**Keywords:** Au/P25; CO oxidation; atmospheric-pressure cold plasma; working gas

---

## 1. Introduction

Supported gold catalysts have attracted increasing research interest after the pioneering work of Haruta [1,2]. Catalytic oxidation of carbon monoxide (CO) over supported gold catalysts has been widely used in indoor air purification, polymer electrolyte membrane fuel cells, low-temperature CO sensors and gas mark [3–6]. Catalytic performance of the supported gold catalyst is closely related with the size of gold nanoparticles, and the support and preparation method. It is generally accepted that the optimal size of gold nanoparticles is in the range of 2–4 nm [7], and $TiO_2$ is the most popular support due to its nontoxicity, low cost, high chemical stability, as well as the strong metal–support interaction formed in CO oxidation [8]. The most adopted method for synthesizing supported gold catalysts is the deposition–precipitation (DP) method. Generally adopted gold precursor ($HAuCl_4$) hydrolyzes in solution to form $Au(OH)_xCl_{4-x}{}^-$. Selection of the support is confined to the isoelectric point [6] to make the positive charged support to absorb the $Au(OH)_xCl_{4-x}{}^-$ species at a proper rate. Consequently, a simple modified impregnation method is developed and employed to synthesize supported gold catalysts [9]. Both the deposition–precipitation method and modified impregnation method need a post thermal activation process, which may result in the side effect of irreversible gold particles aggregation [10,11]. Therefore, it is urgently necessary to develop an alternative low temperature technique to activate the gold catalysts.

Cold plasma can be operated close to room temperature with high energy electrons, and has been proven to be highly suitable to synthesize supported metal catalysts [12–14]. The fast and low-temperature preparation process, and the Coulomb interaction between the charged species and metal precursors ions in cold plasma are conductive to fabrication of metal nanoparticles of small sizes with high dispersion [12]. In addition, the non-thermal equilibrium property of cold plasma is beneficial to enhanced metal–support interaction, amorphous metal nanoparticles [15,16], as well as metal alloy with specific structure [17,18], and metal nanoparticles with specific crystal facet [14]. Thanks to these properties, cold plasma has been successfully employed to prepare and activate gold catalysts, and it has been efficient to enhance the catalytic performance of the gold catalysts [19–25]. Deng et al. [26] used AP oxygen cold plasma to activate Au/P25 catalysts, and found the prepared catalysts exhibited enhanced activity for visible-light photocatalytic oxidation of CO. In previous work, we adopted AP hydrogen and oxygen cold plasma to synthesize $Au/TiO_2$ catalysts, and obtained high performance gold catalysts [27,28]. The effect of discharge time and discharge voltage on the structure and property of the $Au/TiO_2$ catalysts are also investigated and discussed [29]. The results indicate that the small size of gold nanoparticles and the high concentration of active surface oxygen species are the main reasons for the high performance. In spite of this, no system work has been carried out to investigate the influence of the cold plasma working atmosphere on the structure and performance of the supported gold catalysts.

In this work, AP cold plasma is adopted to activate the Au/P25-As catalyst prepared by a simple modified impregnation method. The influence of cold plasma working gas (oxygen, argon, hydrogen, and air) on the structure and performance of the obtained Au/P25catalysts was investigated, and the influence mechanism is discussed.

## 2. Materials and Methods

### 2.1. Catalysts Preparation

Chloroauric acid ($HAuCl_4 \cdot 4H_2O$, ≥99%) purchased from Tianjin Kemiou Chemical Reagent Co. Ltd. was used as gold precursor. Commercial Degussa P25 $TiO_2$ obtained from Germany Degussa Corporation was used as support. Aqueous ammonia solution ($NH_3 \cdot H_2O$, 25%) was acquired from Liaoning Xinxing Chemical Reagent Co. Ltd. All the chemicals were used as received without any further purification.

Au/P25 catalysts with 1 wt% theoretical gold loading were synthesized according to a modified impregnation method reported in a previous study [9]. First, 2.9 mL deionized water and 1.1 mL $HAuCl_4 \cdot 4H_2O$ solution ($C_{Au}$ = 0.01912 g mL$^{-1}$) were sequentially added into a 5 mL measuring cylinder under continuous stirring. Then, the mixed solution was transferred into a 50 mL beaker with 2 g P25 $TiO_2$. The mixture was stirred for a few minutes until it got light yellow and was subsequently aged at room temperature for 12 h. The colloid was rinsed three times with 30 mL aqueous ammonia solution (pH = 11) and three times with 30 mL deionized water in sequence. The product was collected by centrifuge at 10500 rpm for 5 min after each rinse. Finally, Au/P25 catalyst was obtained by drying the product at room temperature for 24 h in a vacuum oven, and was designated as Au/P25-As. The Au content in Au/P25-As is determined to be ca. 0.90 wt% by an Optima 2000DV ICP-AES (Perkin-Elmer, Boston, MA, USA).

Atmospheric-pressure (AP) dielectric barrier discharge (DBD) cold plasma with various working gases was adopted to activate the Au/P25-As catalyst, and the effect of working gas was investigated. Typically, 0.12 g Au/P25-As was activated by AP cold plasma of oxygen, hydrogen, argon, and air at the discharge voltage of 29 kV for one minute, and the obtained catalysts were denoted as Au/P25-$O_2$P, Au/P25-$H_2$P, Au/P25-ArP and Au/P25-AirP, respectively. Flow rate of all the working gases were kept at 100 mL·min$^{-1}$. A schematic diagram of the AP DBD cold plasma device is shown in Figure 1.

The quartz reactor located between high-voltage electrode and ground electrode consists of an upper quartz plate of 90 mm in diameter and 1 mm in thickness and a quartz circular groove of 4.5 mm in inner depth and 70 mm in inner diameter. Both of the electrodes are made of stainless steel. The discharge frequency and discharge voltage were observed by an oscilloscope (DPO2014, Tektronix, Beaverton, OR, USA) with a 1000:1 high voltage probe (Tektronix, P6015A, Beaverton, OR, USA).

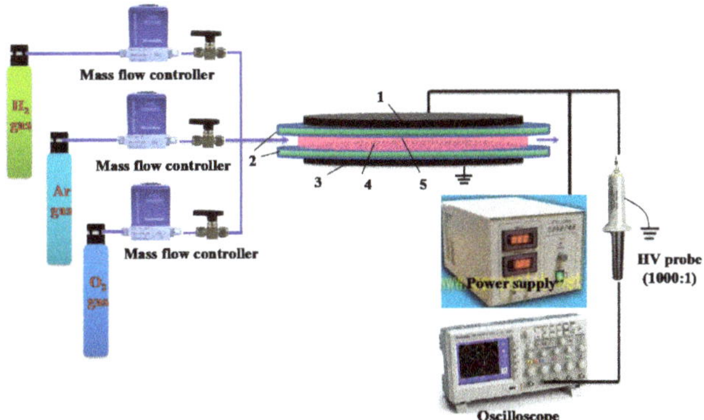


**Figure 1.** Schematic of the atmospheric-pressure (AP) dielectric barrier discharge (DBD) cold plasma device for activating Au/P25 catalysts. 1-discharge electrode, 2-quartz reactor, 3-ground electrode, 4-sample, 5-cold plasma.

### 2.2. Catalysts Characterization

The crystal structures of the synthesized Au/P25 catalysts were analyzed by X-ray diffraction (XRD) on a DX-2700 X-ray power diffractometer (Dandong Haoyuan, Dandong, China) with Cu K$\alpha$ radiation ($\lambda$ = 0.154 nm) at 40 kV and 30 mA. Ultraviolet-Visible (UV-Vis) diffuse reflectance spectroscopy (DRS) was adopted to measure the absorption property of the samples using a U3900 spectrophotometer (Hitachi, Tokyo, Japan). Before testing, the baseline was calibrated using two pieces of BaSO$_4$ white plates at the mode of R%. Transmission electron microscopy (TEM) images of the samples were collected on a HT7700 transmission electron microscope (Hitachi, Tokyo, Japan) with an accelerating voltage of 120 kV. The mean sizes of gold nanoparticles and corresponding size distribution were calculated by selecting more than 120 gold nanoparticles from TEM images. Surface chemical analyses of the samples were performed by X-ray photoelectron spectroscopy (XPS) using an ESCALAN250 X-ray photoelectron spectrometer (Thermo VG, Waltham, MA, USA) equipped with a monochromatic Al K$\alpha$ X-ray source (1486.6 eV photon energy, 150 W). The binding energy of each element in Au/P25 catalysts was calibrated by comparing the standard XPS peak of C1s at 284.6 eV.

### 2.3. Catalytic Activity Evaluation

Catalytic activity of the Au/P25 catalyst was evaluated by CO oxidation in a temperature programmed quartz tube controlled by an electric furnace in the range 30–150 °C. 50 mg Au/P25 catalyst (40–60 mesh) was filled in the middle of a quartz tube with an inner diameter of 4 mm. The catalyst was purged with argon for 15 min prior to reaction. During CO oxidation reaction, the synthetic gas containing 1 vol.% CO, 20 vol.% O$_2$ and balance N$_2$ was fed into the quartz tube at a flow rate of 20 mL·min$^{-1}$. CO concentration was dynamically monitored by a S710 CO$_x$ analyzer (SICK-MAIHAK, Waldkirch, Germany). CO conversion ($X_{CO}$) is defined using the following equation:

$$X_{CO} = \frac{C_{CO}^{in} - C_{CO}^{out}}{C_{CO}^{in}} \times 100\% \tag{1}$$

where $C_{CO}^{in}$ and $C_{CO}^{out}$ represent the volume concentrations of CO before and after reaction at a certain temperature, respectively.

## 3. Results and Discussion

Figure 2 presents the XRD patterns of Au/P25 catalysts as prepared and activated by AP cold plasma with different working gases, as well as P25 support. All the diffraction peaks in the samples can be well indexed as anatase $TiO_2$ (JCPDS no.21-1272) and rutile $TiO_2$ (JCPDS no. 21-1276). The diffraction peaks at 39.15° and 44.05° can be detected for all the samples including the pure P25 $TiO_2$ support, which corresponds to the anatase $TiO_2$ (200) and (210) planes (JCPDS no.21-1272). They are very close to the diffraction peaks of Au (111) and (220) planes in the positions of 38.18° and 44.39° (JCPDS no. 04-0784), respectively. Compared to the pure P25 $TiO_2$ support, the intensity of these diffraction peaks are decreased for the Au/P25 samples ascribing to the supporting of gold species. Therefore, these peaks should be attributed to anatase $TiO_2$ rather than metallic gold. In addition, the nominal loading amount of gold is 1 wt%. The gold species are not detected in the XRD patterns which also indicates that small size of gold species with high dispersion are synthesized, which is consistent with the TEM analysis ($D_{Au}$ = 3–4 nm) (Figure 3). Because cold plasma is generated by high-voltage discharge, many researchers are afraid that it may change or destroy the treated materials. Therefore, the influence of AP cold plasma on the structure of P25 $TiO_2$ support is discussed based on the XRD data. Positions of the strongest characteristic diffraction peaks of the samples, anatase (101) and rutile (110), were summarized in Table 1. There is no obvious difference among them. To further investigate the influence of supporting gold and AP cold plasma activation on the structure of the P25, the weight fraction and average crystallite size of anatase $TiO_2$ and rutile $TiO_2$ were also determined, as summarized in Table 1. The weight fraction of rutile $TiO_2$ ($W_{rutile}$) was obtained according to the following formula [30]:

$$W_{rutile} = \frac{I_{rutile}}{0.884 I_{anatase} + I_{rutile}} \tag{2}$$

where $I_{anatase}$ and $I_{rutile}$ represent the diffraction intensity of anatase (101) and rutile (110), respectively. The average weight fraction of rutile is 18.2% according to the data listed in Table 1. The crystallite size of anatase $TiO_2$ ($D_{anatase}$) and rutile $TiO_2$ ($D_{anatase}$) were obtained according to the Scherrer equation using the characteristic data of anatase (101) and rutile (110). The average crystallite sizes of anatase and rutile $TiO_2$ are 21.3 nm and 29.2 nm, respectively. These results indicate that supporting gold and AP cold plasma activation have little effect on the structure of the P25 support.

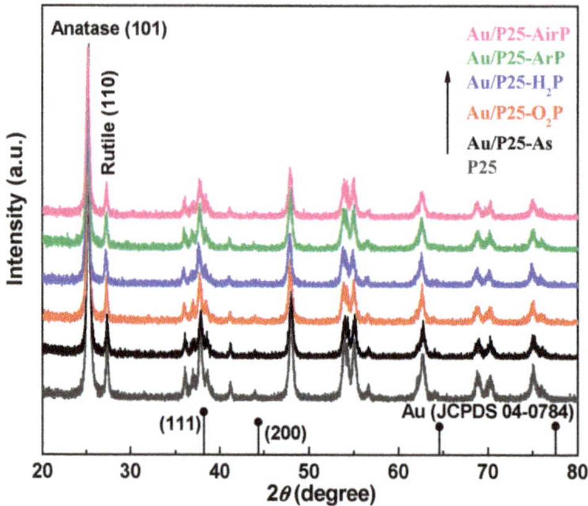


**Figure 2.** X-ray diffraction (XRD) patterns of the Au/P25 catalysts as prepared and activated by AP cold plasma using various working gases, as well as P25 support.

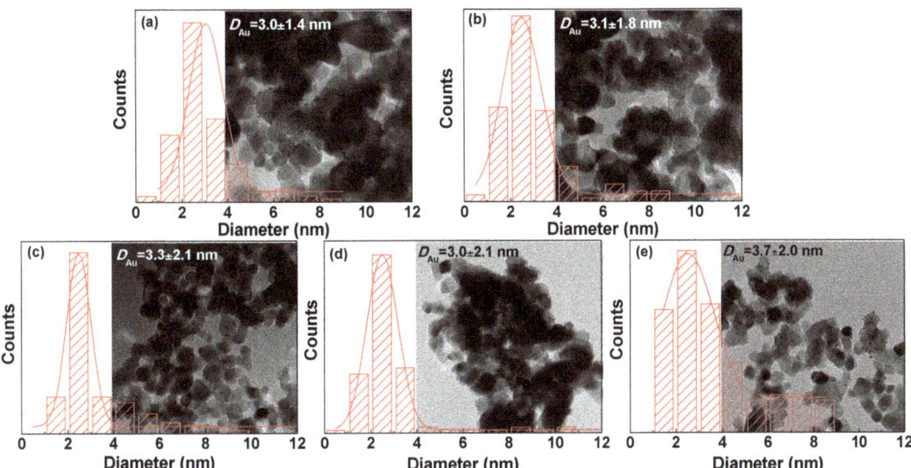

**Figure 3.** Typical transmission electron microscopy (TEM) images of (**a**) Au/P25-As; (**b**) Au/P25-O$_2$P; (**c**) Au/P25-H$_2$P; (**d**) Au/P25-ArP; (**e**) Au/P25-AirP, and the corresponding size distribution histograms of gold nanoparticles.

**Table 1.** Structure properties of the Au/P25 catalysts as prepared and activated by AP cold plasma, as well as P25 support.

| Samples | 2$\theta$ (degree) | | $W_{rutile}$ (%) | $D_{anatase}$ (nm) | $D_{rutile}$ (nm) |
|---|---|---|---|---|---|
| | Anatase (101) | Rutile (110) | | | |
| P25 | 25.3 | 27.5 | 17.9 | 21.1 | 31.0 |
| Au/P25-As | 25.2 | 27.4 | 17.3 | 22.3 | 28.4 |
| Au/P25-O$_2$P | 25.3 | 27.4 | 19.1 | 21.0 | 27.9 |
| Au/P25-H$_2$P | 25.2 | 27.4 | 20.1 | 20.1 | 29.1 |
| Au/P25-ArP | 25.3 | 27.4 | 17.2 | 21.5 | 28.0 |
| Au/P25-AirP | 25.2 | 27.4 | 17.4 | 21.6 | 31.0 |

To investigate the optical properties of the samples, UV-Vis DRS spectra of the Au/P25 catalysts as prepared and activated by AP cold plasma with different working gases, as well as P25 support were measured, as shown in Figure 4. For all the samples, the absorption bands at shorter than 400 nm were ascribed to the P25 support [31], while the absorption bands in the visible region were attributed to gold species [32,33]. It is well known that metallic gold nanoparticles irradiated by visible light can lead to Localized Surface Plasmon Resonance (LSPR). A weak LSPR absorption peak for the as-prepared Au/P25-As is also observed, which is consistent with the XPS results (Table 2). The formation of the metallic gold species can be ascribed to the dissociation of the gold precursors due to their photosensitive property. Obviously, compared to Au/P25-As, the absorption for the Au/P25 catalysts activated by AP cold plasma was dramatically enhanced in visible region, and LSPR absorption bands at ca 560 nm were observed due to the high content of metallic gold (Table 2). It confirms that cationic gold species can be reduced into their metallic state by AP cold plasma using various working gases. However, different LSPR absorption signals for the Au/P25 samples are observed, and the intensity of the LSPR peak follows the order: Au/P25-H$_2$P > Au/P25-ArP > Au/P25-AirP > Au/P25-O$_2$P, which is consistent with the proportion of metallic gold (Table 2) according to the data taken from the result of XPS (Figure 5). The active ground and excited hydrogen atoms generated in cold plasma can not only reduce metal ions with positive standard potential, but also some with negative values [12]. Therefore, the Au/P25-H$_2$P demonstrates the most intense LSPR signal. The weakest LSPR absorption peak was observed for Au/P25-O$_2$P, which may be ascribed to the strong quenching effect of electronegative oxygen gas on the energetic electrons. It can be verified by the intensity order of the LSPR absorption peak: Au/P25-ArP > Au/P25-AirP > Au/P25-$O_2$P.

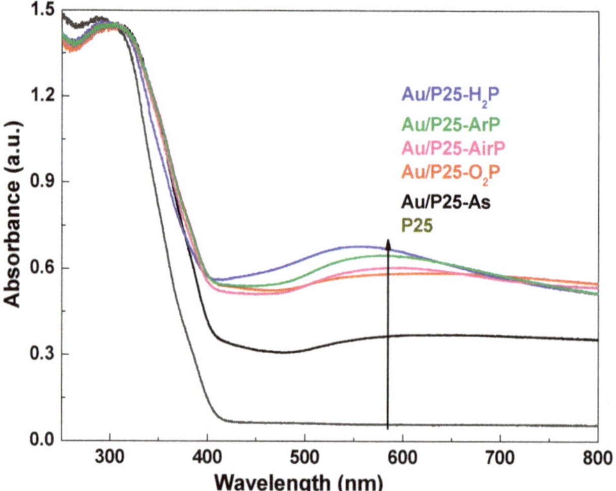


**Figure 4.** UV-Vis diffuse reflectance spectroscopy (DRS) spectra of the Au/P25 catalysts as prepared and activated by AP cold plasma using various working gases, as well as P25 support.

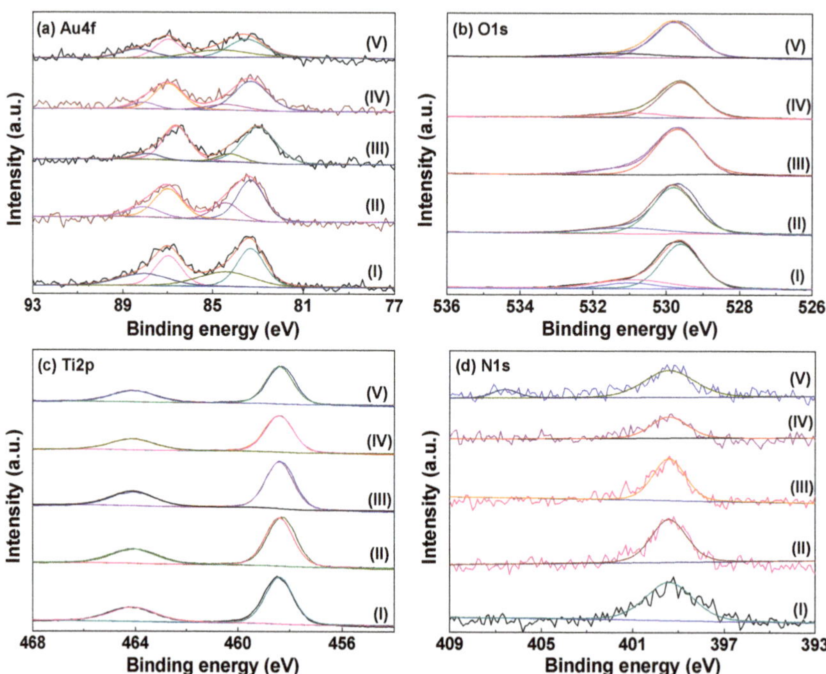


**Figure 5.** X-ray spectroscopy (XPS) spectra of (**a**) Au4f; (**b**) O1s; (**c**) Ti2p; and (**d**) N1s in (I) Au/P25-As; (II) Au/P25-$O_2$P; (III) Au/P25-$H_2$P; (IV) Au/P25-ArP; and (V) Au/P25-AirP.

**Table 2.** Gold nanoparticles diameter and XPS data of the Au/P25 catalysts.

| Samples | $D_{Au}$ [a] (nm) | Binding Energy (eV) | | Proportion (at%) | | Au/Ti Atomic Ratios |
|---|---|---|---|---|---|---|
| | | $Au^0 4f_{7/2}$ | $Au^+ 4f_{7/2}$ | $Au^0/(Au^0+Au^+)$ | $[O]_s/([O]_s+[O]_l)$ | |
| Au/P25-As | $3.0 \pm 1.4$ | 83.3 | 84.4 | 56.9 | 27.4 | 0.023 |
| Au/P25-O$_2$P | $3.1 \pm 1.8$ | 83.3 | 84.4 | 71.4 | 18.9 | 0.019 |
| Au/P25-H$_2$P | $3.3 \pm 2.1$ | 82.9 | 84.1 | 86.2 | 15.4 | 0.018 |
| Au/P25-ArP | $3.0 \pm 2.1$ | 83.3 | 84.4 | 81.9 | 17.3 | 0.020 |
| Au/P25-AirP | $3.7 \pm 2.0$ | 83.3 | 84.5 | 78.3 | 16.5 | 0.018 |

[a] The average size of gold nanoparticles was obtained according to the TEM results.

TEM measurements were carried out to acquire average size and size distribution of gold nanoparticles, which are crucial during the process of gold catalysis [10,34]. Typical TEM images of Au/P25-As, Au/P25-O$_2$P, Au/P25-H$_2$P, Au/P25-ArP, Au/P25-AirP, and the corresponding size distribution histograms of gold nanoparticles are illustrated in Figure 3. Gold nanoparticles are finely dispersed on the P25 support in the Au/P25 samples. It has to be specified that the gold species are the mixture of oxidized and metallic gold species according to the XPS results (Table 2). Either metallic gold or oxidized gold can be distinguished from the P25 TiO$_2$ support in the TEM images due to their contrast ratio. The average gold diameters for Au/P25-As, Au/P25-O$_2$P, Au/P25-H$_2$P, Au/P25-ArP, and Au/P25-AirP are $3.0 \pm 1.4$, $3.1 \pm 1.8$, $3.3 \pm 2.1$, $3.0 \pm 2.1$, and $3.7 \pm 2.0$ nm, respectively (as summarized in Table 2). It was obvious that there is little change in average diameter of the gold nanoparticles for Au/P25-O$_2$P, Au/P25-H$_2$P, Au/P25-ArP after AP cold plasma treatment. Meantime, the size distribution of gold nanoparticles became a little broader than that of Au/P25-As. These indicate that AP cold plasma did not significantly alter the size and size distribution of gold nanoparticles. However, larger particle size of gold nanoparticles is obtained for Au/P25-AirP after cold plasma treatment. Taking the weak influence of oxygen cold plasma on the size of gold nanoparticles into consideration, larger sizes of gold nanoparticles in Au/P25-AirP may result from the poisoning species [NO$_y$]$_s$ during nitrogen and oxygen discharge in air cold plasma [24]. This can be confirmed by the XPS spectrum of N1s for Au/P25-AirP (Figure 5d).

The chemical state of gold species, surface oxygen, and other species play important roles in the activity for supported gold catalysts [23,35]. To further investigate the influence of AP cold plasma activation, XPS spectra of Au/P25-As, Au/P25-O$_2$P, Au/P25-H$_2$P, Au/P25-ArP, and Au/P25-AirP are recorded, as shown in Figure 5. No Cl ions can be detected from the XPS spectra of Cl1s in the Au/P25 samples (not shown here) due to the rinsing of the aqueous ammonia solution and deionized water. In Figure 5a, the XPS spectra of Au4f in these samples can be fitted with two peaks corresponding to metallic Au$^0$ and Au$^+$ [26], revealing that AP cold plasma can reduce the gold precursor species into metallic state gold. The proportion of metallic Au$^0$ and the binding energies of Au4f$_{7/2}$ for these samples are summarized in Table 2. The proportion of the metallic Au$^0$ in the samples follows the order: Au/P25-H$_2$P > Au/P25-ArP > Au/P25-AirP > Au/P25-O$_2$P, which is in line with the intensity sequence of the LSPR peak in UV-Vis DRS spectra (Figure 4). Interestingly, 0.4 eV redshift in the binding energy of Au4f$_{7/2}$ for Au/P25-H$_2$P is observed, which can be explained by the following reasons. One the one hand, AP hydrogen cold plasma for synthesizing supported metal catalysts generally may lead to redshift of the binding energy due to the enhanced strong metal–support interaction [18,36]. On the other hand, Au/P25-H$_2$P has the highest proportion of metallic Au$^0$, which may also enhance the redshift of the binding energy [36].

Surface oxygen species are beneficial to the formation of active intermediates during CO oxidation over supported gold catalysts [6,37]. In Figure 5b, O1s spectra for the Au/P25 samples are illustrated, which can be deconvoluted into two peaks at 529.6 and 530.9 eV, ascribed to crystal lattice oxygen [O]$_l$ and surface oxygen species [O]$_s$, respectively [38]. Based on individual peak area, [O]$_s$ concentration in the oxygen species were calculated and listed in Table 2. The order of [O]$_s$ concentration for all the Au/P25 catalysts was Au/P25-As > Au/P25-O$_2$P > Au/P25-ArP > Au/P25-AirP > Au/P25-H$_2$P, indicating that cold plasma activation can lead to the decline of $[O]_s$ concentration and working gas play important roles in $[O]_s$ concentration. The significant decrease in $[O]_s$ concentration for Au/P25-H$_2$P may result from the consumption of $[O]_s$ by the hydrogen species [27]. Combined with the results of the proportions of metallic Au$^0$ and $[O]_s$ (Table 2), it can be concluded that AP cold plasma can not only decompose gold precursor into metallic Au$^0$ but also form active $[O]_s$ on the P25 surface.

Figure 5c presents the Ti2p XPS spectra of the Au/P25 samples. The peaks at 458.4 and 464.1 eV are attributed to Ti$^{4+}$ of P25 support for the Au/P25 catalysts, confirming that the chemical environment of support didn't vary after cold plasma activation [23,39]. In Figure 5d, N1s spectra for the Au/P25 catalysts are also investigated. The peaks at 399.4 eV for all the Au/P25 catalysts were ascribed to chemisorbed γ-N$_2$ [40]. Interestingly, a new and weak peak at 406.6 eV appeared for Au/P25-AirP, which can be attributed to $[NO_x]_s$ due to the air plasma treatment [24,25]. The atomic ratios of Au/Ti for the Au/P25 samples are determined according to the XPS results and summarized in Table 2. The Au/Ti atomic ratio for Au/P25-As is 0.023. However, they are decreased after AP cold plasma activation due to the dissociation of the gold precursor species. In spite of this, there is no obvious difference for the Au/P25 samples activated by AP cold plasma using different working gases.

Figure 6 presents CO conversion versus reaction temperature over the Au/P25 catalysts prepared by AP cold plasma activation using various working gases, as well as the as-prepared Au/P25-As. All of the Au/P25 catalysts exhibit high CO oxidation activity. CO oxidation activity over the Au/P25 catalysts follows the order: Au/P25-O$_2$P > Au/P25-As > Au/P25-ArP ≈ Au/P25-H$_2$P > Au/P25-AirP. Catalytic performance of the Au/P25 catalysts is closely related with the working atmosphere of AP cold plasma. Au/P25-ArP, Au/P25-H$_2$P, and Au/P25-AirP obtained by AP argon, hydrogen, and air cold plasma activation show poorer CO oxidation activity than the as-prepared Au/P25-As catalyst, and Au/P25-AirP presents the poorest catalytic activity. Interestingly, AP oxygen cold plasma activation can significantly enhance the catalytic performance of the Au/P25 catalyst. 100% CO conversion temperatures for Au/P25-O$_2$P is 40 °C, which is 30 °C lower than that over the as-prepared Au/P25-As catalyst.

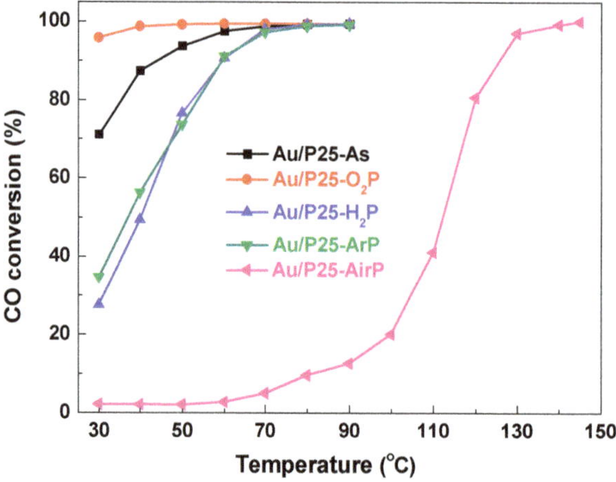


**Figure 6.** CO conversion over the Au/P25 catalysts activated by AP cold plasma using various working gases, as well as the as-prepared Au/P25-As.

AP cold plasma activation with oxygen, argon, and hydrogen as working gas has no significance influence on the size of gold nanoparticles ($D_{Au}$ = 3.0–3.3 nm). AP cold plasma activation of the Au/P25-As catalysts may not only lead to the decomposition of the gold precursor species and formation of active $[O]_s$ species, but also can result in the removal of active $[O]_s$ species after longer AP cold plasma treatment time [29]. The decomposed gold species containing oxidized and metallic gold species can be transformed into active metallic gold species [26]. As a consequence, $[O]_s$ species play important roles in CO oxidation. The highest $[O]_s$ is observed for Au/P25-As (Table 2). Therefore, it seems that it should exhibit the highest CO oxidation activity. In previous work [26], it has been proved that some $[O]_s$ in the as-prepared Au/P25-As sample will be rapidly consumed during the reaction, and lower CO oxidation activity will be obtained. In addition, the high Au/Ti atomic ratio in Au/P25-As is also beneficial to the high CO oxidation activity. For Au/P25-$O_2$P prepared by AP oxygen cold plasma activation, the gold precursor species can be well decomposed, and high concentration of $[O]_s$ species are obtained. Therefore, Au/P25-$O_2$P presents the highest catalytic activity among the Au/P25 catalysts. While, Au/P25-ArP and Au/P25-$H_2$P with less concentration of $[O]_s$ species show poorer CO oxidation performance. Au/P25-AirP presents the poorest CO oxidation catalytic activity among the Au/P25 catalysts, which may be ascribed to the larger size of gold nanoparticles ($D_{Au}$ = 3.7 nm), low concentration of active $[O]_s$, as well as the poisoning $[NO_x]_s$ [24,25].

## 4. Conclusions

AP cold plasma was adopted to activate the Au/P25-As catalyst prepared by a modified impregnation method, and the influence of working gas on the structure and performance of the obtained Au/P25 catalysts was investigated. XRD analyses confirm that supporting gold and AP cold plasma activation have little effect on the P25 support. All of the Au/P25 catalysts exhibit high CO catalytic oxidation activity. Catalytic performance of the Au/P25 catalysts is closely related with the working atmosphere of AP cold plasma. CO oxidation activity over the Au/P25 catalysts follows the order: Au/P25-$O_2$P > Au/P25-As > Au/P25-ArP $\approx$ Au/P25-$H_2$P > Au/P25-AirP. The poor catalytic performance of Au/P25-ArP and Au/P25-$H_2$P is ascribed to the lower concentration of $[O]_s$ species. Au/P25-AirP presents the poorest CO oxidation catalytic activity among the Au/P25 catalysts, which may be ascribed to the larger size of gold nanoparticles, low concentration of active $[O]_s$, as well as the poisoning $[NO_x]_s$. 100% CO conversion temperatures for Au/P25-$O_2$P is 40 °C, which is 30 °C lower than that over the as-prepared Au/P25-As catalyst. The excellent CO oxidation activity over Au/P25-$O_2$P is mainly attributed to the efficient decomposition of gold precursor species, small size of gold nanoparticles, and the high concentration of $[O]_s$ species. AP oxygen cold plasma activation is found to be an efficient method for synthesizing high performance Au/P25 catalysts.

**Author Contributions:** Conceptualization, L.D. and F.Y.; Data curation, J.Z. and D.D.; Investigation, J.Z. and D.D.; Project administration, L.D.; Supervision, L.D. and X.Z.; Writing—original draft, L.D.; Writing—review & editing, X.Z.

**Acknowledgments:** This research was funded by National Natural Science Foundation of China (Grant No. 21773020, 11505019, 21673026), Liaoning Innovative Talents in University (Grant No. LR2017025), Liaoning Natural Science Foundation (Grant No. 20180550085).

**Conflicts of Interest:** The authors declare no conflict of interest.

## References

1. Haruta, M.; Kobayashi, T.; Sano, H.; Yamada, N. Novel gold catalysts for the oxidation of carbon monoxide at a temperature far below 0 °C. *Chem. Lett.* **1987**, *16*, 405–408. [CrossRef]
2. Haruta, M.; Yamada, N.; Kobayashi, T.; Iijima, S. Gold catalysts prepared by coprecipitation for low-temperature oxidation of hydrogen and of carbon monoxide. *J. Catal.* **1989**, *115*, 301–309. [CrossRef]
3. Christensen, C.H.; Nørskov, J.K. Green gold catalysis. *Science* **2010**, *327*, 278–279. [CrossRef] [PubMed]
4. Saavedra, J.; Doan, H.A.; Pursell, C.J.; Grabow, L.C.; Chandler, B.D. The critical role of water at the gold-titania interface in catalytic CO oxidation. *Science* **2014**, *345*, 1599–1602. [CrossRef] [PubMed]
5. Wang, Y.; Widmann, D.; Behm, R.J. Influence of $TiO_2$ bulk defects on CO adsorption and CO oxidation on Au/$TiO_2$: Electronic metal-support interactions (EMSIs) in supported Au catalysts. *ACS Catal.* **2017**, *7*, 2339–2345. [CrossRef]
6. Kung, H.H.; Kung, M.C.; Costello, C.K. Supported Au catalysts for low temperature CO oxidation. *J. Catal.* **2003**, *216*, 425–432. [CrossRef]

7.   Valden, M.; Lai, X.; Goodman, D.W. Onset of catalytic activity of gold clusters on Titania with the appearance of nonmetallic properties. *Science* **1998**, *281*, 1647–1650. [CrossRef] [PubMed]

8.   Widmann, D.; Liu, Y.; Schüth, F.; Behm, R.J. Support effects in the Au-catalyzed CO oxidation-correlation between activity, oxygen storage capacity, and support reducibility. *J. Catal.* **2010**, *276*, 292–305. [CrossRef]

9.   Delannoy, L.; El, H.N.; Musi, A.; To, N.N.L.; Krafft, J.M.; Louis, C. Preparation of supported gold nanoparticles by a modified incipient wetness impregnation method. *J. Phys. Chem. B* **2006**, *110*, 22471–22478. [CrossRef] [PubMed]

10.  Akita, T.; Lu, P.; Ichikawa, S.; Tanaka, K.; Haruta, M. Analytical TEM study on the dispersion of Au nanoparticles in Au/P25 catalyst prepared under various temperatures. *Surf. Interface Anal.* **2001**, *31*, 73–78. [CrossRef]

11.  Overbury, S.H.; Schwartz, V.; Mullins, D.R.; Yan, W.; Dai, S. Evaluation of the Au size effect: CO oxidation catalyzed by Au/P25. *J. Catal.* **2006**, *241*, 56–65. [CrossRef]

12.  Di, L.B.; Zhang, J.S.; Zhang, X.L. A review on the recent progress, challenges and perspectives of atmospheric-pressure cold plasma for preparation of supported metal catalysts. *Plasma Process. Polym.* **2018**, *15*, 1700234. [CrossRef]

13.  Liu, C.J.; Zhao, Y.; Li, Y.; Zhang, D.S.; Chang, Z.; Bu, X.H. Perspectives on electron-assisted reduction for preparation of highly dispersed noble metal catalysts. *ACS Sustain. Chem. Eng.* **2014**, *2*, 3–13. [CrossRef]

14.  Wang, Z.; Zhang, Y.; Neyts, E.C.; Cao, X.; Zhang, X.; Jang, B.W.L.; Liu, C.J. Catalyst Preparation with Plasmas: How Does It Work? *ACS Catal.* **2018**, *8*, 2093–2110. [CrossRef]

15.  Xu, Z.J.; Qi, B.; Di, L.B.; Zhang, X.L. Partially crystallized Pd nanoparticles decorated TiO$_2$ prepared by atmospheric-pressure cold plasma and its enhanced photocatalytic performance. *J. Energy Chem.* **2014**, *23*, 679–683. [CrossRef]

16.  Zou, J.J.; Zhang, Y.P.; Liu, C.J. Reduction of supported noble-metal ions using glow discharge plasma. *Langmuir* **2006**, *22*, 11388–11394. [CrossRef] [PubMed]

17.  Wang, W.; Wang, Z.; Wang, J.; Zhong, C.J.; Liu, C.J. Highly active and stable Pt-Pd alloy catalysts synthesized by room-temperature electron reduction for oxygen reduction reaction. *Adv. Sci.* **2017**, *4*, 1600486. [CrossRef] [PubMed]

18.  Di, L.B.; Duan, D.Z.; Park, D.W.; Ahn, W.S.; Lee, B.J.; Zhang, X.L. Cold plasma for synthesizing high performance bimetallic PdCu catalysts: Effect of reduction sequence and Pd/Cu atomic ratios. *Top. Catal.* **2017**, *60*, 925–933. [CrossRef]

19.  Furusho, H.; Kitano, K.; Hamaguchi, S.; Nagasaki, Y. Preparation of stable water-dispersible PEGylated gold nanoparticles assisted by nonequilibrium atmospheric-pressure plasma jets. *Chem. Mater.* **2009**, *21*, 3526–3535. [CrossRef]

20.  Sharma, R.; Rimmer, R.D.; Gunamgari, J.; Shekhawat, R.S.; Davis, B.J.; Mazumder, M.K.; Lindquist, D.A. Plasma-assisted activation of supported Au and Pd catalysts for CO oxidation. *IEEE Trans. Ind. Appl.* **2005**, *41*, 1373–1376. [CrossRef]

21.  Liu, X.; Mou, C.Y.; Lee, S.; Li, Y.; Secrest, J.; Jang, B.W.-L. Room temperature O$_2$ plasma treatment of SiO$_2$ supported Au catalysts for selective hydrogenation of acetylene in the presence of large excess of ethylene. *J. Catal.* **2012**, *285*, 152–159. [CrossRef]

22.  Wei, Z.; Liu, C. Synthesis of monodisperse gold nanoparticles in ionic liquid by applying room temperature plasma. *Mater. Lett.* **2011**, *65*, 353–355. [CrossRef]

23.  Zhang, S.; Li, X.S.; Zhu, B.; Liu, J.L.; Zhu, X.; Zhu, A.M.; Jang, B.W.-L. Atmospheric-pressure O$_2$ plasma treatment of Au/P25 catalysts for CO oxidation. *Catal. Today* **2015**, *256*, 142–147. [CrossRef]

24.  Fan, H.Y.; Shi, C.; Li, X.S.; Zhang, S.; Liu, J.L.; Zhu, A.M. In-situ plasma regeneration of deactivated Au/P25 nanocatalysts during CO oxidation and effect of N$_2$ content. *Appl. Catal. B Environ.* **2012**, *119*, 49–55. [CrossRef]

25.  Zhu, B.; Li, X.S.; Liu, J.L.; Liu, J.B.; Zhu, X.; Zhu, A.M. In-situ regeneration of Au nanocatalysts by atmospheric-pressure air plasma: Significant contribution of water vapor. *Appl. Catal. B Environ.* **2015**, *179*, 69–77. [CrossRef]

26.  Deng, X.Q.; Zhu, B.; Li, X.S.; Liu, J.L.; Zhu, X.; Zhu, A.M. Visible-light photocatalytic oxidation of CO over plasmonic Au/P25: Unusual features of oxygen plasma activation. *Appl. Catal. B Environ.* **2016**, *188*, 48–55. [CrossRef]

27. Di, L.B.; Zhan, Z.B.; Zhang, X.L.; Qi, B.; Xu, W.J. Atmospheric-Pressure DBD Cold Plasma for Preparation of High Active Au/P25 Catalysts for Low-Temperature CO Oxidation. *Plasma Sci. Technol.* **2016**, *18*, 544–548. [CrossRef]
28. Di, L.B.; Duan, D.Z.; Zhang, X.L.; Qi, B.; Zhan, Z.B. Effect of $TiO_2$ Crystal Phase and Preparation Method on the Catalytic Performance of Au/$TiO_2$ for CO Oxidation. *IEEE Trans. Plasma Sci.* **2016**, *44*, 2692–2698. [CrossRef]
29. Zhang, X.L.; Xu, W.W.; Duan, D.Z.; Park, D.-W.; Di, L.B. Atmospheric-pressure oxygen cold plasma for synthesizing Au/$TiO_2$ catalysts: Effect of discharge voltage and discharge time. *IEEE Trans. Plasma Sci.* **2018**, *46*, 2776–2781. [CrossRef]
30. Zhang, H.; Banfield, J.F. Understanding polymorphic phase transformation behavior during growth of nanocrystalline aggregates: Insights from $TiO_2$. *J. Phys. Chem. B* **2000**, *104*, 3481–3487. [CrossRef]
31. Di, L.B.; Xu, Z.J.; Zhang, X.L. Atmospheric-pressure cold plasma for synthesizing Ag modified Degussa P25 with visible light activity using dielectric barrier discharge. *Catal. Today* **2013**, *211*, 143–146. [CrossRef]
32. Zanella, R.; Giorgio, S.; Shin, C.H.; Henry, C.R.; Louisa, C. Characterization and reactivity in CO oxidation of gold nanoparticles supported on $TiO_2$ prepared by deposition-precipitation with NaOH and urea. *J. Catal.* **2004**, *222*, 357–367. [CrossRef]
33. Orendorff, C.J.; Sau, T.K.; Murphy, C.J. Shape-dependent plasmon-resonant gold nanoparticles. *Small* **2006**, *2*, 636–639. [CrossRef] [PubMed]
34. Haruta, M. Size-and support-dependency in the catalysis of gold. *Catal. Today* **1997**, *36*, 153–166. [CrossRef]
35. Delannoy, L.; Weiher, N.; Tsapatsaris, N.; Beesley, A.M.; Nchari, L.; Schroeder, S.L.M.; Louis, C. Reducibility of supported gold (III) precursors: Influence of the metal oxide support and consequences for CO oxidation activity. *Top. Catal.* **2007**, *44*, 263–273. [CrossRef]
36. Di, L.B.; Zhang, X.L.; Xu, Z.J.; Wang, K. Atmospheric-pressure cold plasma for preparation of high performance Pt/$TiO_2$ photocatalyst and its mechanism. *Plasma Chem. Plasma Process.* **2014**, *34*, 301–311. [CrossRef]
37. Daniells, S.T.; Overweg, A.R.; Makkee, M.; Moulijn, J.A. The mechanism of low-temperature CO oxidation with Au/$Fe_2O_3$ catalysts: A combined Mössbauer, FT-IR, and TAP reactor study. *J. Catal.* **2005**, *230*, 52–65. [CrossRef]
38. Liu, L.; Gu, X.; Cao, Y.; Yao, X.; Zhang, L.; Tang, C.; Gao, F.; Dong, L. Crystal-plane effects on the catalytic properties of Au/$TiO_2$. *ACS Catal.* **2013**, *3*, 2768–2775. [CrossRef]
39. Yang, Q.Y.; Zhu, Y.; Tian, L.; Pei, Y.; Qiao, M.H.; Fan, K.N. Influence of Au/$TiO_2$ Catalyst Preparation Parameters on the Selective Hydrogenation of Crotonaldehyde. *Acta Phys. Chim. Sin.* **2009**, *25*, 1853–1860. [CrossRef]
40. Di, L.B.; Shi, C.; Li, X.S.; Liu, J.L.; Zhu, A.M. Uniformity, structure, and photocatalytic activity of $TiO_2$ films deposited by atmospheric-pressure linear cold plasma. *Chem. Vap. Depos.* **2012**, *18*, 309–314. [CrossRef]

 *nanomaterials* 

*Article*

# Effective Oxygen Reduction Reaction Performance of FeCo Alloys In Situ Anchored on Nitrogen-Doped Carbon by the Microwave-Assistant Carbon Bath Method and Subsequent Plasma Etching

**Mincong Liu [1], Feng Yu [1,*], Cunhua Ma [1], Xueyan Xue [1], Haihai Fu [1], Huifang Yuan [1], Shengchao Yang [1], Gang Wang [1], Xuhong Guo [1,2,*] and Lili Zhang [3,*]**

[1]  Key Laboratory for Green Processing of Chemical Engineering of Xinjiang Bingtuan, School of Chemistry and Chemical Engineering, Shihezi University, Shihezi 832003, China
[2]  State Key Laboratory of Chemical Engineering, East China University of Science and Technology, Shanghai 200237, China
[3]  Institute of Chemical and Engineering Sciences, Agency for Science, Technology and Research, Jurong Island 627833, Singapore
*  Correspondence: yufeng05@mails.ucas.ac.cn (F.Y.); guoxuhong@ecust.edu.cn (X.G.); zhang_lili@ices.a-star.edu.sg (L.Z.); Tel.: +86-993-205-7272 (F.Y.)

Received: 17 August 2019; Accepted: 5 September 2019; Published: 8 September 2019

**Abstract:** Electrocatalysts with strong stability and high electrocatalytic activity have received increasing interest for oxygen reduction reactions (ORRs) in the cathodes of energy storage and conversion devices, such as fuel cells and metal-air batteries. However, there are still several bottleneck problems concerning stability, efficiency, and cost, which prevent the development of ORR catalysts. Herein, we prepared bimetal FeCo alloy nanoparticles wrapped in Nitrogen (N)-doped graphitic carbon, using Co-Fe Prussian blue analogs ($Co_3[Fe(CN)_6]_2$, Co-Fe PBA) by the microwave-assisted carbon bath method (MW-CBM) as a precursor, followed by dielectric barrier discharge (DBD) plasma treatment. This novel preparation strategy not only possessed a fast synthesis rate by MW-CBM, but also caused an increase in defect sites by DBD plasma treatment. It is believed that the co-existence of Fe/Co-N sites, rich active sites, core-shell structure, and FeCo alloys could jointly enhance the catalytic activity of ORRs. The obtained catalyst exhibited a positive half-wave potential of 0.88 V vs. reversible hydrogen electrode (RHE) and an onset potential of 0.95 V vs. RHE for ORRs. The catalyst showed a higher selectivity and long-term stability than Pt/C towards ORR in alkaline media.

**Keywords:** FeCo alloy; oxygen reduction reaction; microwave-assisted carbon bath method; plasma; defect sites

---

## 1. Introduction

Fuel cells and rechargeable zinc-air batteries are the most promising clean auto power for the next generation due to their low cost and high energy density [1–4]. It is already widely known that the oxygen reduction reaction (ORR) is a slow kinetic process in cathodic reactions [5,6]. Precious metal Pt-based electrocatalysts usually possess high catalytic activity for ORRs, while their application for ORRs is not satisfactory, because of the problems, such as high cost and poor long-term durability [7,8]. Therefore, it is essential to develop electrocatalysts with high activity, long life, and low cost to substitute for the conventional catalysts of ORRs [9–11].

In order to reduce the loading of Pt and Pt-M (M = Fe, Co, Ni, Cu etc.), alloy catalysts have been extensively discussed [12–14]. The second element alloyed with Pt can adjust the catalytic ability of the surface of the Pt-based catalyst and improve the catalytic performance [15]. For example, Zhang et al. [16]

synthesized bimetallic PtNi/C with hollow structures through a facile solution-based approach and observed distance lattice contraction of Pt due to the presence of Ni-improved ORR performance. It was found that PtNi/C showed excellent half-wave potential ($E_{1/2}$) of 0.88 V vs. reversible hydrogen electrode (RHE) and a stable 4-electron pathway. At the meantime, the morphology and size of Pt-M alloys also have an important influence on the electrocatalytic performance. Ma et al. [17] used a simple method to synthesize a hexagonal nanosheet PtFe alloy, a hexagonal nanosheet PtFe alloy with uniform distribution and ultra-small (ca. 2.6 nm) particle size, as a high efficiency electrocatalyst. The PtFe alloy showed the highest initial potential of 0.95 V vs. RHE. Up till now, non-Pt of the $M_1$-$M_2$ ($M_1$, $M_2$ = Fe, Co, Ni etc.) system has been reported, which is beneficial to improve the conductivity of catalysts and activate each other's active sites by the doping of $M_1$/$M_2$ [18,19]. Wen et al. [20] reported FeCo alloy nanoparticles embedded in Nitrogen (N)-doped carbon with excellent ORR performance, which was attributed to the core-shell nanostructure, the large specific surface area, and the synergetic effect of the mutual element for FeCo@NC. Wang et al. [21] prepared FeNi nanoparticles wrapped in N-doped carbon nanotubes (NCNTS,) in which NCNTS effectively prevented the oxidation and aggregation of FeNi nanoparticles, showing an initial potential of 0.95 V vs. RHE. All the previous research showed that the catalysts with bimetallic active sites exhibited optimal performance with large surface areas, porous nanostructures, and rich active centers.

Herein, we prepared FeCo alloy nanoparticles wrapped with N-doped carbon (FeCo@NC) using Co-Fe PBA as a precursor via the microwave-assisted carbon bath method (MW-CBM). Dielectric barrier discharge (DBD) plasma is then used to produce the catalyst with more defect sites (DBD-FeCo@NC). PBA have different bimetallic compositions, uniform sizes, morphology, and structure, which are considered to be ideal precursors for the synthesis of hollow and porous electrocatalysts [22]. In addition, MW-CBM has been successfully used in our previous work of $LiFePO_4$/C [23], $LFePO_4$/MEGO [24], Fe/C [25], and Ni/VMT [26]. The MW-CBD has rapid heating efficiency due to the high efficiency of microwave absorption of columnar carbon. due to its advantages of rapid heating efficiency and low side reactions [27]. The MW-CBD prevents the reaction between air and catalyst precursor due to the air reacting with the columnar carbon during the heating process. Moreover, the plasma-assisted preparation method is used in the synthesis and modification of electrocatalyst materials, such as making active sites or exfoliating catalysts [28,29]. The as-prepared DBD-FeCo@NC exhibited good electrochemical ORR performance, e.g., a positive half-wave potential of 0.88 V vs. RHE and an onset potential of 0.95 V vs. RHE. We believe that this strategy provides potential for the preparation of similar superstructures of other effective catalysts with much active sites.

## 2. Materials and Methods

### 2.1. Synthesis of Samples

Synthesis of $Co_3[Fe(CN)_6]_2$ (Co-Fe PBA): 2 mmol of $K_3[Fe(CN)_6]$ was dissolved in 100 mL deionized (DI) water and is labelled as solution A. A total of 3 mmol of $Co(NO)_2$ was dissolved in 100 mL DI water and is labeled as solution B. Solution B was slowly added to solution A under continuous magnetic stirring and the mixture was left for stirring for 3.5 h at room temperature. After aging for 24 h without stirring at room temperature, the precipitate was centrifuged for several times with DI water and absolute ethanol. The final product of Co-Fe PBA was obtained after drying at 80 °C for 8 h under vacuum.

Synthesis of FeCo@NC catalysts: 1.0 g of Co-Fe PBA was put into a small graphite crucible (d = 1 cm, h = 1.5 cm). The small graphite crucible containing Co-Fe PBA was then put inside a 150 mL crucible and the graphite crucible was embedded in commercial columnar carbon material. The 150 mL crucible with a small graphite crucible inside was then placed in a commercial microwave oven under microwave irradiation with 900 W for 10 min. The obtained product was further treated by 0.5 M $H_2SO_4$ with ultrasound for 1 h to remove impurities. The product was washed by centrifugation using DI water. The final product was obtained after drying at 80 °C in a vacuum and is denoted as FeCo@NC.

Synthesis of DBD-FeCo@NC catalysts: 50 mg of FeCo@NC was treated in Ar atmosphere under a DBD plasma reactor at an input power of 50 V × 1.5 A AC (alternating current) for 30 min to prepare DBD-FeCo@NC.

### 2.2. Characterizations

The field emission Tecnai G2 F20 electron (Hillsboro, OR, USA) microscope was used to analyze transmission electron microscopy (TEM). X-ray diffraction (XRD, D8 Advance, Bruker, Karlsruhe, Germany) with Cu-K radiation was used to characterize the structures of crystallographic phases for the products. The Raman spectra was analyzed by a Laser Confocal Micro-Raman Spectroscope (LabRAM HR800, Horiba Jobin Yvon, French) with a laser wavelength of 532 nm. The surface chemical compositions were tested by using an X-ray photoelectron spectroscope (XPS, ESCALAB 250Xi, Thermo Fisher Scientific, MA, USA).

### 2.3. Electrochemical Measurements

#### 2.3.1. ORR Text

The electrochemical performance was tested by using a CHI760D electrochemical station with a three-electrode cell system at room temperature in 0.1 M KOH (Potassium hydroxide) as an electrode solution. The reference electrode and the counter electrode was the Ag/AgCl electrode and a Pt wire, respectively. To prepare the catalyst ink for electrochemical analysis, 5 mg of the catalyst was dispersed into 0.5 mL of ethanol containing a Nafion solution (5 wt%, DuPont) with the aid of ultrasonication. A total of 10 μL of the catalyst ink was then coated on the glassy carbon disc electrode (3 mm in diameter) and dried at 60 °C. The catalyst loading was controlled at 0.0142 mg/cm$^2$. The catalyst on the glassy carbon rotating disk electrode was used as a working electrode, with a rotating rate varying from 625 to 2500 rpm at a scan rate of 10 mV/s. Ag/AgCl and platinum plate were used as the reference and counter electrodes, respectively. The cyclic voltammetry (CV) curves were obtained at a sweep speed of 50 mV/s in the potential range between −0.8 and 0.2 V after purging $O_2$ or $N_2$ for 20 min.

In a typical ORR program, the following equation can be used to estimate the number of electron transfers based on the slope of the Koutecky–Levich (K–L) graph:

$$\frac{1}{J} = \frac{1}{J_k} + \frac{1}{B\omega^{1/2}} \tag{1}$$

where $J$ represents the current density measured on a rotating disk electrode (RDE), $J_k$ is the kinetic current density, and $\omega$ acts as the electrode rotation speed. $B$ is derived from the following equation:

$$B = 0.2nFC_0D_0^{2/3}v^{-1/6} \tag{2}$$

among which $n$ represents the electron transfer number of each $O_2$ molecule in the ORR process, $F$ is 96,485 C/mol (Faraday constant), $C_0$ is $1.2 \times 10^{-3}$ mol/L (the dissolved $O_2$ concentration), $D_0$ is $1.9 \times 10^{-5}$ cm$^2$/s (the $O_2$ diffusion coefficient), and $v$ is 0.01 cm$^2$/s (the electrolyte kinematic viscosity).

#### 2.3.2. OER (Oxygen Evolution Reaction) Text

The electrochemical performances of the prepared products were measured in an $O_2$-saturated 1 M KOH at a 10 mV/s scanning rate with a three-electrode system using a CHI760D electrochemical station. The catalyst ink coated on Ni-foam was used as the working electrode. Pt foil and Ag/AgCl were used as the counter electrode and reference electrode, respectively. To prepare the catalyst ink, 2 mg of samples were dispersed in 500 μL water and 500 μL ethanol mixture, which contained 30 mL 60 wt% polytetrafluoroethylene (PTFE) solution with the aid of ultrasonication. Subsequently, 100 μL of as-prepared catalyst ink was loaded on surface of the Nickel foam surface ($1 \times 1$ cm$^2$) and then dried at room temperature.

All potentials by measuring were called reversible hydrogen electrode (RHE) by RHE calibration, as follows:

$$E_{RHE} = E_{Ag/AgCl} + 0.197 + 0.059 \, pH \tag{3}$$

For the whole polarization curve, linear sweep voltammetry (LSV) was performed at a 1.0 mV/s scanning rate, which were IR (deviation caused by I-current and R-resistance) corrected. Calculating the overpotential ($\eta$) is expressed as follows: $\eta = E_{RHE} - 1.23$ V.

## 3. Results

As shown in Figure 1a, it was observed from the XRD pattern of Co-Fe PBA that the peaks of XRD are related to the Co$_3$[Fe(CN)$_6$]$_2$(H$_2$O)$_{10}$ (JCPDS No. 46-907) [20]. At the same time, Figure 1b showed the FTIR spectrum of CoFe-PBA. It can be seen that Fe$^{III}$-CN-Co$^{II}$ and Fe$^{II}$-CN-Co$^{III}$ appeared at positions of 2111 cm$^{-1}$ and 2158 cm$^{-1}$, respectively. Additionally, two peaks at 1609 cm$^{-1}$ and 3416 cm$^{-1}$ corresponding to the bending vibration and stretching vibration absorption peak of O-H in water molecules were obviously detected, indicating that some water molecules entered the PBA lattice. Combined with the XRD spectrum and FTIR spectrum, it was shown that CoFe-PBA was successfully synthesized.

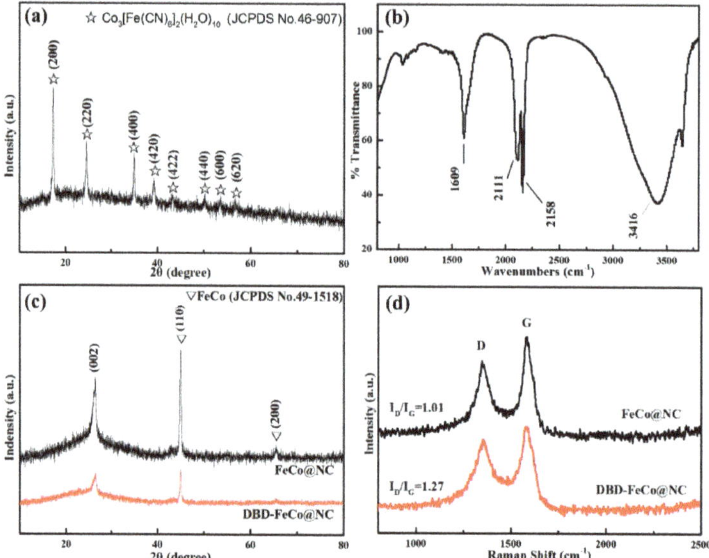

**Figure 1.** (**a**) XRD pattern, (**b**) FTIR spectra of CoFe-PBA, (**c**) XRD pattern, and (**d**) Raman spectra of FeCo@NC and DBD-FeCo@NC.

In order to determine the structural feature of the as-prepared products, the XRD pattern (Figure 1c) of FeCo@NC and DBD-FeCo@NC were presented in Figure 1c. The XRD patterns of the two showed a sharp peaks at 44.8° and 65.3°, which can be indexed to the diffraction from (110) and (200) planes of FeCo alloy (JCPDS No. 49-1567) [30]. Additionally, the two catalysts had broad peaks at 2θ = 26°, which formed on the (002) plane of graphite. FeCo@NC prepared by MW-CBM exhibited a high degree of graphitization. However, the peak intensity of graphite carbon was reduced for DBD-FeCo@NC after plasma treatment, indicating that the crystallinity of graphite carbon was decreased. The molecular structures of FeCo@NC and DBD-FeCo@NC were investigated using the Raman spectra. As can be seen from Figure 1d thatthe G band and the D band were around 1580 cm$^{-1}$ and 1350 cm$^{-1}$, respectively. The value of $I_D/I_G$ can be used to estimate the defect levels in the Raman spectra of carbon-based materials. It was found that the $I_D/I_G$ values were 1.01 for FeCo@NC, and 1.27 for DBD-FeCo@NC, respectively, indicating that the plasma treatment promoted the defective sites for DBD-FeCo@NC [31,32]. This conclusion was consistent with our XRD results and high-resolution TEM (HRTEM) images. The catalysts exposed more active sites due to increased defect sites, which was beneficial for ORR performance.

XPS was further employed to survey the surface composition of the FeCo@NC and DBD-FeCo@NC. Figure 2a showed the survey XPS spectrum of FeCo@NC and DBD-FeCo@NC. Figure 2b showed that C 1s spectra can be divided into three peaks, corresponding to sp$^2$ hybridized C (284.2 eV), C-O/C-N (285.5 eV), and O-C=O (288.0 eV) [33]. Figure 2c displayed the N 1s spectra of FeCo@NC and DBD-FeCo@NC, which could be deconvoluted into four peaks relevant to pyridinic N (398.2 eV), pyrrolic N (400.0 eV), graphitic N (401.1 eV), and oxidized N (406.6 eV). All of these N species were reported to show advantages for ORR, apart from the uncertain contribution of the oxidized N. As shown in Table 1, the different nitrogen type contents of FeCo@NC and DBD-FeCo@NC were 0.26 vs 0.58 at.% (pyridinic N), 0.30 vs 0.33 at.% (pyrrolic N), 0.26 vs 0.26 at.% (graphitic N), and 0.98 vs 0.50 at.% (oxidized N), respectively. It was found that DBD-FeCo@NC expressed an increased in the content of pyridinic N after plasma treatment, which was good for ORR activity [34,35]. The O 1s XPS spectra of all samples were given in Figure 2c. The peak at 531.2 eV and 532.0 eV belonged to oxygen defects and O-H sites from surface-absorbed water, respectively. Moreover, the amount of oxygen defects for FeCo@NC and DBD-FeCo@NC was 72.98% and 85.57%, respectively. The increase of oxygen vacancies after plasma treatment was beneficial to the adsorption and reduction of oxygen [36,37]. Figure 2e showed the Fe 2p spectra of FeCo@NC and DBD-FeCo@NC. The peak at the binding energy of 711.2 eV corresponded to Fe 2p$_{3/2}$, the peak at 725.2 eV was relevant to Fe 2p$_{1/2}$, and the peak at 718.4 eV was a satellite peak, confirming the existence of Fe-N-C structure [38]. The high-resolution XPS spectra of Co 2p (Figure 2f) revealed that there was a weak pair of doublets for the Co 2p$_{3/2}$ and Co 2p$_{1/2}$ signals at 780.6 eV and 796.2 eV, and the peak at 719.0 eV was a satellite peak, indicating the presence of Co-N-C species [39].

**Table 1.** Atomic content of FeCo@NC and DBD-FeCo@NC.

| Sample | Content (at.%) | | | | | Content of N Species (at.%) | | | |
|---|---|---|---|---|---|---|---|---|---|
| | C | O | N | Fe | Co | Pyridinic | Pyrrolic | Graphitic | Oxidized |
| FeCo@NC | 50.42 | 32.89 | 1.8 | 9.44 | 5.45 | 0.26 | 0.30 | 0.26 | 0.98 |
| DBD-FeCo@NC | 60.05 | 24.65 | 1.67 | 8.64 | 4.99 | 0.58 | 0.33 | 0.26 | 0.50 |

The morphology of FeCo@NC and DBD-FeCo@NC was investigated by transmission electron microscopy (TEM). A closer view of the TEM image displayed that FeCo@NC existed in a core-shell structure with a diameter of 30–50 nm (Figure 3a). The high-resolution TEM (HRTEM) in Figure 3b of FeCo@NC further revealed that the FoCo alloy nanoparticles were wrapped by graphitic carbon layers. The well-defined crystalline lattice gaps were 0.201 nm (core) and 0.35 nm (shell), which associated with the (110) plane of the FeCo phase and (002) the plane of the graphitic carbon, respectively. The HRTEM image in Figure 3c disclosed that DBD-FeCo@NC maintained a core-shell structure well after plasma treatment, and the lattice fringe spacing (0.201 nm) of the FeCo phase can also be observed. It was worth noting that the thickness of the graphite carbon layer of DBD-FeCo@NC was stripped from 2.68 nm to 1.06 nm after plasma treatment, compared to FeCo@NC. The thin layer structure of DBD-FeCo@NC reduced the surface reaction resistance of the catalyst, which was beneficial to ion transport/transfer [40]. More importantly, there were a few defects in DBD-FeCo@NC, which indicated that the interface active site was enhanced by plasma treatment and was beneficial for ORR [41,42]. HAADF-STEM (High-Angle Annular Dark Field- Scanning transmission electron microscopy) images (Figure 3e) and relevant elemental mapping of Fe, Co, C, and N revealed the core-shell morphology with C, N, Fe, and Co and the uniform distribution of Fe and Co elements in the core-shell structure, proving that the active centers were doped and distributed uniformly on the catalyst.

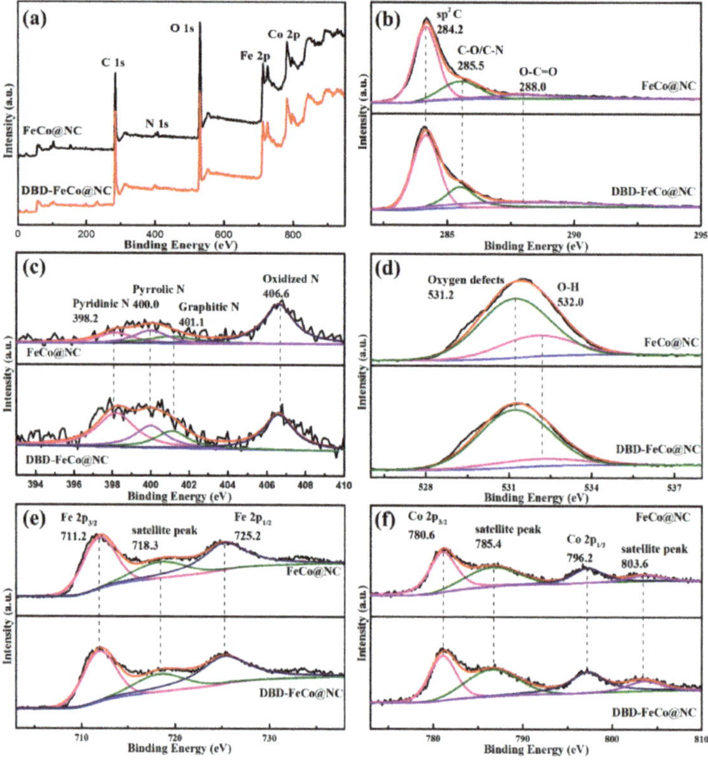

**Figure 2.** (**a**) XPS spectrum, (**b**) C 1s spectra, (**c**) N 1s spectra, (**d**) O 1s spectra, (**e**) Fe 2p spectra, (**f**) Co 2p spectra of FeCo@NC, and DBD-FeCo@NC.

The catalytic performance of ORR was tested using RDE with the electrolyte of 0.1 M KOH. LSV curves (Figure 4a) for the FeCo@NC, DBD-FeCo@NC, and 20 wt% Pt/C were examined at a 10 mV/s scanning rate 1600 rpm rotation speed in $O_2$-saturated electrolytes. Not surprisingly, DBD-FeCo@NC disclosed the limited current density of 5.66 mA/cm$^2$ and an onset potential of 0.96 V vs. RHE, which was greater than FeCo@NC (4.42 mA/cm$^2$, 0.88 V vs. RHE) and 20 wt.% Pt/C (5.01 mA/cm$^2$, 0.93 V vs. RHE). It was observed that DBD-FeCo@NC showed the half-wave potential at 0.88 V vs. RHE, which was 100 mV higher than that of FeCo@NC without plasma treatment. The ORR performance was comparable to other alloy catalysts (Table 2). Figure 4b displayed that CV showed a noticeable reduction peak at a 50 mV s$^{-1}$ scanning rate in $O_2$-saturated electrolytes, but no peak in $N_2$-saturated electrolytes. These results indicated the obvious ORR activity toward DBD-FeCo@NC. The reaction kinetics of DBD-FeCo@NC was measured by an ORR polarization technique at 625–2500 rpm (Figure 4c). The Koutecky–Levich (K–L) graphs of the DBD-FeCo@NC exhibited an approximately linear relationship between $\omega^{-1/2}$ and $j^{-1}$ in Figure 4d. In the range of 0.3–0.6 V, the average electron transfer number n value for DBD-FeCo@NC was calculated to be about 3.8 from K–L graphs, indexing a near four-electron transfer mechanism. Durability tests were performed for 20 wt% Pt/C and DBD-FeCo@NC by *i-t* measurement (Figure 4e). After 32,000 s of continuous operation, the current density of DBD-FeCo@NC can still be maintained at 75.6% (vs 35.6%, 20 wt% Pt/C), manifesting that DBD-FeCo@NC exhibited good durability. The methanol tolerance of DBD-FeCo@ NC and 20 wt% Pt/C was examined by adding

3 M methanol to 0.1 M KOH solution after 200 s. The chronoamperometric response (Figure 4f) showed that DBD-FeCo@NC only decreased slightly, and its current remained at 89.2% after 1200 s, while the 20 wt% Pt/C was only 43.1% of its maximum current. The results displayed that the stability and selectivity of the DBD-FeCo@ NC catalyst for ORR was higher than 20 wt% Pt/C.

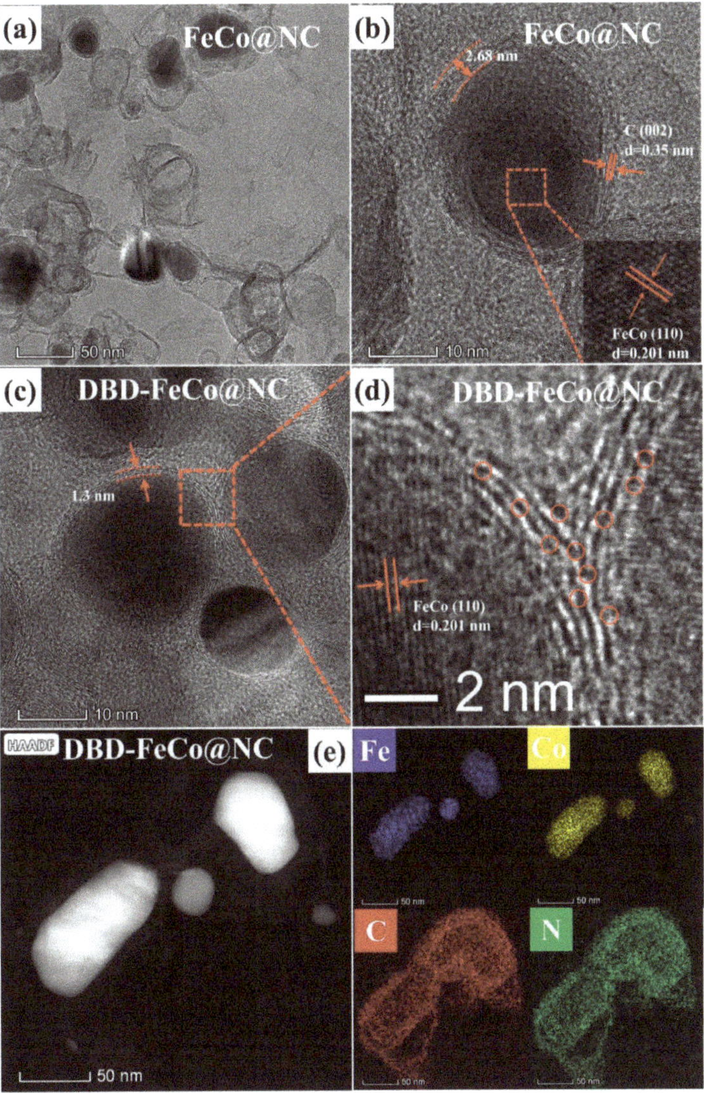


**Figure 3.** (**a**) Transmission electron microscopy (TEM) image for FeCo@NC, (**b**) high-resolution TEM (HRTEM) image for FeCo@NC, (**c**,**d**) HRTEM images for dielectric barrier discharge (DBD)-FeCo@NC, (**e**) HAADF-STEM image and the corresponding elemental mapping of DBD-FeCo@NC.

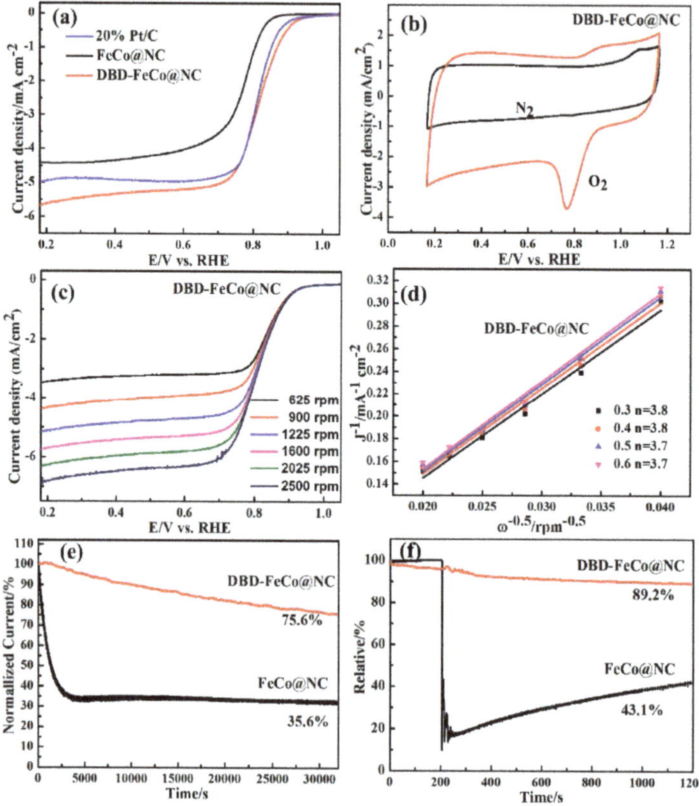

**Figure 4.** (**a**) Linear sweep voltammetry (LSV) curves of FeCo@NC, DBD-FeCo@NC, and 20 wt.% Pt/C in $O_2$-saturated 0.1 M KOH, with a speed of 1600 rpm at a sweep rate of 10 mV/s; (**b**) cyclic voltammetry (CV) curves of DBD-FeCo@NC at a scan rate of 50 mV/s in $N_2$-saturated or $O_2$-saturated 0.1 M KOH electrolyte; (**c**) LSVs of DBD-FeCo@NC at various rotation speeds and corresponding K–L plots (**d**); (**e**) long-term stability tests; and (**f**) tolerance to alcohol poisoning tests of DBD-FeCo@NC and 20 wt.% Pt/C via the oxygen reduction reaction (ORR) cathodic current-time (*i-t*) method.

**Table 2.** FeCo@NC and DBD-FeCo@NC compared with other alloy catalysts. Reversible hydrogen electrode (RHE); microwave-assisted carbon bath method (MW-CBM).

| Catalysts | Preparation Method | Onset Potential (V vs. RHE) | Half-Wave Potential (V vs. RHE) | Limiting-Current Density (mA/cm$^2$) | Ref. |
|---|---|---|---|---|---|
| PtNi/C | Solution synthesis | - | 0.88 | - | [16] |
| PtFe alloy | Solution synthesis | 0.95 | 0.88 | 5.83 | [17] |
| FeCo@NC-750 | Furnace heating | 0.94 | 0.80 | 4.82 | [20] |
| FeNi@NCNTs | Furnace heating | 0.95 | 0.77 | 4.70 | [21] |
| FeCo@NC | MW-CBM | 0.88 | 0.78 | 4.42 | this work |
| DBD-FeCo@NC | MW-CBM | 0.96 | 0.88 | 5.66 | this work |

The electrocatalytic properties of FeCo@NC and DBD-FeCo@NC for the OER activity were also studied. As displayed in Figure 5a, the initial potential of DBD-FeCo@NC was 1.49 V vs. RHE, which was more negative than that of FeCo@NC (1.55 V vs. RHE), and the initial potential of nickel foam displayed 1.58 V vs. RHE, suggesting an enhanced OER activity in DBD-FeCo@NC. It is well known that the potential demanded to provide the current density of 10 mA/cm$^2$ is the key benchmark for OER [43–45]. Under the current density of 10.0 mA/cm$^2$, overpotential of 386 mV and 335 mV were estimated for FeCo@NC and DBD-FeCo@NC, respectively. As illustrated in Figure 5b, the DBD-FeCo@NC displayed a Tafel slope (111 mV/dec), which was much less than that of FeCo@NC (209 mV/dec), demonstrating the most favorable OER kinetics and highly active DBD-FeCo@NC. The current *i-t* curve (Figure 5c) showed excellent stability of DBD-FeCo@NC under a current density of 10 mA/cm$^2$, which retained 91.3% of the initial catalytic current after 14 h of continuous testing.

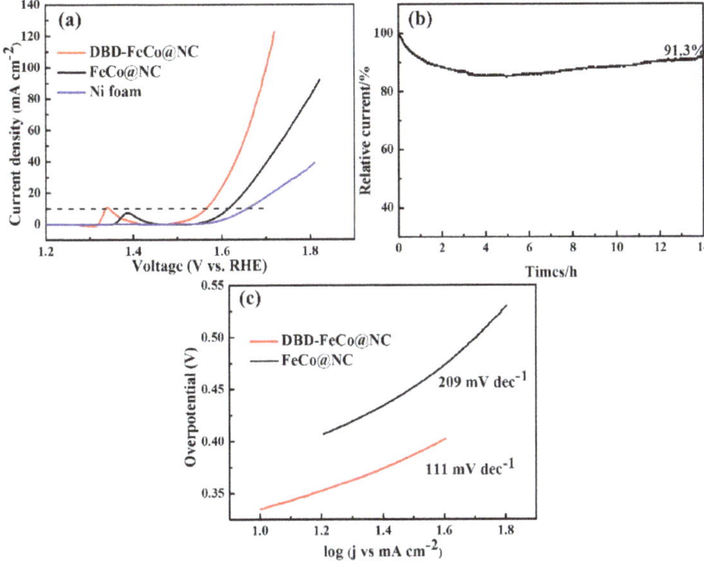


**Figure 5.** (**a**) The LSV polarization curves and (**b**) Tafel slope of OER for FeCo@NC and DBD-FeCo@NC, (**c**) Galvanostatic measurement of OER of DBD-FeCo@NC at a current density of 10 mA cm$^{-2}$.

Based on the structure and morphology of the catalysts, the excellent electrocatalytic activities for both ORR and OER can be attributed to several factors: (1) The active sites of Fe/Co-N$_x$-C can adjust the electronic polarities and surface properties, which can improve the activity of the catalyst; (2) the unique core-shell structure offered more active sites and effectively prevented the coagulation and dissolution/redeposition of FeCo alloy nanoparticles, thus maintaining high electrocatalytic stability; (3) the plasma treatment made the catalyst possess more defect sites, which played an important role in improving catalytic performance; (4) the Co-Fe system enhanced the conductivity of the catalysts.

## 4. Conclusions

In summary, we synthesized N-doped graphite carbon-coated FeCo alloy core-shell nanoparticles via the microwave-assisted carbon bath method and further treatment with DBD plasma as bi-functional ORR/OER catalysts. The materials exhibited unique core-shell structure, Fe/Co-N-C sites, the existence of FeCo alloys, and enriched defect active sites. Because of these characteristics, DBD-FeCo@NC displayed excellent ORR performance, as well as good OER activity. This synthetic route is facile and scalable to prepare catalytic materials with different metal-doped and abundant defect active sites, giving a wide possibility for large-scale application in practice.

**Author Contributions:** F.Y., X.G., and L.Z. designed the experiments. C.M., S.Y., G.W., and X.X. administered the experiments. M.L., H.Y., and H.F. performed experiments. M.L. and F.Y. collected data. X.G. and L.Z. gave conceptual advice. All authors analyzed, discussed the data, and wrote the manuscript.

**Funding:** This work was supported by the National Natural Science Foundation of China (21663022, 21865025).

**Conflicts of Interest:** The authors declare no conflict of interest.

## References


1. Su, C.-Y.; Cheng, H.; Li, W.; Liu, Z.-Q.; Li, N.; Hou, Z.; Bai, F.-Q.; Zhang, H.-X.; Ma, T.-Y. Atomic Modulation of FeCo-Nitrogen-Carbon Bifunctional Oxygen Electrodes for Rechargeable and Flexible All-Solid-State Zinc-Air Battery. *Adv. Energy Mater.* **2017**, *7*, 1602420. [CrossRef]
2. Zhang, X.; Lyu, D.; Mollamahale, Y.B.; Yu, F.; Qing, M.; Yin, S.; Zhang, X.; Tian, Z.Q.; Shen, P.K. Critical role of iron carbide nanodots on 3D graphene based nonprecious metal catalysts for enhancing oxygen reduction reaction. *Electrochim. Acta* **2018**, *281*, 502–509. [CrossRef]
3. Xu, N.; Zhang, Y.; Zhang, T.; Liu, Y.; Qiao, J. Efficient quantum dots anchored nanocomposite for highly active ORR/OER electrocatalyst of advanced metal-air batteries. *Nano Energy* **2019**, *57*, 176–185. [CrossRef]
4. Zhao, Q.; Katyal, N.; Seymour, I.D.; Henkelman, G.; Ma, T. Vanadium(III) Acetylacetonate as an Efficient Soluble Catalyst for Lithium-Oxygen Batteries. *Angew. Chem. Int. Ed.* **2019**, *58*, 12553–12557. [CrossRef]
5. Liu, S.; Wang, M.; Sun, X.; Xu, N.; Liu, J.; Wang, Y.; Qian, T.; Yan, C. Zinc-Air Batteries: Facilitated Oxygen Chemisorption in Heteroatom-Doped Carbon for Improved Oxygen Reaction Activity in All-Solid-State Zinc–Air Batteries (Adv. Mater. 4/2018). *Adv. Mater.* **2018**, *30*, 1870028. [CrossRef]
6. Liu, M.; Guo, X.; Hu, L.; Yuan, H.; Wang, G.; Dai, B.; Zhang, L.; Yu, F. Fe3O4/Fe3C@nitrogen-doped carbon for enhancing oxygen reduction reaction. *Chem. Nano Mat.* **2019**, *5*, 187–193. [CrossRef]
7. Li, Y.; Zhong, C.; Liu, J.; Zeng, X.; Qu, S.; Han, X.; Deng, Y.; Hu, W.; Lu, J. Atomically Thin Mesoporous Co3 O4 Layers Strongly Coupled with N-rGO Nanosheets as High-Performance Bifunctional Catalysts for 1D Knittable Zinc-Air Batteries. *Adv. Mat.* **2017**, *30*, 1703657. [CrossRef]
8. Hu, L.; Yu, F.; Yuan, H.; Wang, G.; Liu, M.; Wang, L.; Xue, X.; Peng, B.; Tian, Z.; Dai, B. Improved oxygen reduction reaction via a partially oxidized Co-CoO catalyst on N-doped carbon synthesized by a facile sand-bath method. *Chin. Chem. Lett.* **2019**, *30*, 624–629. [CrossRef]
9. Han, J.; Meng, X.; Lu, L.; Bian, J.; Li, Z.; Sun, C. Single-Atom Fe-Nx-C as an Efficient Electrocatalyst for Zinc-Air Batteries. *Adv. Funct. Mat.* **2019**. [CrossRef]
10. Wang, Y.; Zhu, M.; Wang, G.; Dai, B.; Yu, F.; Tian, Z.; Guo, X. Enhanced Oxygen Reduction Reaction by In Situ Anchoring Fe2N Nanoparticles on Nitrogen-Doped Pomelo Peel-Derived Carbon. *Nanomaterials* **2017**, *7*, 404. [CrossRef]
11. Shi, L.; Wu, T.; Wang, Y.; Zhang, J.; Wang, G.; Zhang, J.; Dai, B.; Yu, F. Nitrogen-Doped Carbon Nanoparticles for Oxygen Reduction Prepared via a Crushing Method Involving a High Shear Mixer. *Materials* **2017**, *10*, 1030. [CrossRef] [PubMed]
12. Du, X.X.; He, Y.; Wang, X.X.; Wang, J.N. Fine-grained and fully ordered intermetallic PtFe catalysts with largely enhanced catalytic activity and durability. *Energy Environ. Sci.* **2016**, *9*, 2623–2632. [CrossRef]
13. Chong, L.; Wen, J.; Kubal, J.; Sen, F.G.; Zou, J.; Greeley, J.; Chan, M.; Barkholtz, H.; Ding, W.; Liu, D.-J. Ultralow-loading platinum-cobalt fuel cell catalysts derived from imidazolate frameworks. *Science* **2018**, *362*, 1276–1281. [CrossRef] [PubMed]
14. Jia, Q.; Zhao, Z.; Cao, L.; Li, J.; Ghoshal, S.; Davies, V.; Stavitski, E.; Attenkofer, K.; Liu, Z.; Li, M.; et al. Roles of Mo Surface Dopants in Enhancing the ORR Performance of Octahedral PtNi Nanoparticles. *Nano Lett.* **2018**, *18*, 798–804. [CrossRef] [PubMed]
15. Chen, Q.; Yang, Y.; Cao, Z.; Kuang, Q.; Du, G.; Jiang, Y.; Xie, Z.; Zheng, L. Excavated Cubic Platinum-Tin Alloy Nanocrystals Constructed from Ultrathin Nanosheets with Enhanced Electrocatalytic Activity. *Angew. Chem. Int. Ed.* **2016**, *55*, 9021–9025. [CrossRef] [PubMed]
16. Zhang, G.-R.; Wöllner, S. Hollowed structured PtNi bifunctional electrocatalyst with record low total overpotential for oxygen reduction and oxygen evolution reactions. *Appl. Catal. B Environ.* **2018**, *222*, 26–34. [CrossRef]
17. Wang, N.; Li, Y.; Guo, Z.; Li, H.; Hayase, S.; Ma, T. Minute quantities of hexagonal nanoplates PtFe alloy with facile operating conditions enhanced electrocatalytic activity and durability for oxygen reduction reaction. *J. Alloy. Compd.* **2018**, *752*, 23–31. [CrossRef]
18. Zeng, M.; Liu, Y.; Zhao, F.; Nie, K.; Han, N.; Wang, X.; Huang, W.; Song, X.; Zhong, J.; Li, Y. Metallic Cobalt Nanoparticles Encapsulated in Nitrogen-Enriched Graphene Shells: Its Bifunctional Electrocatalysis and Application in Zinc-Air Batteries. *Adv. Funct. Mater.* **2016**, *26*, 4397–4404. [CrossRef]
19. Bates, M.K.; Jia, Q.; Doan, H.; Liang, W.; Mukerjee, S. Charge-Transfer Effects in Ni–Fe and Ni–Fe–Co Mixed-Metal Oxides for the Alkaline Oxygen Evolution Reaction. *ACS Catal.* **2015**, *6*, 155–161. [CrossRef]

20. Cai, P.; Ci, S.; Zhang, E.; Shao, P.; Cao, C.; Wen, Z. FeCo Alloy Nanoparticles Confined in Carbon Layers as High-activity and Robust Cathode Catalyst for Zn-Air Battery. *Electrochim. Acta* **2016**, *220*, 354–362. [CrossRef]

21. Zhao, X.; Abbas, S.C.; Huang, Y.; Lv, J.; Wu, M.; Wang, Y. Robust and Highly Active FeNi@NCNT Nanowire Arrays as Integrated Air Electrode for Flexible Solid-State Rechargeable Zn-Air Batteries. *Adv. Mater. Interfaces* **2018**, *5*, 1701448. [CrossRef]

22. Wei, L.; Karahan, H.E.; Zhai, S.; Liu, H.; Chen, X.; Zhou, Z.; Lei, Y.; Liu, Z.; Chen, Y. Amorphous Bimetallic Oxide-Graphene Hybrids as Bifunctional Oxygen Electrocatalysts for Rechargeable Zn-Air Batteries. *Adv. Mater.* **2017**, *29*, 1701410. [CrossRef] [PubMed]

23. Yu, F.; Lim, S.H.; Zhen, Y.; An, Y.; Lin, J. Optimized electrochemical performance of three-dimensional porous LiFePO 4 /C microspheres via microwave irradiation assisted synthesis. *J. Power Sources* **2014**, *271*, 223–230. [CrossRef]

24. Yu, F.; Zhang, L.; Lai, L.; Zhu, M.; Guo, Y.; Xia, L.; Qi, P.; Wang, G.; Dai, B. High Electrochemical Performance of LiFePO4 Cathode Material via In-Situ Microwave Exfoliated Graphene Oxide. *Electrochim. Acta* **2015**, *151*, 240–248. [CrossRef]

25. Liu, M.; Yin, X.; Guo, X.; Hu, L.; Yuan, H.; Wang, G.; Wang, F.; Chen, L.; Zhang, L.; Yu, F. High efficient oxygen reduction performance of Fe/Fe3C nanoparticles in situ encapsulated in nitrogen-doped carbon via a novel microwave-assisted carbon bath method. *Nano Mater. Sci.* **2019**, *1*, 131–136. [CrossRef]

26. Li, P.; Wen, B.; Yu, F.; Zhu, M.; Guo, X.; Han, Y.; Kang, L.; Huang, X.; Dan, J.; Ouyang, F.; et al. High efficient nickel/vermiculite catalyst prepared via microwave irradiation-assisted synthesis for carbon monoxide methanation. *Fuel* **2016**, *171*, 263–269. [CrossRef]

27. Yu, F.; Zhang, L.; Zhu, M.; An, Y.; Xia, L.; Wang, X.; Dai, B. Overwhelming microwave irradiation assisted synthesis of olivine-structured LiMPO4 (M=Fe, Mn, Co and Ni) for Li-ion batteries. *Nano Energy* **2014**, *3*, 64–79. [CrossRef]

28. Ding, D.; Song, Z.-L.; Cheng, Z.-Q.; Liu, W.-N.; Nie, X.-K.; Bian, X.; Chen, Z.; Tan, W. Plasma-assisted nitrogen doping of graphene-encapsulated Pt nanocrystals as efficient fuel cell catalysts. *J. Mater. Chem. A* **2014**, *2*, 472–477. [CrossRef]

29. Wang, Y.; Yu, F.; Zhu, M.; Ma, C.; Zhao, D.; Wang, C.; Zhou, A.; Dai, B.; Ji, J.; Guo, X. N-Doping of plasma exfoliated graphene oxide via dielectric barrier discharge plasma treatment for the oxygen reduction reaction. *J. Mater. Chem. A* **2018**, *6*, 2011–2017. [CrossRef]

30. Lin, H.; Ruirui, Z.; Lingzhi, W.; Fapei, Z.; Qianwang, C. Synthesis of FeCo nanocrystals encapsulated in nitrogen-doped graphene layers for use as highly efficient catalysts for reduction reactions. *Nanoscale* **2014**, *7*, 450–454.

31. Wang, X.; Wang, J.; Wang, D.; Dou, S.; Ma, Z.; Wu, J.; Tao, L.; Shen, A.; Ouyang, C.; Liu, Q.; et al. One-pot synthesis of nitrogen and sulfur co-doped graphene as efficient metal-free electrocatalysts for the oxygen reduction reaction. *Chem. Commun.* **2014**, *50*, 4839–4842. [CrossRef] [PubMed]

32. Dou, S.; Shen, A.; Tao, L.; Wang, S. Molecular doping of graphene as metal-free electrocatalyst for oxygen reduction reaction. *Chem. Commun.* **2014**, *50*, 10672–10675. [CrossRef] [PubMed]

33. Chen, H.; Wang, G.; Chen, L.; Dai, B.; Yu, F. Three-Dimensional Honeycomb-Like Porous Carbon with Both Interconnected Hierarchical Porosity and Nitrogen Self-Doping from Cotton Seed Husk for Supercapacitor Electrode. *Nanomaterials* **2018**, *8*, 412. [CrossRef] [PubMed]

34. Wu, Z.Y.; Xu, X.X.; Hu, B.C.; Liang, H.W.; Lin, Y.; Chen, L.F.; Yu, S.H. Iron Carbide Nanoparticles Encapsulated in Mesoporous Fe-N-Doped Carbon Nanofibers for Efficient Electrocatalysis. *Angew. Chem. Int. Ed. Engl.* **2015**, *54*, 8179–8183. [CrossRef] [PubMed]

35. Li, J.S.; Li, S.L.; Tang, Y.J.; Han, M.; Dai, Z.H.; Bao, J.C.; Lan, Y.Q. Nitrogen-doped Fe/Fe3C@graphitic layer/carbon nanotube hybrids derived from MOFs: efficient bifunctional electrocatalysts for ORR and OER. *Chem. Commun.* **2015**, *51*, 2710–2713. [CrossRef] [PubMed]

36. Lima, F.H.B.; Calegaro, M.L.; Ticianelli, E.A. Electrocatalytic activity of dispersed platinum and silver alloys and manganese oxides for the oxygen reduction in alkaline electrolyte. *Russ. J. Electrochem.* **2006**, *42*, 1283–1290. [CrossRef]

37. Zeng, Y.; Lai, Z.; Han, Y.; Zhang, H.; Xie, S.; Lu, X. Oxygen-Vacancy and Surface Modulation of Ultrathin Nickel Cobaltite Nanosheets as a High-Energy Cathode for Advanced Zn-Ion Batteries. *Adv. Mater.* **2018**, 1802396. [CrossRef]

38. Lin, Q.; Bu, X.; Kong, A.; Mao, C.; Zhao, X.; Bu, F.; Feng, P. New Heterometallic Zirconium Metalloporphyrin Frameworks and Their Heteroatom-Activated High-Surface-Area Carbon Derivatives. *J. Am. Chem. Soc.* **2015**, *137*, 2235–2238. [CrossRef]

39. Palaniselvam, T.; Kashyap, V.; Bhange, S.N.; Baek, J.-B.; Kurungot, S. Nanoporous Graphene Enriched with Fe/Co-N Active Sites as a Promising Oxygen Reduction Electrocatalyst for Anion Exchange Membrane Fuel Cells. *Adv. Funct. Mater.* **2016**, *26*, 2150–2162. [CrossRef]

40. Liao, C.; Xu, Q.; Wu, C.; Fang, D.; Chen, S.; Chen, S.; Luo, J.; Li, L. Core–shell nano-structured carbon composites based on tannic acid for lithium-ion batteries. *J. Mater. Chem. A* **2016**, *4*, 17215–17224. [CrossRef]

41. Shen, A.; Zou, Y.; Wang, Q.; Dryfe, R.A.W.; Huang, X.; Dou, S.; Dai, L.; Wang, S. Oxygen Reduction Reaction in a Droplet on Graphite: Direct Evidence that the Edge Is More Active than the Basal Plane. *Angew. Chem. Int. Ed.* **2014**, *53*, 10804–10808. [CrossRef] [PubMed]

42. Zhang, M.; Li, P.; Tian, Z.; Zhu, M.; Wang, F.; Li, J.; Dai, B.; Yu, F.; Qiu, H.; Gao, H. Clarification of Active Sites at Interfaces between Silica Support and Nickel Active Components for Carbon Monoxide Methanation. *Catalysts* **2018**, *8*, 293. [CrossRef]

43. Fu, H.; Liu, Y.; Chen, L.; Shi, Y.; Kong, W.; Hou, J.; Yu, F.; Wei, T.; Wang, H.; Guo, X. Designed formation of NiCo2O4 with different morphologies self-assembled from nanoparticles for asymmetric supercapacitors and electrocatalysts for oxygen evolution reaction. *Electrochim. Acta* **2019**, *296*, 719–729. [CrossRef]

44. Xue, X.; Yu, F.; Peng, B.; Wang, G.; Lv, Y.; Chen, L.; Yao, Y.; Dai, B.; Shi, Y.; Guo, X. One-step synthesis of nickel–iron layered double hydroxides with tungstate acid anions via flash nano-precipitation for the oxygen evolution reaction. *Sustain. Energy Fuels* **2019**, *3*, 237–244. [CrossRef]

45. Yan, J.; Kong, L.; Ji, Y.; White, J.; Li, Y.; Zhang, J.; An, P.; Liu, S.; Lee, S.-T.; Ma, T. Single atom tungsten doped ultrathin $\alpha$-Ni(OH)$_2$ for enhanced electrocatalytic water oxidation. *Nat. Commun.* **2019**, *10*. [CrossRef] [PubMed]

*Article*

# Enhanced Breakdown Strength and Thermal Conductivity of BN/EP Nanocomposites with Bipolar Nanosecond Pulse DBD Plasma Modified BNNSs

**Yan Mi *, Jiaxi Gou, Lulu Liu, Xin Ge, Hui Wan and Quan Liu**

State Key Laboratory of Power Transmission Equipment & System Security and New Technology, School of Electrical Engineering, Chongqing University, Chongqing 400044, China; cqugjx@163.com (J.G.); xz_liulu@163.com (L.L.); 20181102038t@cqu.edu.cn (X.G.); 13320248214@163.com (H.W.); 20171102050t@cqu.edu.cn (Q.L.)

* Correspondence: miyan@cqu.edu.cn; Tel.: +86-2365111172

Received: 20 August 2019; Accepted: 26 September 2019; Published: 30 September 2019

**Abstract:** Filling epoxy resin (EP) with boron nitride (BN) nanosheets (BNNSs) can effectively improve the thermal conductivity of BN/EP nanocomposites. However, due to the few hydroxyl groups on the surface of BNNSs, silane coupling agent (SCA) cannot effectively modify BNNSs. The agglomeration of BNNSs is severe, which significantly reduces the AC breakdown strength of the composites. Therefore, this paper uses atmospheric pressure bipolar nanosecond pulse dielectric barrier discharge (DBD) Ar+H$_2$O low temperature plasma to hydroxylate BNNSs to improve the AC breakdown strength and thermal conductivity of the composites. X-ray photoelectron spectroscopy (XPS) shows that the hydroxyl content of the BNNSs surface increases nearly two fold after plasma modification. Fourier transform infrared spectroscopy (FTIR) and thermogravimetric analysis (TGA) show that plasma modification enhances the dehydration condensation reaction of BNNSs with SCA, and the coating amount of SCA on the BNNSs surface increases by 45%. The breakdown test shows that the AC breakdown strength of the composites after plasma modification is improved under different filling contents. With the filling content of BNNSs increasing from 10% to 20%, the composites can maintain a certain insulation strength. Meanwhile, the thermal conductivity of the composites increases by 67% as the filling content increases from 10% (SCA treated) to 20% (plasma and SCA treated). Therefore, the plasma hydroxylation modification method used in this paper can provide a basis for the preparation of high thermal conductivity insulating materials.

**Keywords:** bipolar nanosecond pulse; low temperature plasma; hydroxylation modification; BN nanosheets; nanocomposites; breakdown strength; thermal conductivity

---

## 1. Introduction

Breakdown strength and thermal conductivity of insulating materials are important factors affecting the safe and stable operation of electrical equipments. With the development of UHV transmission technology, higher and higher requirements for the performance of insulating materials are put forward. Epoxy resin has good insulation properties and is often used in the electronics and power industries, such as power electronics packaging materials, dry-type transformers, bushings and so on. However, the thermal conductivity of epoxy resin is low, and the heat generated during operation of a device cannot be dissipated in time, resulting in a decrease in the dielectric strength and insulation life of epoxy resin [1,2]. At present, a common method to improve the thermal conductivity of epoxy resin is to add thermal conductive insulating nanoparticles, such as Al$_2$O$_3$, Si$_3$N$_4$, AlN, and BN, to the epoxy resin [3–5]. Studies have shown that BN has a higher thermal conductivity than other particles, and it has a better effect on the thermal conductivity of epoxy resin. The higher the content of

BN, the higher the thermal conductivity of the epoxy resin composite [6,7]. However, when the BN content exceeds a certain amount, serious agglomeration will occur, which will lead to a decrease in the AC breakdown strength of the composite, thus limiting the improvement in the thermal conductivity and practical application of epoxy resin composites [8,9]. Therefore, it is important to improve the breakdown strength of high thermal conductivity nanocomposites.

Further research shows that improving the dispersion of nanoparticles in matrix materials is an effective method to improve the insulation properties of nanocomposites, and this is also a research hotspot and challenge for nanomaterials [10–12]. At present, the main method to improve the dispersion of nanoparticles is surface modification of nanoparticles, and silane coupling agent (SCA) is a commonly used surface modifier [13]. However, the surface of BN has a small content of -OH groups [14], and the bonding degree between BN and SCA is low, so it is necessary to hydroxylate the surface first. The methods for surface hydroxylation modification of BN mainly include high temperature annealing [15], NaOH heat treatment [16], $H_2O_2$ heat treatment [17], and concentrated mixed acid treatment [18]. However, these methods take a long time, and the strong oxidizing chemical reagents used in these methods are not environmentally friendly.

In recent years, some scholars have begun to modify nanoparticles by plasma. Hydroxyl groups are bonded on the surface of nanoparticles by the collisions of high-energy particles in plasma, and this method has the advantages of high efficiency and environmental friendliness. Guangning Wu et al. [19] modified nano-$Al_2O_3$ with atmospheric pressure air plasma excited by high frequency AC power, and effectively increased the hydroxyl content on the surface of $Al_2O_3$. Yeongseon Kim et al. [14] used atmospheric pressure RF plasma to modify BN and successfully introduced hydroxyl groups on the BN surface, which enhanced the binding degree of BN and SCA. Thus, a high-frequency AC source is currently used to hydroxylate nanoparticles. Some studies show that nanosecond pulse discharge can produce more uniform plasma and has a higher energy efficiency than high-frequency AC power because nanosecond pulse power has a shorter rising edge than AC power, which can trigger overvoltage breakdown and enhance the ionization and electronic excitation processes [20–22]. Moreover, pulse discharge plasma has a lower temperature and less influence on the material [23].

In addition to the above advantages, bipolar nanosecond pulse dielectric barrier discharge (DBD) has a reverse electric field generated by the charge accumulated on the surface of the dielectric after the last pulse discharge, and this field is in the same direction as the electric field generated by the next pulse. The electric field is enhanced after superposition, thereby promoting the development of discharge and the generation of active particles. Yunfei Liu et al. [24] used atmospheric pressure bipolar nanosecond pulse discharge to generate plasma in air to modify the surface of PET films. Experiments showed that uniform plasma can be generated when the air gap is less than 1.2 mm; the surface of the modified film had no obvious morphological change, and oxygen-containing polar groups were successfully introduced. Dezheng Yang et al. [25] studied the atmospheric pressure air DBD emission spectra driven by different polarity pulses. The results showed that the $N_2$ ($C_3\Pi_u \rightarrow B_3\Pi_g$) spectral emission intensity produced by bipolar pulse discharge was approximately 4–5 times that of unipolar pulse discharge, which indicated that bipolar pulse discharge was more conducive to exciting active substances.

In view of the current problems existing in BN/EP composites and the advantages of plasma modification by bipolar nanosecond pulse discharge, this paper uses atmospheric pressure bipolar nanosecond pulse DBD Ar+$H_2O$ low temperature plasma for hydroxylation modification of BNNSs to enhance the combination of BNNSs and SCA, increase the coating amount of SCA, and improve the AC breakdown strength and thermal conductivity of BN/EP nanocomposites, which provides a basis for the preparation of high thermal conductivity insulating materials for power systems.

## 2. Materials and Methods

### *2.1. Materials*

The average particle size of the hexagonal BNNSs was 50 nm, and the BNNSs were purchased from Chaowei Nanotechnology Co., Ltd. (Shanghai, China). The scanning electron microscopy (SEM) image is shown in Figure 1. Ar (purity ≥ 99.999%) was purchased from Chaoyang Gas Co., Ltd. (Chongqing, China). The SCA KH560 and 0.1 M oxalic acid titration solution were purchased from Aladdin Biochemical Technology Co., Ltd. (Shanghai, China). Bisphenol A epoxy resin E51, methyl tetrahydrophthalic anhydride (MeTHPA) curing agent HKR-0719, and the accelerant DMP-30 were purchased from Huakai Resin Co., Ltd. (Jining, Shandong, China). Anhydrous ethanol was purchased from Cologne Chemicals Co., Ltd. (Chengdu, Sichuan, China). Pure water was purchased from Chuandong Chemical (Group) Co., Ltd. (Chongqing, China).

**Figure 1.** SEM image of pristine epoxy resin (EP) with BN nanosheets (BNNSs).

### *2.2. Plasma Modification of Epoxy Resin (EP) with BN Nanosheets (BNNSs)*

A schematic diagram of the plasma modification is shown in Figure 2. The power supply is a laboratory-made bipolar nanosecond pulse generator [26]. The electrode plates are stainless-steel circular electrodes with a diameter of 100 mm. The barrier dielectric is a circular quartz glass with a diameter of 110 mm and a thickness of 1 mm. A 500 Ω noninductive resistor is also connected in series to DBD as a protective resistor. The whole DBD device is placed in a transparent glass box, and two gas channels are placed the box. One channel is directly connected to high-purity Ar, while through the other channel, Ar is passed into a gas-washing bottle containing pure water. The flow ratio of the two is 4:1 and the total flow rate is 2 L/min. The flow rate is controlled by a D07-7 gas mass flow controller (Qixing Huachuang Electronics Co., Ltd., Beijing, China). In the experiment, 0.5 g BNNSs is spread on the quartz glass and then placed on the ground electrode plate. The distance between the plates is 2 mm. The glass box is placed over the DBD device and filled with Ar and $H_2O$ mixed gas for 3 min. Then the generator is activated to produce plasma for 30 s. BNNSs are removed, stirred well and placed into the glass box for another 30 s treatment. This process is repeated three times to complete the hydroxylation modification. The voltage and current waveforms of DBD are measured by a high voltage probe P6015A (Tektronix. Inc., Beaverton, OR, USA), Pearson current sensor 2877 (Pearson Electronics. Inc., Palo Alto, CA, USA) and an oscilloscope DPO 4054 (Tektronix. Inc., Beaverton, OR, USA). The waveforms are shown in Figure 3. The voltage amplitude is 4 kV, the pulse width is 300 ns, the frequency is 1 kHz, and the positive and negative pulse interval is

2 µs. The current exhibits the typical characteristic of pulse DBD. There is a pulse current at both the rising and falling edges, and the former is larger than the latter. This difference is because the charge accumulated on the dielectrics at the beginning of discharge is still very small, so the previous current is mainly affected by the applied electric field, and the latter current is generated by the inverse electric field generated by the charge accumulated on the dielectrics minus the applied electric field [21]. The plasma image is taken by a Canon EOS 750D camera (Canon Inc., Tokyo, Japan) with an exposure time of 100 ms, an aperture value of f/10, and an ISO speed of 6400. The image is shown in Figure 3. The plasma fills the entire air gap uniformly, which is advantageous for surface modification of BNNSs.

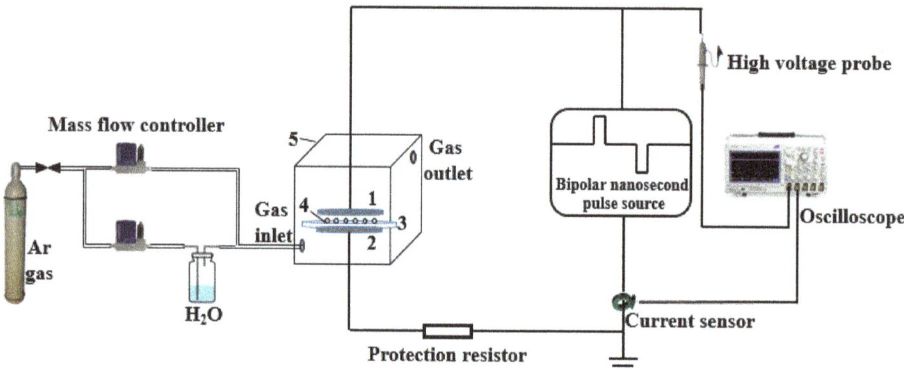


**Figure 2.** Schematic diagram of plasma modification of BNNSs. 1: High voltage electrode, 2: ground electrode, 3: silica glass, 4: BNNSs, 5: glass box.

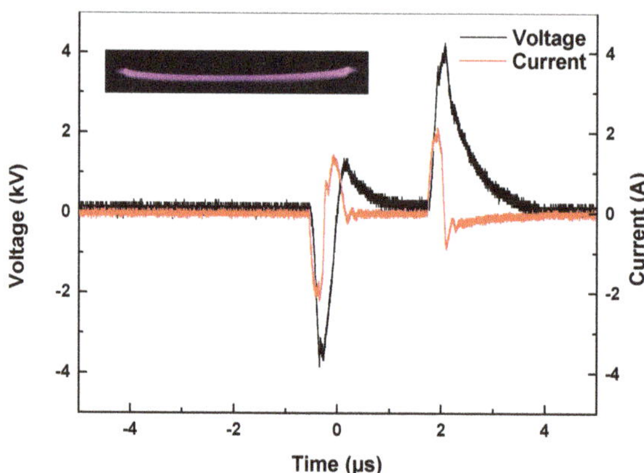


**Figure 3.** Voltage and current waveforms of dielectric barrier discharge (DBD) and image of plasma.

## 2.3. SCA Treated BNNSs

The treatment process of the SCA is shown in Figure 4. A certain amount of SCA KH560 is incorporated into a mixed solution of absolute ethanol and pure water. The mass ratio of SCA, ethanol and pure water is 1:144:16. The amount of SCA is estimated by the following formula [27]:

$$m_0 = m.s/s_0, \tag{1}$$

where $m_0$ is the amount of SCA, $m$ is the mass of the nanoparticles, $s$ is the specific surface area of the nanoparticles (34.318 m²·g⁻¹), $s_0$ is the minimum coating area of SCA (322 m²·g⁻¹). In general, the actual amount is less than the calculated amount. Therefore, the amount of SCA is 10% of the mass of the nanoparticles. The pH of the mixed solution is adjusted to 4 with oxalic acid, and then hydrolyzed in a water bath at 40 °C for 2 h. Quantitative BNNSs are then added to the mixed solution and sonicated for 30 min. The solution is stirred with a magnetic stirrer at 800 r/min for 2 h. Then, the suspension is centrifuged for 3 times with absolute ethanol for 5 min each time at a rotation speed of 3000 r/min to remove SCA and other impurities without grafting. The product is then dried in a vacuum oven at 60 °C for 12 h. Finally, the dried BNNSs are ground into powder by a planetary ball mill.

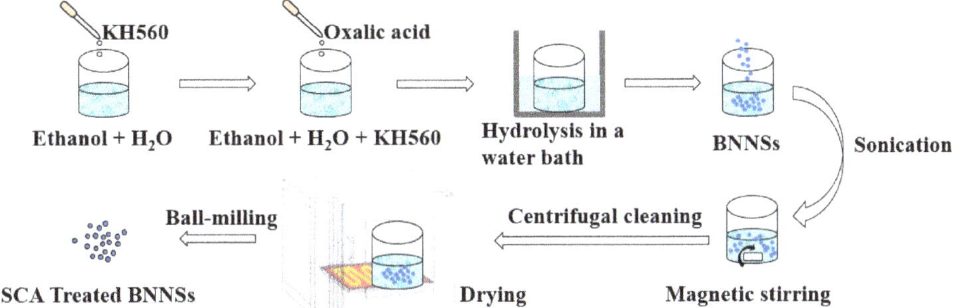


**Figure 4.** Silane coupling agent (SCA) treatment process.

*2.4. Preparation of the Epoxy Composites*

The sample preparation process of the BN/EP nanocomposites is shown in Figure 5. A certain proportion of epoxy resin, curing agent and BNNSs are magnetically stirred in a water bath at 70 °C for 1 h at a speed of 800 r/min. Then, the solution is ultrasonically dispersed in a water bath at 70 °C for 30 min. A certain amount of accelerant (the mass ratio of epoxy resin, curing agent and accelerant is 10:8:0.2) is added, and magnetic stirring is continued for 15 min. The resulting solution is then placed in an oven and degassed under vacuum at 60 °C for 30 min. Finally, the mixture is poured into a mold, and cured in an oven at 90 °C for 2 h and then at 110 °C for 2 h.

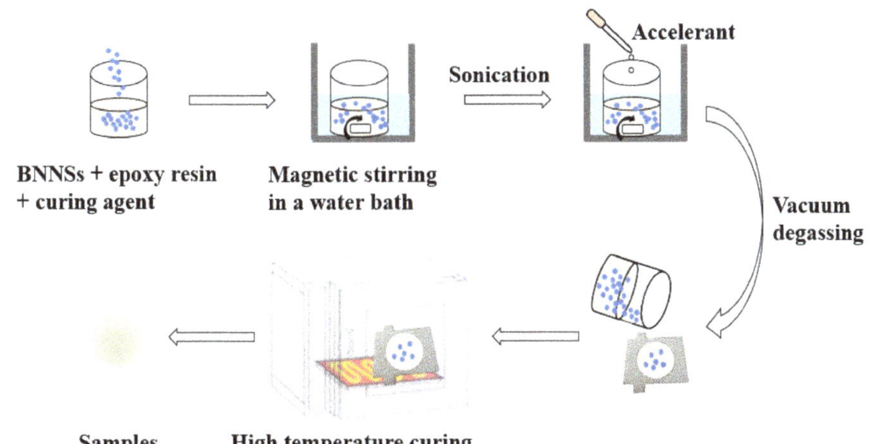


**Figure 5.** Preparation process of the epoxy composites.

*2.5. Characterization*

The elements and hydroxyl groups on the surface of the BNNSs before and after plasma modification were analyzed by X-ray photoelectron spectroscopy (XPS, ESCALAB 250Xi, Thermo

Fisher Scientific Co., Ltd., MA, USA). The binding energy was calibrated with reference to the C1s peak of 284.8 eV. Fourier transform infrared spectroscopy (FTIR, Nicolet iS50, Thermo Fisher Scientific Co., Ltd., MA, USA) was used to test whether SCA was grafted onto the BNNSs surface, and the spectrum was recorded from 4000 cm$^{-1}$ to 500 cm$^{-1}$ with a resolution of 4 cm$^{-1}$. TGA of pristine BNNSs and SCA treated BNNSs was carried out using a thermogravimetric differential thermal analyzer (TGA/DSC1/1600LF, METTLER TOLEDO Group, Zurich, Switzerland) at a heating rate of 10 °C/min under a nitrogen atmosphere from 25 °C to 800 °C to evaluate the amount of coating of SCA on BNNSs. The HCDJC-100kV voltage breakdown tester (Huace Instrument Co., Ltd., Beijing, China) was used to test the AC breakdown strength of the BN/EP nanocomposites at different contents. The electrodes are stainless-steel ball electrodes with a diameter of 20 mm. The thickness of the samples is 1mm, and the diameter is 40 mm. To prevent flashover, the sample is immersed in silicone oil. It is uniformly pressurized at a speed of 2 kV/s until breakdown. The thermal conductivity of the nanocomposites was measured by a laser thermal conductivity meter (LFA467HT, Netzsch. Ltd., Selb, Germany).

*2.6. Statistical Analysis*

The experimental data were statistically analyzed using OriginPro software, and the breakdown strength data were fitted using MATLAB software. One-way ANOVA was used to assess statistically significant differences in the experimental data ($p < 0.05$ was considered statistically significant).

## 3. Results and Discussion

*3.1. X-Ray Photoelectron Spectroscopy (XPS)*

The XPS spectra of BNNSs before and after plasma modification are shown in Figure 6. It can be seen that the BNNSs mainly have B, N, C and O on the surface, and the O peak of the BNNSs is obviously enhanced after plasma modification. To further analyze the effect of plasma hydroxylation modification, the peak of B1s was fitted, and the peak results are shown in Figure 7. The peak at 190.6 eV represents the B-N bond, and the peak at 191.5 eV represents the B-O bond, which forms –OH groups on the BNNSs [28]. Therefore, the effect of the hydroxylation modification can be measured by the content of the B-O bond. After plasma modification, the content of the B-O bond increased from 3.06% to 9.01%, indicating that the plasma modification significantly increased the hydroxyl content of the BNNSs surface.

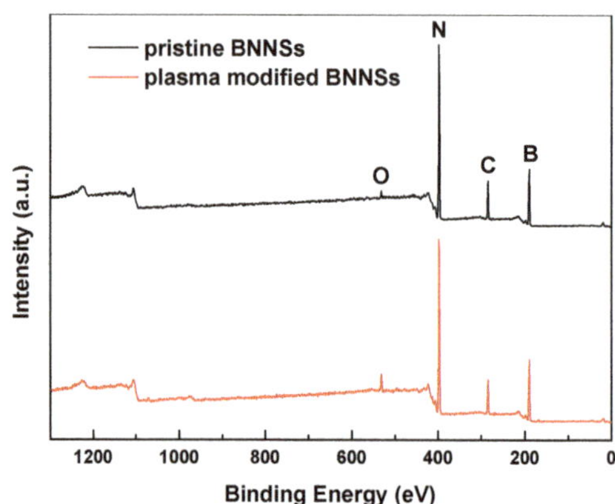


**Figure 6.** XPS survey spectra of pristine BNNSs and plasma modified BNNSs.

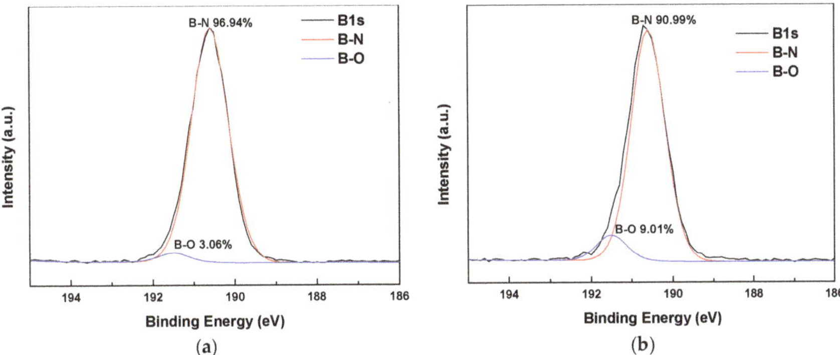

**Figure 7.** X-ray photoelectron spectroscopy (XPS) spectra of B1s of (**a**) pristine BNNSs, and (**b**) plasma modified BNNSs.

This increase is mainly due to the physical and chemical reactions initiated by the collision of high-energy particles in the plasma. First, the high-energy electrons generated by the discharge collide with the ground-state argon atoms, as the reactions reported in the article [29].

Since the ionization energy of argon is relatively high, in addition to direct ionization, it is excited to generate excited state argon. Then excited state argon atoms are ionized to generate more charged particles or returns to the ground state to release photons, further promoting the development of discharge. High-energy particles, such as electrons, excited state argon atoms and photons, have higher energy to break the B-N bond and create active sites on the B atom, which provides a basis for hydroxyl groups bonding.

Hydroxyl groups are mainly produced by ionization decomposition of water. The main reactions are reported in the article [30].

There are two main ways for water to decompose to produce hydroxyl groups. One is the direct collision between electrons and water molecules, and the other is the collision between excited argon atoms and water molecules. The generated hydroxyl group has excess electrons, and can interact with the electron-deficient B atom to form the covalent bond B-OH through Lewis acid-base interactions [28], thereby increasing the hydroxyl content of the BNNSs.

### 3.2. Fourier Transform Infrared Spectroscopy (FTIR)

Figure 8 is the Fourier transform infrared spectra of pristine BNNSs, SCA treated BNNSs, plasma and SCA treated BNNSs. All three particles have three absorption peaks at 3394, 1324, and 765 cm$^{-1}$, which represent the -OH vibration peak, the B-N stretching vibration and bending vibration peak, respectively [5]. After SCA treatment, weak -CH2- antisymmetric and symmetric stretching vibration peaks appear at 2932 and 2871 cm$^{-1}$, which represent the carbon chain in the organic functional group of SCA [31]. A Si-O-Si stretching vibration peak appears at 1100 cm$^{-1}$, which is formed by the dehydration condensation reaction of Si-OH in the SCA. The Si-O-C, which is a hydrolyzable functional group of SCA, stretching vibration peak appears at 1020 cm$^{-1}$, indicating that some SCA is not sufficiently hydrolyzed. 918 cm$^{-1}$ is the vibration peak of B-O-Si [32], which is formed by the dehydration condensation reaction between B-OH on the surface of BNNSs and Si-OH of SCA. Moreover, the BNNSs treated with plasma and SCA have a stronger B-O-Si vibration peak than those treated with SCA only. This difference is because the plasma modification increases the hydroxyl content of the BNNSs, thereby enhancing the degree of binding of the BNNSs to SCA. A schematic diagram of the reaction between SCA and BNNSs is shown in Figure 9.

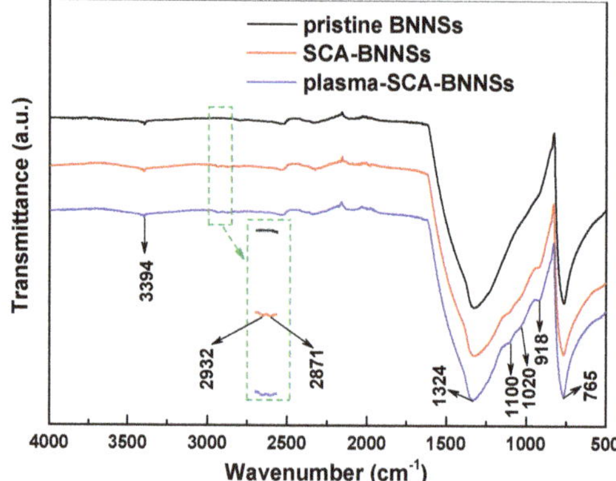

**Figure 8.** Fourier transform infrared spectroscopy (FTIR) spectra of pristine BNNSs, SCA treated BNNSs (SCA-BNNSs), and plasma and SCA treated BNNSs (plasma-SCA-BNNSs).

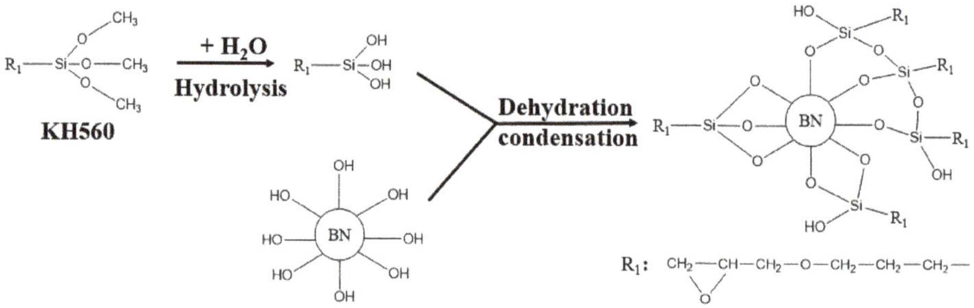

**Figure 9.** Schematic diagram of the reaction between SCA and BNNSs.

### 3.3. Thermogravimetric Analysis (TGA)

Figure 10 shows the thermogravimetric curves of pristine BNNSs, SCA treated BNNSs, and plasma and SCA treated BNNSs. It can be seen that pristine BNNSs are very stable, and their weight does not change substantially as the temperature increases. The BNNSs treated by the two methods show different degrees of weight loss. The weight loss at approximately 200 °C is most likely caused by the thermal desorption of water [32]. It can be seen that plasma and SCA treated BNNSs have more dehydration weight loss, which may be because some of the hydroxyl groups on the surface of the BNNSs do not react with SCA, and there are more redundant hydroxyl groups after plasma modification that adsorb more water. In addition, a portion of Si-OH does not react with B-OH and may adsorb more water. The weight loss from 200 °C to 800 °C is caused by the decomposition of SCA; thus, the coating rate of SCA can be estimated by subtracting the mass fractions of 200 °C and 800 °C. To estimate more accurately, the TGA tests of SCA-BNNSs and plasma-SCA-BNNSs were carried out 3 times, respectively. The interval between each treatment and test is less than 1 h. The results are shown in Table 1 and Figure 11. After the plasma modification, the average SCA coating rate increases from 1.47% to 2.24%, which is 52.4% higher and consistent with the experimental results in Section 3.2.

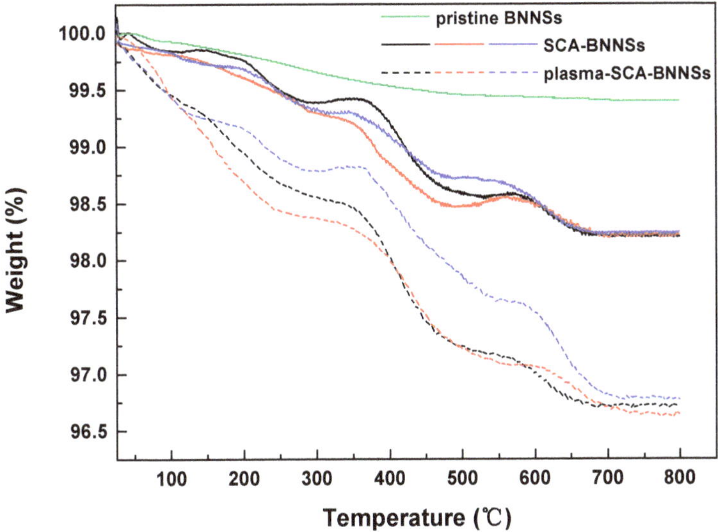

**Figure 10.** Thermal degradation of pristine BNNSs, SCA treated BNNSs (SCA-BNNSs), and plasma and SCA treated BNNSs (plasma-SCA-BNNSs). The interval between each treatment and test is less than 1 h.

**Table 1.** Coating rate of SCA treated BNNSs (SCA-BNNSs), and plasma and SCA treated BNNSs (plasma-SCA-BNNSs).

| Sample Number | SCA-BNNSs | Plasma-SCA-BNNSs |
| --- | --- | --- |
| 1 | 1.45% | 2.39% |
| 2 | 1.56% | 2.26% |
| 3 | 1.39% | 2.06% |
| Average | 1.47% | 2.24% |

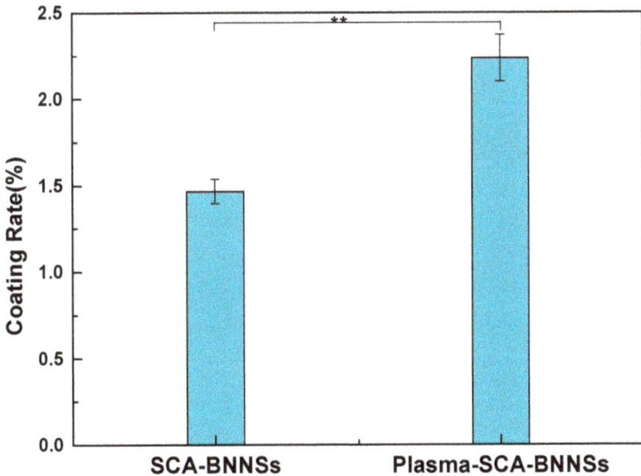

**Figure 11.** Coating rate of SCA treated BNNSs (SCA-BNNSs), and plasma and SCA treated BNNSs (plasma-SCA-BNNSs). ** $p < 0.01$. The interval between each treatment and test is less than 1 h.

### 3.4. Breakdown Strength

In the breakdown test, 10 samples were prepared for each kind of composites, and then analyzed using two-parameter Weibull distribution [33]:

$$P = 1 - \exp[-(\frac{E}{\alpha})^{\beta}], \tag{2}$$

where $P$ is the breakdown probability when the field strength is $E$, $E$ is the breakdown field strength of the test, $\alpha$ is the size parameter or characteristic breakdown field strength, indicating the field strength value when the breakdown probability is 63.2%, $\beta$ is the shape parameter, indicating the dispersion of the experimental data; the smaller the value, the greater the dispersion. For the breakdown field strength measured for each sample, the breakdown probability can be calculated using the following formula:

$$P_i = \frac{i - 0.5}{n + 0.25} \times 100\%, \tag{3}$$

where $i$ represents the ordinal number of the $E$ value in ascending order, and $n$ represents the total number of test samples for each material. Perform two logarithmic operations on Equation (2), and it can be transformed into a linear Equation:

$$\lg\left(\ln\left(\frac{1}{1-P}\right)\right) = \beta\lg E - \beta\lg\alpha, \tag{4}$$

Therefore, $\lg(\ln(1/(1-P)))$ can be linearly fitted to $\lg E$ to calculate the characteristic breakdown strength $\alpha$.

Figure 12 is the Weibull distribution of the AC breakdown strength of the composites at different contents, the parameters of the Weibull analysis are shown in Table 2. $\beta$ is greater than 14, even close to 70, indicating that the dispersion of experimental data of breakdown strength is small. $R^2$ is greater than 0.8, indicating that the degree of fit is better. Figure 13 is the characteristic breakdown strength of the composites at different contents. The breakdown strength of both kinds of composites decreases with increasing BNNSs content, but the plasma and SCA treated samples have higher strength than the samples treated only with SCA. This difference may be due to the increase in the hydroxyl groups on the surface of the BNNSs after plasma modification, which improves the coating rate of SCA. The curing reaction of epoxy resin is shown in Figure 14. Since the KH560 SCA has an epoxy group, it can participate in the curing reaction of the epoxy resin, thereby enhancing the degree of bonding between BNNSs and EP matrix. The dispersion of BNNSs in the epoxy resin is improved and the area of the interface bonding region increases. Therefore, there are more deep traps introduced in the nanocomposites [34,35]. On the one hand, deep traps reduce the mobility and average free path of carriers. On the other hand, the trapped carriers near the electrode form space charges of the same polarity, weakening the electric field near the electrode and suppressing the charge injection. The polymer chain fracture becomes difficult and the breakdown strength increases [36]. If 90% of the breakdown strength of pure epoxy resin is acceptable for insulation, the content of BNNSs can increase from approximately 10% to 20% after plasma modification, which will help to improve the thermal conductivity of the composites.

### 3.5. Thermal Conductivity

Figure 15 shows the thermal conductivity of the two kinds of composites at different contents. The values of three samples for each material were averaged and analyzed for significance. The higher the BNNSs content, the higher the thermal conductivity. Moreover, the thermal conductivity of the plasma and SCA treated samples is higher than that of samples treated only with SCA, which may be because the plasma modification increases the hydroxyl content of the BNNSs, thereby increasing the coating rate of SCA. Increasing the coating enhances the interface bonding ability of the BNNSs and epoxy resin, improving the compatibility of the two materials, reducing phonon scattering and interface thermal resistance, and improving the thermal conductivity [37,38]. In addition, combined with the breakdown strength in Section 3.4, when the BNNSs content is increased from 10% (SCA treated) to 20% (plasma and SCA treated), the thermal conductivity of the composites will increase by 67%.

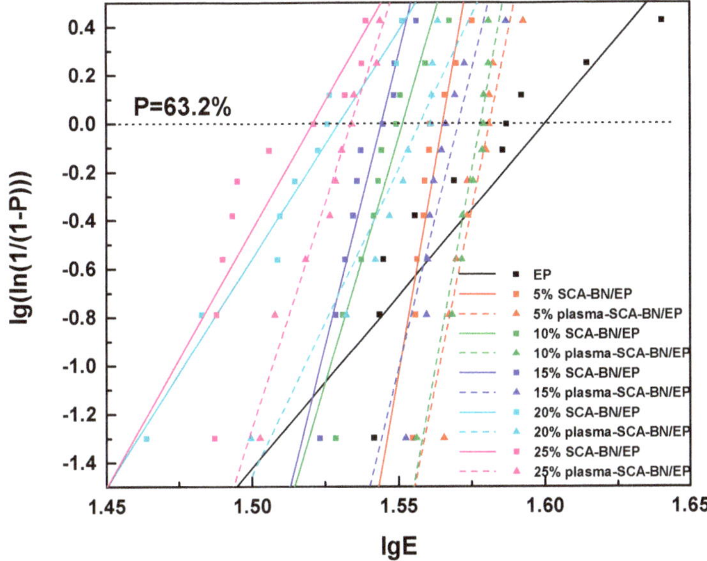


**Figure 12.** Weibull plots of the breakdown strength of epoxy resin (EP) and BN/EP nanocomposites with SCA treated BNNSs (SCA-BN/EP) and plasma and SCA treated BNNSs (plasma-SCA-BN/EP) at different contents (5%, 10%, 15%, 20%, 25%).

**Table 2.** Parameters of the Weibull analysis.

| Samples | $\alpha$ (kV/mm) | $\beta$ | $R^2$ |
| --- | --- | --- | --- |
| EP | 39.81 | 14.23 | 0.8134 |
| 5% SCA-BN/EP | 36.75 | 67.53 | 0.8493 |
| 5% plasma-SCA-BN/EP | 38.11 | 58.70 | 0.8690 |
| 10% SCA-BN/EP | 35.60 | 40.55 | 0.8885 |
| 10% plasma-SCA-BN/EP | 37.85 | 65.46 | 0.9345 |
| 15% SCA-BN/EP | 35.00 | 48.10 | 0.9172 |
| 15% plasma-SCA-BN/EP | 37.21 | 48.93 | 0.8715 |
| 20% SCA-BN/EP | 33.86 | 18.84 | 0.9612 |
| 20% plasma-SCA-BN/EP | 36.10 | 25.31 | 0.9174 |
| 25% SCA-BN/EP | 33.33 | 21.24 | 0.8079 |
| 25% plasma-SCA-BN/EP | 34.19 | 37.20 | 0.9622 |

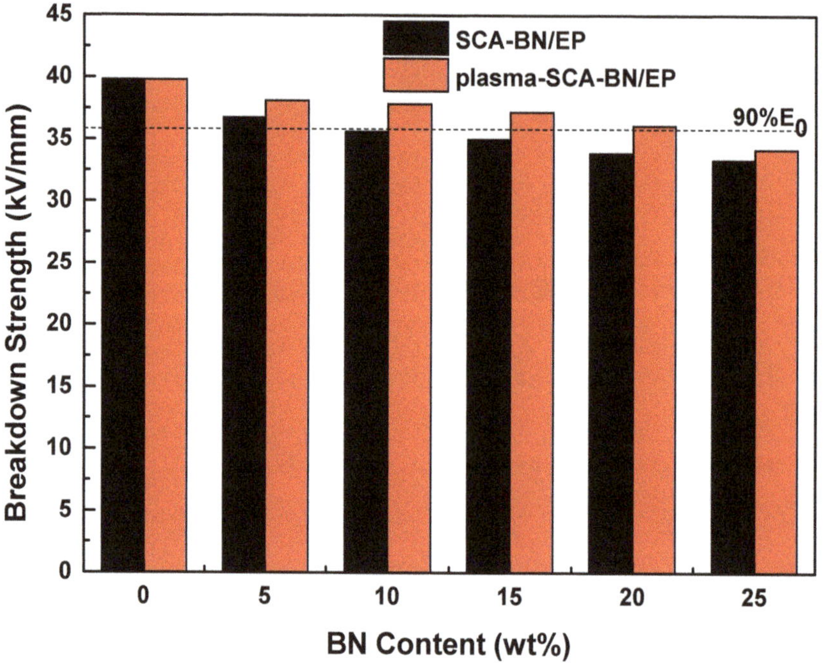


**Figure 13.** Breakdown strength of the EP and BN/EP nanocomposites with SCA treated BNNSs (SCA-BN/EP) and plasma and SCA treated BNNSs (plasma-SCA-BN/EP) at different contents.

**Figure 14.** Schematic diagram of curing reaction of epoxy resin.

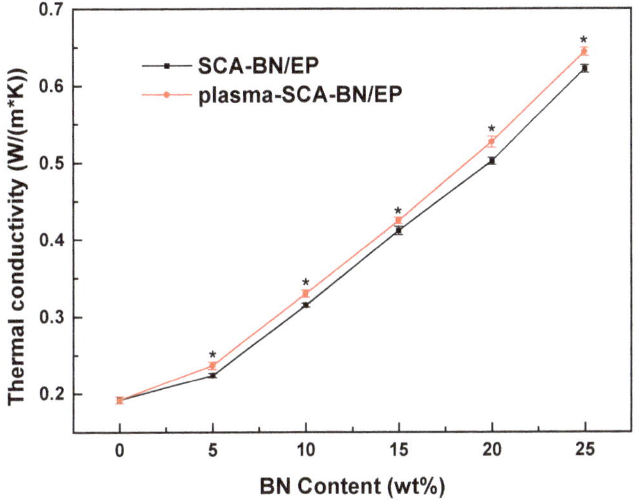


**Figure 15.** Thermal conductivity of BN/EP nanocomposites with SCA treated BNNSs (SCA-BN/EP) and plasma and SCA treated BNNSs (plasma-SCA-BN/EP) at different contents. * $p < 0.05$.

## 4. Conclusions

To simultaneously improve the AC breakdown strength and thermal conductivity of BN/EP nanocomposites, the hydroxylation of BNNSs was carried out by atmospheric pressure bipolar nanosecond pulse DBD Ar+$H_2$O low temperature plasma. XPS results indicate that plasma modification significantly increases the hydroxyl content of the BNNSs. The FTIR and TGA results show that more hydroxyl groups and SCA are combined after plasma modification, which improves the coating rate of SCA on the BNNSs surface. The breakdown test and thermal conductivity results show that the plasma and SCA treated nanocomposites have higher breakdown strength and thermal conductivity than those treated with only SCA, although the increased amplitude is not very large. At the same breakdown strength (90% of the breakdown strength of pure epoxy resin), the BNNSs content can increase to 20% after plasma modification, whereby the thermal conductivity of the composites will increase by 67%. Therefore, this paper provides a simple, efficient and environmentally friendly plasma hydroxylation modification method, which also provides effective guidance for the preparation of high thermal conductivity insulating composites. Of course, this method needs further improvement to achieve higher breakdown strength for practical applications.

**Author Contributions:** Conceptualization, Y.M. and J.G.; Methodology, Y.M. and J.G.; Validation, Q.L., J.G. and L.L.; Formal analysis, J.G. and L.L.; Investigation, J.G., L.L. and X.G.; Resources, H.W. and X.G.; Writing—original draft preparation, J.G.; Writing—review and editing, J.G., L.L.; Supervision, Y.M.

**Funding:** This research was funded by the National "111" Project of China (B08036).

**Conflicts of Interest:** The authors declare no conflict of interest.

## References


1. Huang, X.Y.; Jiang, P.K.; Tanaka, T. A review of dielectric polymer composites with high thermal conductivity. *IEEE Electr. Insul. Mag.* **2011**, *27*, 8–16. [CrossRef]
2. Zhi, C.Y.; Bando, Y.; Terao, T.; Tang, C.C.; Kuwahara, H.; Golberg, D. Towards Thermoconductive, Electrically Insulating Polymeric Composites with Boron Nitride Nanotubes as Fillers. *Adv. Funct. Mater.* **2009**, *19*, 1857–1862. [CrossRef]
3. Xu, Y.S.; Chung, D.D.L.; Mroz, C. Thermally conducting aluminum nitride polymer-matrix composites. *Compos. Part A Appl. Sci. Manuf.* **2001**, *32*, 1749–1757. [CrossRef]

4. He, H.; Fu, R.; Han, Y.; Shen, Y.; Wang, D. High thermal conductive $Si_3N_4$ particle filled epoxy composites with a novel structure. *J. Electron. Packag.* **2007**, *129*, 469–472. [CrossRef]

5. Yu, J.H.; Huang, X.Y.; Wu, C.; Wu, X.F.; Wang, G.L.; Jiang, P.K. Interfacial modification of boron nitride nanoplatelets for epoxy composites with improved thermal properties. *Polymer* **2012**, *53*, 471–480. [CrossRef]

6. Zhu, B.L.; Ma, J.; Wu, J.; Yung, K.C.; Xie, C.S. Study on the Properties of the Epoxy-Matrix Composites Filled with Thermally Conductive AlN and BN Ceramic Particles. *J. Appl. Polym. Sci.* **2010**, *118*, 2754–2764. [CrossRef]

7. Chiang, T.H.; Hsieh, T.E. A study of encapsulation resin containing hexagonal boron nitride (hBN) as inorganic filler. *J. Inorg. Organomet. Polym. Mater.* **2006**, *16*, 175–183. [CrossRef]

8. Marx, P.; Wanner, A.J.; Zhang, Z.C.; Jin, H.F.; Tsekmes, I.A.; Smit, J.J.; Kern, W.; Wiesbrock, F. Effect of Interfacial Polarization and Water Absorption on the Dielectric Properties of Epoxy-Nanocomposites. *Polymers* **2017**, *9*, 195. [CrossRef]

9. Wang, Z.B.; Iizuka, T.; Kozako, M.; Ohki, Y.; Tanaka, T. Development of Epoxy/BN Composites with High Thermal Conductivity and Sufficient Dielectric Breakdown Strength Part I—Sample Preparations and Thermal Conductivity. *IEEE Trans. Dielectr. Electr. Insul.* **2011**, *18*, 1963–1972. [CrossRef]

10. Liu, D.; Hoang, A.T.; Pourrahimi, A.M.; Pallon, L.K.H.; Nilsson, F.; Gubanski, S.M.; Olsson, T.; Hedenqvist, M.S.; Gedde, U.W. Influence of Nanoparticle Surface Coating on Electrical Conductivity of $LDPE/Al_2O_3$ Nanocomposites for HVDC Cable Insulations. *IEEE Trans. Dielectr. Electr. Insul.* **2017**, *24*, 1396–1404. [CrossRef]

11. Gao, M.Z.; Zhang, P.H. Relationship between dielectric properties and nanoparticle dispersion of nano-$SiO_2$/epoxy composite. *Acta Phys. Sin.* **2016**, *65*, 192–199. [CrossRef]

12. Yang, G.Q.; Cui, J.; Ohki, Y.; Wang, D.Y.; Li, Y.; Tao, K. Dielectric and relaxation properties of composites of epoxy resin and hyperbranched-polyester-treated nanosilica. *RSC Adv.* **2018**, *8*, 30669–30677. [CrossRef]

13. Jesionowski, T.; Krysztafkiewicz, A. Influence of silane coupling agents on surface properties of precipitated silicas. *Appl. Surf. Sci.* **2001**, *172*, 18–32. [CrossRef]

14. Kim, Y.; Hwang, S.; So, J.I.; Kim, C.L.; Kim, M.; Sang, E.S. Treatment of Atmospheric-Pressure Radio Frequency Plasma on Boron Nitride for Improving Thermal Conductivity of Polydimethylsiloxane Composites. *Macromol. Res.* **2018**, *26*, 864–867. [CrossRef]

15. Yu, B.; Xing, W.; Guo, W.; Qiu, S.; Xin, W.; Lo, S.; Yuan, H. Thermal exfoliation of hexagonal boron nitride for effective enhancements on thermal stability, flame retardancy and smoke suppression of epoxy resin nanocomposites via sol–gel process. *J. Mater. Chem. A* **2016**, *4*, 7330–7340. [CrossRef]

16. Kim, K.; Kim, M.; Hwang, Y.; Kim, J. Chemically modified boron nitride-epoxy terminated dimethylsiloxane composite for improving the thermal conductivity. *Ceram. Int.* **2014**, *40*, 2047–2056. [CrossRef]

17. Huang, X.Y.; Zhi, C.Y.; Jiang, P.K.; Golberg, D.; Bando, Y.; Tanaka, T. Polyhedral Oligosilsesquioxane-Modified Boron Nitride Nanotube Based Epoxy Nanocomposites: An Ideal Dielectric Material with High Thermal Conductivity. *Adv. Funct. Mater.* **2013**, *23*, 1824–1831. [CrossRef]

18. Hou, J.; Li, G.H.; Yang, N.; Qin, L.L.; Grami, M.E.; Zhang, Q.X.; Wang, N.Y.; Qu, X.W. Preparation and characterization of surface modified boron nitride epoxy composites with enhanced thermal conductivity. *RSC Adv.* **2014**, *4*, 44282–44290. [CrossRef]

19. Wu, X.H.; Wu, G.N.; Yang, Y.; Zhang, X.T.; Lei, Y.X.; Zhong, X.; Zhu, J. Influence of Nanoparticle Plasma Modification on Trap Properties of Polyimide Composite Films. *Proc. CSEE* **2018**, *38*, 3410–3418. [CrossRef]

20. Shao, T.; Zhang, C.; Long, K.H.; Zhang, D.D.; Wang, J.; Yan, P.; Zhou, Y.X. Surface modification of polyimide films using unipolar nanosecond-pulse DBD in atmospheric air. *Appl. Surf. Sci.* **2010**, *256*, 3888–3894. [CrossRef]

21. Shao, T.; Yu, Y.; Zhang, C.; Zhang, D.D.; Niu, Z.; Wang, J.; Yan, P.; Zhou, Y.X. Excitation of Atmospheric Pressure Uniform Dielectric Barrier Discharge Using Repetitive Unipolar Nanosecond-pulse Generator. *IEEE Trans. Dielectr. Electr. Insul.* **2010**, *17*, 1830–1837. [CrossRef]

22. Yuan, H.; Wang, W.C.; Yang, D.Z.; Zhao, Z.L.; Zhang, L.; Wang, S. Atmospheric air dielectric barrier discharge excited by nanosecond pulse and AC used for improving the hydrophilicity of aramid fibers. *Plasma Sci. Technol.* **2017**, *19*, 125401. [CrossRef]

23. Wu, S.; Xu, H.; Lu, X.; Pan, Y. Effect of Pulse Rising Time of Pulse dc Voltage on Atmospheric Pressure Non-Equilibrium Plasma. *Plasma Process Polym.* **2013**, *10*, 136–140. [CrossRef]

24. Liu, Y.; Chunqiang, S.U.; Xiang, R.; Fan, C.; Zhou, W.; Feng, W.; Ding, W. Experimental study on surface modification of PET films under bipolar nanosecond-pulse dielectric barrier discharge in atmospheric air. *Appl. Surf. Sci.* **2014**, *313*, 53–59. [CrossRef]

25. Yang, D.Z.; Yang, Y.; Li, S.Z.; Nie, D.X.; Zhang, S.; Wang, W.C. A homogeneous dielectric barrier discharge plasma excited by a bipolar nanosecond pulse in nitrogen and air. *Plasma Sources Sci. Technol.* **2012**, *21*, 035004. [CrossRef]

26. Mi, Y.; Wan, H.; Bian, C.H.; Peng, W.C.; Gui, L. An MMC-based Modular Unipolar/Bipolar High-voltage Nanosecond Pulse Generator with Adjustable Rise/Fall Time. *IEEE Trans. Dielectr. Electr. Insul.* **2019**, *26*, 515–522. [CrossRef]

27. Ma, X.K.; Lee, N.H.; Oh, H.J.; Jung, S.C.; Lee, W.J.; Kim, S.J. Morphology control of hexagonal boron nitride by a silane coupling agent. *J. Cryst. Growth* **2011**, *316*, 185–190. [CrossRef]

28. Pakdel, A.; Bando, Y.; Golberg, D. Plasma-Assisted Interface Engineering of Boron Nitride Nanostructure Films. *ACS Nano* **2014**, *8*, 10631–10639. [CrossRef]

29. Park, G.; Lee, H.; Kim, G.; Lee, J.K. Global Model of He/$O_2$ and Ar/$O_2$ Atmospheric Pressure Glow Discharges. *Plasma Process. Polym.* **2010**, *5*, 569–576. [CrossRef]

30. Liu, D.X.; Sun, B.W.; Iza, F.; Xu, D.H.; Wang, X.H.; Rong, M.Z.; Kong, M.G. Main species and chemical pathways in cold atmospheric-pressure $Ar^+$ $H_2O$ plasmas. *Plasma Sources Sci. Technol.* **2017**, *26*, 045009. [CrossRef]

31. Joni, I.M.; Balgis, R.; Ogi, T.; Iwaki, T.; Okuyama, K. Surface functionalization for dispersing and stabilizing hexagonal boron nitride nanoparticle by bead milling. *Colloids Surf. A Physicochem. Eng. Asp.* **2011**, *388*, 49–58. [CrossRef]

32. Seyhan, A.T.; Goncu, Y.; Durukan, O.; Akay, A.; Ay, N. Silanization of boron nitride nanosheets (BNNSs) through microfluidization and their use for producing thermally conductive and electrically insulating polymer nanocomposites. *J. Solid State Chem.* **2017**, *249*, 98–107. [CrossRef]

33. Tuncer, E.; James, D.R.; Sauers, I.; Ellis, A.R.; Pace, M.O. On dielectric breakdown statistics. *J. Phys. D Appl. Phys.* **2006**, *39*, 4257–4268. [CrossRef]

34. Tanaka, T. Dielectric nanocomposites with insulating properties. *IEEE Trans. Dielectr. Electr. Insul.* **2005**, *12*, 914–928. [CrossRef]

35. Wang, X.Y.; Andritsch, T.; Chen, G.; Virtanen, S. The Role of the Filler Surface Chemistry on the Dielectric and Thermal Properties of Polypropylene Aluminium Nitride Nanocomposites. *IEEE Trans. Dielectr. Electr. Insul.* **2019**, *26*, 1009–1017. [CrossRef]

36. Li, S.T.; Min, D.M.; Wang, W.W.; Chen, G. Linking Traps to Dielectric Breakdown through Charge Dynamics for Polymer Nanocomposites. *IEEE Trans. Dielectr. Electr. Insul.* **2016**, *23*, 2777–2785. [CrossRef]

37. Gu, J.W.; Zhang, Q.Y.; Dang, J.; Xie, C. Thermal conductivity epoxy resin composites filled with boron nitride. *Polym. Adv. Technol.* **2012**, *23*, 1025–1028. [CrossRef]

38. Gu, J.W.; Liang, C.B.; Dang, J.; Dong, W.C.; Zhang, Q.Y. Ideal dielectric thermally conductive bismaleimide nanocomposites filled with polyhedral oligomeric silsesquioxane functionalized nanosized boron nitride. *RSC Adv.* **2016**, *6*, 35809–35814. [CrossRef]

 *nanomaterials*

*Article*

# A Novel Route to Manufacture 2D Layer $MoS_2$ and $g\text{-}C_3N_4$ by Atmospheric Plasma with Enhanced Visible-Light-Driven Photocatalysis

**Bo Zhang, Zhenhai Wang, Xiangfeng Peng, Zhao Wang *, Ling Zhou and QiuXiang Yin**

National Engineering Research Center of Industry Crystallization Technology, School of Chemical Engineering and Technology, Tianjin University, Tianjin 300072, China

* Correspondence: wangzhao@tju.edu.cn; Tel.: +86-138-2052-5018

Received: 12 July 2019; Accepted: 6 August 2019; Published: 8 August 2019

**Abstract:** An atmospheric plasma treatment strategy was developed to prepare two-dimensional (2D) molybdenum disulfide ($MoS_2$) and graphitic carbon nitride ($g\text{-}C_3N_4$) nanosheets from $(NH_4)_2MoS_4$ and bulk $g\text{-}C_3N_4$, respectively. The moderate temperature of plasma is beneficial for exfoliating bulk materials to thinner nanosheets. The thicknesses of as-prepared $MoS_2$ and $g\text{-}C_3N_4$ nanosheets are 2–3 nm and 1.2 nm, respectively. They exhibited excellent photocatalytic activity on account of the nanosheet structure, larger surface area, more flexible photophysical properties, and longer charge carrier average lifetime. Under visible light irradiation, the hydrogen production rates of $MoS_2$ and $g\text{-}C_3N_4$ by plasma were 3.3 and 1.5 times higher than the corresponding bulk materials, respectively. And $g\text{-}C_3N_4$ by plasma exhibited 2.5 and 1.3 times degradation rates on bulk that for methyl orange and rhodamine B, respectively. The mechanism of plasma preparation was proposed on account of microstructure characterization and online mass spectroscopy, which indicated that gas etching, gas expansion, and the repulsive force of electron play the key roles in the plasma exfoliation. Plasma as an environmentally benign approach provides a general platform for fabricating ultrathin nanosheet materials with prospective applications as photocatalysts for pollutant degradation and water splitting.

**Keywords:** dielectric barrier discharge plasma; $MoS_2$ nanosheets; $g\text{-}C_3N_4$ nanosheets; photodegradation; water splitting; gas etching; repulsive force

---

## 1. Introduction

Photocatalytic technology is a fascinating strategy in addressing energy shortages and environmental pollution [1–5]. Two-dimensional (2D) materials have a wide range of applications in the field of photocatalysis due to their special structure and excellent optical and electrical properties [6–8]. 2D nanosheets made of a few atomic layers are mainly synthesized from layered structural materials. Graphene is the typical 2D material [9–11], and has been applied in catalysts and electronics. In recent years, other layered structural materials, for example, metal-free materials [12], transitional metal dichalcogenides [13], and transitional metal carbides [14] have also been exfoliated into 2D nanosheets to explore their unique properties and applications. However, the preparation method is an important factor that restricts the performance and production of two-dimensional materials. At present, there are some commonly methods for produce 2D materials, such as ultrasonication [15], hydrothermal method [16] and chemical vapor deposition method [17], etc. However, a fast, high yield, and an environmentally-friendly method for 2D materials manufacture is still urgently needed.

Dielectric barrier discharge (DBD) plasma is a kind of cold plasma used to prepare nano-sized materials in atmosphere. DBD plasma has excellent advantages in the preparation of nanomaterials, which can be attributed to its large amount of active substances, ambient temperature and nonequilibrium state [18–20]. DBD plasma is more remarkably used in a more controlled method for producing structures and in surface induction processes, in comparison with the traditional thermal methods [21,22]. Wang et al. utilized water plasma to prepare 2D layered double hydroxide nanosheets to improve the rate of oxygen evolution reaction [23]. N-doped graphene was exfoliated by DBD plasma for oxygen reduction reaction [24]. Our group developed a way to prepare graphene using atmospheric plasma and proposed the preparation mechanism [25].

Both $MoS_2$ and $g$-$C_3N_4$, typical two-dimensional layered materials, are the most interesting photocatalysts. $MoS_2$, a typical transitional metal dichalcogenides with a sandwich layered structure, has been far and wide exploited in photocatalytic $H_2$ production [26,27]. $g$-$C_3N_4$, a non-metal semiconductor with a grapheme-like layered structure that was successfully used in split water and degrade organic contaminants under visible light irradiation [28]. Nevertheless, their photocatalytic performance was limited for bare bulk structure [29–31]. Hence, much research has been conducted to improve their photolytic activity, such as preparing ultrathin $MoS_2$ nanosheets and exfoliating thick $g$-$C_3N_4$ into few layers nanosheets. Xie et al. synthesized $MoS_2$ nanosheets by hydrothermal method, with a thickness of 5.9 nm by hydrothermal method, corresponding to 9 S-MO-S atomic layers [32]. Niu et al. used direct thermal treatment to prepare $g$-$C_3N_4$ nanosheets with a yield of around 6%, and its thickness is about 2 nm [33]. Zhang et al. utilized bulk $g$-$C_3N_4$ as the precursor and produced the $g$-$C_3N_4$ nanosheets by ultrasonic treatment [34].

Here, we developed a novel process of DBD plasma to the preparation of ultrathin $MoS_2$ nanosheets and $g$-$C_3N_4$ nanosheets from $(NH_4)_2MoS_4$ and bulk $g$-$C_3N_4$, respectively. Compared to the traditional methods, this method has the advantages of high yield and environmentally friendly (without solvent). The structure, morphology, and property of prepared $MoS_2$ nanosheets and $g$-$C_3N_4$ nanosheets were studied. Moreover, the photolytic activity was evaluated by $H_2$ evolution reaction and organic degradation under visible light irradiation. Finally, we believe that the plasma method can be a universal method for two-dimensional material preparation.

## 2. Experimental

### 2.1. Sample Preparation

All the materials were purchased from Shanghai Aladdin Biochemical Technology Co. (Shanghai, China). All chemicals are used directly without further purification.

Synthesis of $MoS_2$ nanosheets. 0.5 g of $(NH_4)_2MoS_4$ powder was treated by DBD plasma in $H_2$/Ar atmosphere, and the process was carried out at moderate temperature. Details of the DBD plasma treatment have been described previously. [25] More details about DBD treatment can be found in the Supplementary Material. The schematic representation of DBD plasma setup can be found in Figure S1. The sample was placed in the quartz ring between the two electrodes of DBD plasma generator. Total DBD treatment time was 1 h. The temperature of the DBD was measured by infrared imaging (Ircon, 100PHT, Everett, WA, USA), indicating that the DBD plasma process was at around 150 °C (Figure S2).

Synthesis of $g$-$C_3N_4$ nanosheets. 0.5 g of bulk $g$-$C_3N_4$ powder was treated by DBD plasma at moderate temperature and air atmosphere. The specific processing is the same as described above.

Images of samples (Figure S3) reveals color change before and after plasma treatment.

### 2.2. Characterizations

The crystalline phase of $g$-$C_3N_4$ and $MoS_2$ were analyzed on D/Max-2500 V diffractometer (Cu Kα α = 0.154 nm, 4°/min, Rigaku, Tokyo, Japan). The field emission scanning electron microscopy (SEM) and high-resolution transmission electron microscopy (HRTEM) images were obtained from Zeiss-Merlin scanning electron microscopy (Jena, Germany) and JEOL-2100F transmission electron microscopy (Tokyo, Japan), respectively. X-ray photoelectron spectroscopy (XPS) was conducted on a Perkin Elmer PHI-1600 system (MA, USA). An atomic force microscope (AFM) was used on Agilent 5500 (CA, USA). The FTIR of the prepared samples were analyzed using a Nicolet-560 (MN, USA).

Optical properties were measured on a UV-vis spectrophotometer (UV-2600, Shimadzu, Kyoto, Japan). The photoluminescence (PL) spectra and time-resolved fluorescence decay spectra was surveyed on a HORIBA Jobin Yvon Fluorolog-3 spectrophotometer (Paris, France) with an excitation wavelength at 330 nm. The $N_2$ Brunauer-Emmett-Teller (BET) surface area was measured on a Nova Automated Gas Sorption System (Quantachrome Corporation, FL, USA) after degassed at 150 °C for 4 h. The gas products were monitored online and analyzed with an SHP8400PMS-L mass spectrometer (SDPTOP, Shanghai, China).

### 2.3. Photocatalytic Activity

The photocatalytic $H_2$ production experiment was carried out by the Perfect Light IIIAG system (Beijing, China). The visible light source used in the experiment was a 300 W Xe lamp (420 nm filter).

Hydrogen production over Eosin Y-sensitized $MoS_2$. 15 mL Triethanolamine (TEOA) was mixed with 85 mL of deionized water, then 25 mg $MoS_2$ and 70 mg Eosin Y were sonicated in a short time to make it evenly dispersed in the solution, and finally the solution was made to drop PH = 7 by adding concentrated hydrochloric acid. The reactor continued to vacuum and circulate cold water to eliminate the influence of other factors on the experiment. Gas chromatography can accurately determine the amount of hydrogen in the system.

Hydrogen production over $Pt/g$-$C_3N_4$. 100 mg powder samples was dispersed in deionized water. 67 µL $H_2PtCl_6$ ($Pt/g$-$C_3N_4$: 0.5 wt %) was dispersed in the suspension. Under visible light, photoreduction and $H_2$ production were conducted on a Perfect Light IIIAG system. 10 vol % triethanolamine was added as sacrificial reagent. The other steps were the same as above.

Degrading pollutants by $g$-$C_3N_4$. Briefly, 50 mg catalyst was dispersed in 100 mL of 20 mg/L RhB or 10 mg/L MO solution, and the suspension was then placed into a 200 mL vessel with continuous stirring. Dark treatment of 30 min was necessary to approach an adsorption/desorption equilibrium, and a 300 W Xe lamp and 420 nm cutoff filter was used to supply visible light. 3 mL of solution was taken at intervals, and then the supernatant was obtained by high-speed centrifugation. The concentration was measured by UV-2600 spectrophotometer (Kyoto, Japan).

### 2.4. Photoelectrochemical Measurements

The Photocurrent density was conducted at the electrochemical analyzer. A three-electrode electrochemical cell was used for the measurements, tin oxide mixed with fluorine (FTO) conductive glass loaded with a sample as an working electrode, and its reference electrode and counter electrode were Hg/HgO electrode and Pt wire, respectively. The 0.1 M $Na_2SO_4$ solution containing 1 mM Eosin-y was used as an electrolyte. A 300 W Xe lamp equipped with 500 nm bandpass filter acts as the solar light source. At the beginning of the test, we fully introduced $N_2$ into the electrolyte to remove dissolved oxygen.

## 3. Results and Discussion

### 3.1. Sample Characterization

Figure 1a shows the dispersion of $g$-$C_3N_4$ in isopropanol before and after DBD plasma treatment. The plasma treated sample was a suspension in the liquid and has a lighter color than un-treated samples. It is in agreement with 2D $g$-$C_3N_4$ nanosheets described in the literature [35]. Figure 1b shows $MoS_2$ were dispersed in deionized water after 10 min ultrasonication and allowed to stand for 24 h. Through observation, the samples by DBD plasma were found to maintain better dispersibility than samples by calcination. This indicates that the sample prepared by DBD has smaller size or thinner nanosheets [32].

**Figure 1.** Images of (**a**) g-C$_3$N$_4$ suspensions with and without DBD plasma treatment, (**b**) MoS$_2$ suspensions by DBD plasma and calcination treatment.

Figure 2a show XRD patterns of two different structures of MoS$_2$ by calcination and DBD plasma. The as-prepared bulk MoS$_2$ and MoS$_2$ nanosheets exhibit similar (002), (100), (103) and (110) planes at 14.1°, 32.9°, 39.5° and 58.7°, respectively, which can be indicated as hexagonal 2H-MoS$_2$ (JCPDS 75-1539) [36]. It shows that the DBD plasma can successfully decompose (NH$_4$)$_2$MoS$_4$ to MoS$_2$. The peaks intensity of MoS$_2$ nanosheets by plasma is weaker than that of bulk MoS$_2$, indicating the crystallinity is low and meet the characteristics of the two-dimensional structure [37]. Figure 2b show the XRD patterns of two different structures of g-C$_3$N$_4$. They both exhibited two peaks at 13.0° and 27.2°, corresponding to the (100) and (002) planes of g-C$_3$N$_4$ (JCPDS 87-1526), respectively [9,37]. It shows the structure of interlayer stacking and conjugated aromatic system stacking. It indicates that both of samples have identical crystal structures. In addition, the intensity of the peak at 13.0° of samples remarkably decreased after DBD plasma treatment, which is consistent with typical XRD patterns of 2D g-C$_3$N$_4$ nanosheets in previous reports [38,39]. The result confirms that DBD plasma can exfoliate bulk g-C$_3$N$_4$ into nanosheets successfully.

The XPS survey spectra of MoS$_2$ and g-C$_3$N$_4$ can be seen from Figure S4 in Supplementary Material, and the atomic concentration of those were shown in Tables S1 and S2. They exhibted that the as-obtained samples matched the theoretical chemical formulas.

Figure 2c,d show the high-resolution Mo 3d and S 2p XPS spectra of MoS$_2$ samples. As can be seen from Figure 2c, there are three peaks located at 226.4 eV, 229.2 eV and 232.5 eV, the first peak can be ascribed to S 2s, and the latter two peaks correspond to Mo 3d$_{5/2}$ and Mo 3d$_{3/2}$, respectively [40]. Figure 2d shows the S 2p spectrum containing two peaks with binding energies of 162.0 eV and 163.5 eV, corresponding to S 2p$_{3/2}$ and S 2p$_{1/2}$, respectively [41]. The results indicate that this samples are MoS$_2$ with Mo$^{4+}$ and S$^{2-}$. Bulk MoS$_2$ and MoS$_2$ nanosheets all exhibited similar XPS spectrum, indicating MoS$_2$ by DBD and calcination have similar chemical structural compositions.

Figure 2e,f show the XPS spectra of g-C$_3$N$_4$ samples. C 1s spectra is shown in Figure 2e. Two symmetrical peaks at 284.8 and 288.4 eV were observed in both g-C$_3$N$_4$ nanosheets and bulk. The peak at 284.8 eV can be ascribed to the inherent adventitious carbon, and the other peak located at 288.4 eV was identified as sp$^2$-hybridized carbon (N–C=N) in g-C$_3$N$_4$ chemical structure [42–44]. Figure 2e shows the N 1s spectra of g-C$_3$N$_4$ samples. To analyze the chemical bonds of the functional groups, the N 1s spectra are deconvoluted into four peaks at 398.6, 399.3, 400.0, and 401.2 eV, respectively. The peak at 398.6 eV is attributed to the sp$^2$-hybridized nitrogen that existed in triazine rings (C–N=C); the peak at 399.3 eV is ascribed to the N atoms bonded to three sp$^2$ carbon atoms (N–(C)$_3$); the peak at 400.0 eV is attributed to the presence of amide (N–C=O); and the peak at 401.3 eV is due to the existence of amino functional groups (C–N–H). The other peak at 404.3 eV is ascribed to π-excitations [35,42–44]. Overall, g-C$_3$N$_4$ nanosheets exhibit the same chemical composition and element coordination as bulk g-C$_3$N$_4$. The result confirms that the DBD plasma process do not change the basic chemical composition of bulk g-C$_3$N$_4$ when g-C$_3$N$_4$ nanosheets is generated.

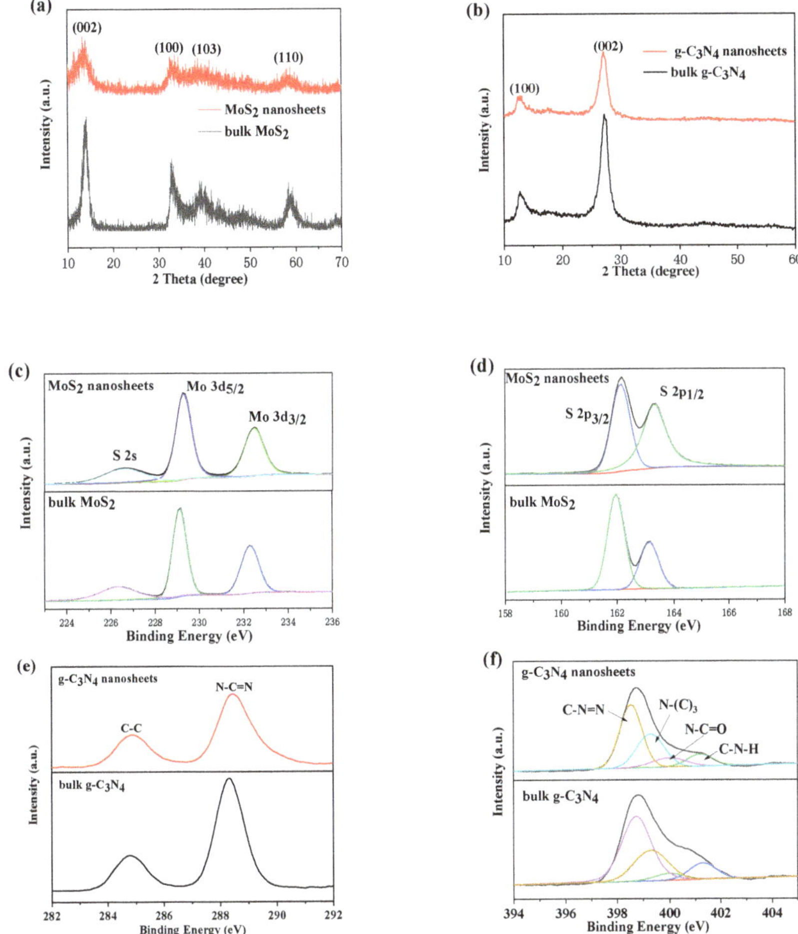

**Figure 2.** XRD patterns of (**a**) bulk $MoS_2$ and $MoS_2$ nanosheets, (**b**) bulk $g$-$C_3N_4$ and $g$-$C_3N_4$ nanosheets. XPS spectra (**c**) Mo 3d and (**d**) S 2p of bulk $MoS_2$ and $MoS_2$ nanosheets. (**e**) C 1s and (**f**) N 1s of bulk $g$-$C_3N_4$ and $g$-$C_3N_4$ nanosheets.

The chemical construction of plasma-prepared $g$-$C_3N_4$ nanosheets was characterized by the FTIR spectra. Figure S5 shows that bulk $g$-$C_3N_4$ have a sharp peak at 810 $cm^{-1}$ is assigned to the heptazine ring system, other peaks in the region of around 1000–1800 $cm^{-1}$ are attributed to bridging C–NH–C units or trigonal C–N–(C)–C units, the broad peak at 3000–3600 $cm^{-1}$ is ascribed to N–H and O–H stretching [45–48]. The spectrum of $g$-$C_3N_4$ nanosheets exhibit almost the same peaks with bulk $g$-$C_3N_4$, confirming ultrathin nanosheets prepared by cold plasma are indeed $g$-$C_3N_4$ nanosheets that possess the uniform chemical structure as the layered bulk $g$-$C_3N_4$. The results are in agreement with XPS and XRD, further confirming that the basic chemical composition have no change by plasma treatment.

Figure 3a shows that UV-vis diffuse reflectance spectra, and it revealed photo absorption properties of bulk $g$-$C_3N_4$ and $g$-$C_3N_4$ nanosheets. After plasma treatment, $g$-$C_3N_4$ nanosheets showed an obvious blue shift, which is typical for 2D materials. Based on the absorption edge, the band gaps were 2.70 and 2.75 eV for bulk and 2D $g$-$C_3N_4$, respectively. Due to the quantum confinement effect, 2D $g$-$C_3N_4$ has a bigger band gap and increases the charge carrier generated ability which will enhance the photocatalytic performance [31,33].

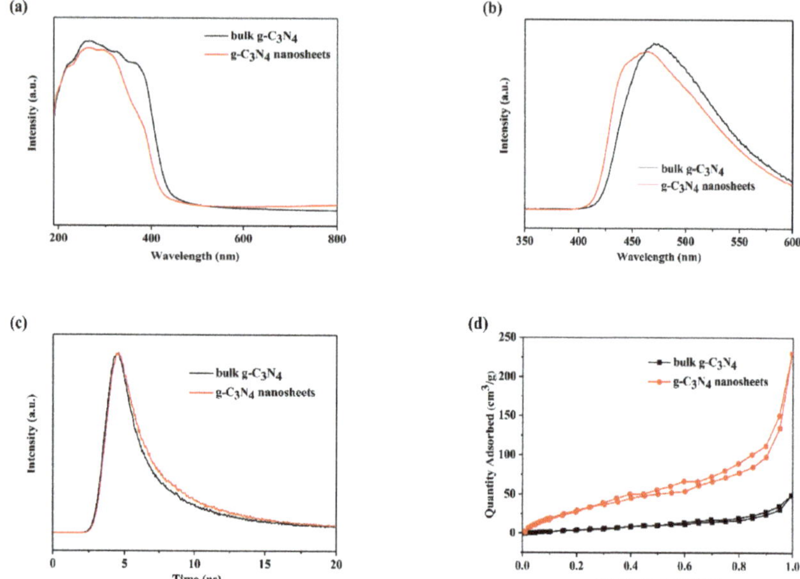

**Figure 3.** (**a**) UV-vis diffuse reflectance spectra (DRS) of bulk g-C$_3$N$_4$ and g-C$_3$N$_4$ nanosheets. (**b**) Photoluminescence (PL) spectra and (**c**) time-resolved fluorescence decay spectra of bulk g-C$_3$N$_4$ and g-C$_3$N$_4$ nanosheets. (**d**) N$_2$ adsorption-desorption isotherms of bulk g-C$_3$N$_4$ and g-C$_3$N$_4$ nanosheets.

Photoluminescence (PL) and time-resolved fluorescence decay spectra are employed to investigate the charge carrier separation and recombination properties of g-C$_3$N$_4$ nanosheets by plasma treatment. Figure 3b depicts the PL spectra of bulk g-C$_3$N$_4$ and g-C$_3$N$_4$ nanosheets, and a blue shift can be obviously observed in the g-C$_3$N$_4$ nanosheets due to the quantum confinement effect [49]. This result is consistent with UV-vis DRS result.

Figure 3c shows the time-resolved fluorescence decay spectra of bulk g-C$_3$N$_4$ and g-C$_3$N$_4$ nanosheet. The fluorescent intensities of both samples decay exponentially. However, the g-C$_3$N$_4$ nanosheets exhibited slower decay kinetics than bulk samples. Hence, the lifetime of the photo-induced charge carriers of plasma-exfoliated g-C$_3$N$_4$ nanosheets was longer than that of bulk g-C$_3$N$_4$. According to the fitting calculation of the spectra data, the average lifetime of charge carriers was 6.7 and 6.3 ns for bulk and 2D g-C$_3$N$_4$, respectively. The recombination of charge carriers was restrained in g-C$_3$N$_4$ nanosheets. It means that 2D g-C$_3$N$_4$ by plasma have a more flexible electron transfer ability than bulk g-C$_3$N$_4$. It is also favorable for photocatalytic performance.

Figure 3d exhibits nitrogen adsorption–desorption isotherms. The isotherms of both g-C$_3$N$_4$ exhibit type-1 curve with H3 hysteresis loops. In addition, the BET surface area of g-C$_3$N$_4$ nanosheets increased from 41.79 to 136.8 m$^2$/g via plasma treatment, which due to the special sheet structure. Hence, plasma-exfoliated g-C$_3$N$_4$ nanosheets have better photocatalytic activity than bulk that due to the presence of more active sites on a larger specific surface area.

Figure 4 shows the photocurrent density of MoS$_2$ and g-C$_3$N$_4$. Figure 4a shows that MoS$_2$ nanosheets exhibit the higher photocurrent density than bulk MoS$_2$, indicating MoS$_2$ nanosheets by plasma have higher photoelectric conversion efficiency. Figure 4b shows the photocurrent density of bulk g-C$_3$N$_4$ and g-C$_3$N$_4$ nanosheets. g-C$_3$N$_4$ nanosheets also have higher photocurrent density. These results are consistent with the above results.

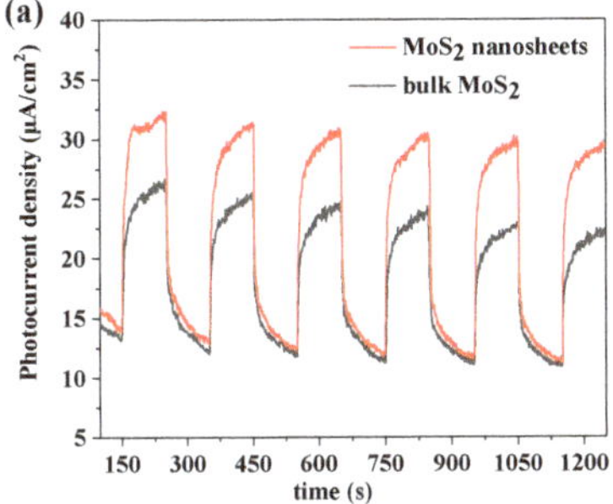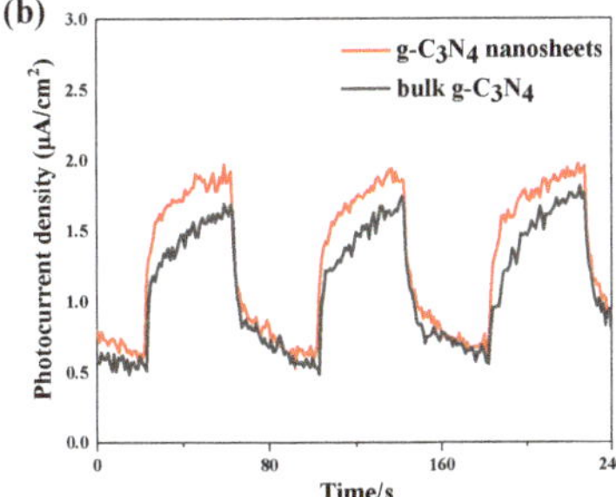

**Figure 4.** Photocurrent density vs time for (**a**) bulk $MoS_2$ and $MoS_2$ nanosheets, (**b**) bulk g-$C_3N_4$ and g-$C_3N_4$ nanosheets.

## 3.2. Morphology

The morphology of g-$C_3N_4$ nanosheets and bulk g-$C_3N_4$ were analyzed by FESEM. Figure S6 show that bulk g-$C_3N_4$ are aggregated particles and layered structures, whereas g-$C_3N_4$ nanosheets significantly exhibited a different microstructure that is more loose than bulk g-$C_3N_4$, further confirming that plasma treatment can change structure of bulk g-$C_3N_4$. Further discussion will be given along with TEM results.

TEM images of $MoS_2$ and g-$C_3N_4$ are shown in Figure 5. Figure 5a,b show that $MoS_2$ prepared by plasma exhibited large area of nanosheet structure with wrinkle-like, indicating that the thickness of the nanosheet was very thin. Figure 5c,d show the microstructure of the bulk $MoS_2$, which can be observed to be the thick flat structure composed of particle packing. Figure 5e,f are TEM images of both g-$C_3N_4$ samples. They show that the samples by plasma are almost transparent. Hence, bulk g-$C_3N_4$ is also successfully exfoliated into thin g-$C_3N_4$ nanosheets by DBD plasma treatment [50–52], which is in agreement with previous results.

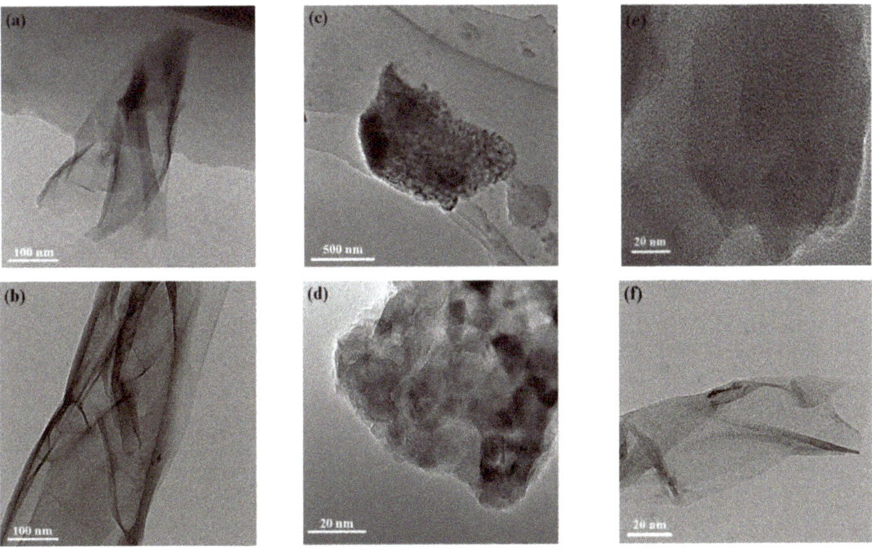

**Figure 5.** TEM images of (**a**,**b**) $MoS_2$ nanosheets, (**c**,**d**) bulk $MoS_2$, (**e**) bulk g-$C_3N_4$ and (**f**) g-$C_3N_4$ nanosheets.

Atomic force microscope (AFM) was conducted measure the thickness of the as-prepared ultrathin sheets. As shown in Figure 6a,b, there are micron-sized nanosheets of $MoS_2$ by plasma, with varying thickness due to the folds and bends of the edges. Heights of 2.04 nm and 3.18 nm of the flakes were observed, corresponding to the green and red lines. It indicates a $MoS_2$ layer less than 5 "S-MO-S" atomic layers [53]. Figure 6c,d show 20 nm–25 nm thickness of bulk $MoS_2$, due to the uneven accumulation of particles on its surface. As shown in Figure 6e,f, g-$C_3N_4$ nanosheets are deposited on the mica and exhibited a uniform thickness of approximately 1.2 nm, which corresponds to a g-$C_3N_4$ layer has 3 single atom layers [54,55]. The results are consistent with the observation from TEM. This further confirms that DBD plasma treatment can successfully prepare ultrathin $MoS_2$ nanosheets and g-$C_3N_4$ nanosheets.

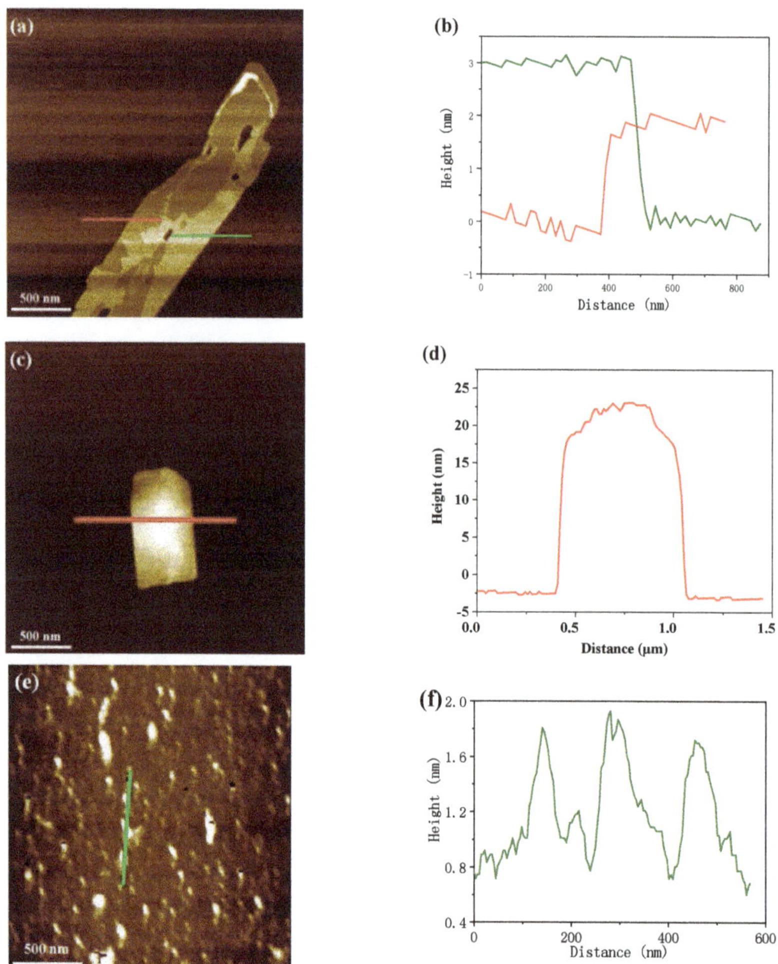

**Figure 6.** AFM image and corresponding cross-section profile of (**a**,**b**) $MoS_2$ nanosheets, (**c**,**d**) bulk $MoS_2$ and (**e**,**f**) g-$C_3N_4$ nanosheets.

### 3.3. Mechanism Analysis

To verify that the mode of action was operative during the DBD plasma treatment process for $(NH_4)_2MoS_4$ to $MoS_2$ nanosheets and bulk g-$C_3N_4$ to 2D g-$C_3N_4$, the DBD plasma reactor was connected to an online mass spectrum. It is used to analyze the intermediate product during the plasma process. $H_2/Ar$ was continuously purged into the plasma reactor. Figure 7a shows the components during the preparation process of $MoS_2$ nanosheets, measured by mass spectrometry. As mentioned in Experimental procedures, the plasma treatment process was a batch operation of 3 min for each experiment. It shows that the sample has been treated at 18 min. M 17 and M 34 have peaks between 18 min and 23 min. The changes of M 17 and M 34 corresponding to $NH_3$ and $H_2S$ were generated in DBD plasma treatment. Moreover, M 2 was decreased that exhibit the opposite variation trend. It confirms that $H_2$ was consumed in DBD plasma treatment. Overall, the mass spectrum results show that $H_2$ served as etching sources to react with $(NH_4)_2MoS_4$ and gaseous ammonia and hydrogen sulfide were generated in plasma process. The characteristic peak of $H_2O$ is not found. The large amount of gas generated is expanded to open the layers of bulk $(NH_4)_2MoS_4$ to $MoS_2$ nanosheets.

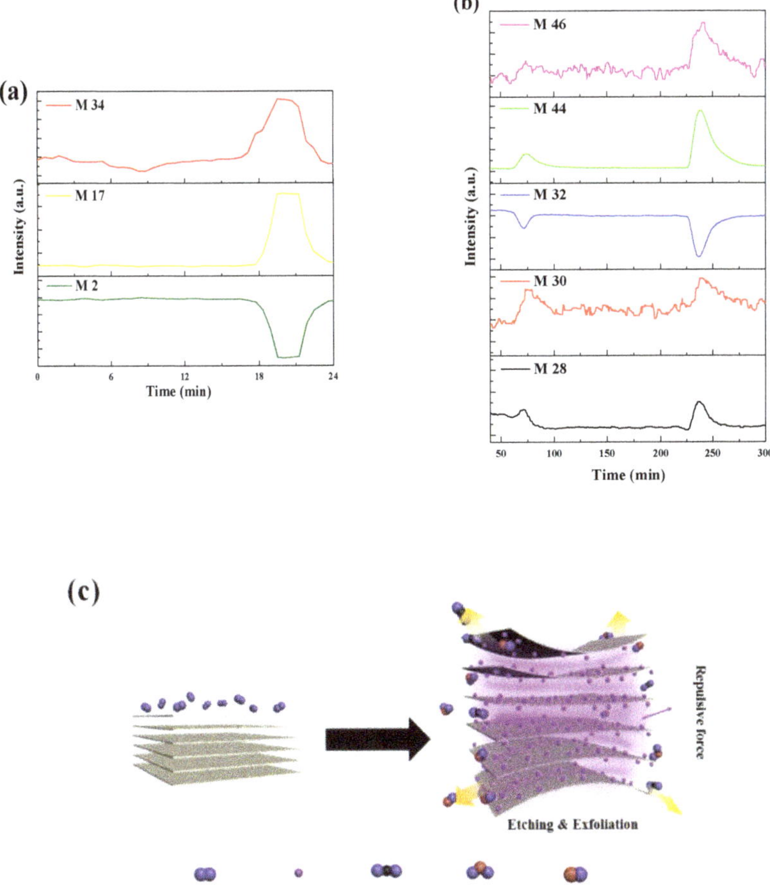



**Figure 7.** Mass spectrum during DBD plasma treatment process of (**a**) $MoS_2$ nanosheets and (**b**) g-$C_3N_4$ nanosheets. (**c**) Brief schematic illustration DBD plasma exfoliation process.

Figure 7b shows the components during the preparation process from g-$C_3N_4$ nanosheets measured by mass spectrometry. It shows that the bulk g-$C_3N_4$ has been treated twice at about 30 min and 50 min, respectively. It indicates that M 28, M 30, M 44, and M 46 increase and then returned to being constant. The changes of M 28, M 30 and M 46 imply that CO, NO and $NO_2$ were generated.

M 44 shows the change in $CO_2$ or $N_2O$ that confirm bulk g-$C_3N_4$ was oxidized into gaseous oxide during DBD plasma treatment. Nevertheless, M 32 decreases indicated $O_2$ was consumed in plasma treatment. The results show that $O_2$ served as oxidant to react with bulk g-$C_3N_4$ and that gaseous oxycarbide and oxynitride were generated in an plasma process.

Figure 7c illustrates the schematic of $MoS_2$ nanosheets and g-$C_3N_4$ nanosheets were generated in DBD plasma. According to the above analysis, etching and gas expansion are the main reasons for exfoliation process. In addition, we propose electrons would adhere to the surface or between the layers of bulk materials in plasma, and the repulsive force would exfoliate bulk materials into two-dimension nanosheets. It also facilitates the exfoliation.

*3.4. Photocatalytic Activity*

The photocatalytic performance of $MoS_2$ nanosheets are evaluated by hydrogen production. Hydrogen production is tested in dye sensitization systems under optical light. Figure 8a shows that the hydrogen production of $MoS_2$ nanosheets reaches 4.88 mmol/g per hour. However, the hydrogen production of bulk $MoS_2$ is only 1.47 mmol/g per hour. The former is about 3.3 times that of the latter. It shows that $MoS_2$ nanosheets by plasma has excellent hydrogen production activity, which is attributed to its thinner sheet structure and better charge transfer efficiency. As can be seen from Figure 8b, there was no remarkable decrease in the hydrogen production of $MoS_2$ nanosheets after three cycles of experiments, indicating that it has superior stability.

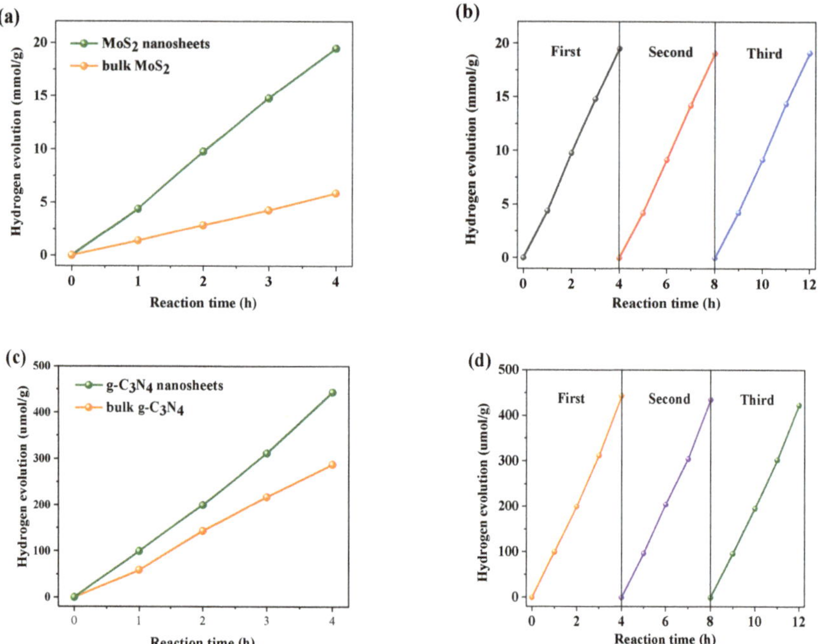

**Figure 8.** (a) Photocatalytic $H_2$ evolution rates and (b) stability of bulk $MoS_2$ and $MoS_2$ nanosheets under visible light irradiation, (c) Photocatalytic $H_2$ evolution rates and (d) stability of bulk g-$C_3N_4$ and g-$C_3N_4$ nanosheets.

In order to evaluate the activity of 2D g-$C_3N_4$ by DBD plasma, $H_2$ evolution and RhB/MO degradation reactions were conducted in optical light irradiation. Figure 8c shows that 0.5% Pt/g-$C_3N_4$ nanosheets demonstrated better catalytic performance than bulk that in water splitting, whose $H_2$ evolution can reached 5.5 μmol/h and was 1.5 times of the latter. Therefore, the 2D nanosheets obtained by plasma treatment have superior properties in photocatalytic reactions. In addition, 2D structure provides faster electron transfer rate and extended lifetime of electrons and holes. The obtained 2D g-C$_3$N$_4$ exhibits a large specific surface area. This is in agreement with the previous characterization results. All these factors contribute to the superior H$_2$ evolution rate. Figure 8d shows the hydrogen production cycle stability experiment of g-C$_3$N$_4$ nanosheets. After 12 h, the hydrogen production amount only slightly decreased, indicating that it has good stability.

Figure 9a shows that the concentration of the MO solution continuously decreased under visible light exposure from a starting concentration of 10 mg/L with both g-C$_3$N$_4$. The 2D g-C$_3$N$_4$ successfully degraded the MO in 2 h, and the rate is 2.5 times faster than the bulk. Figure 9b also shows the degradation reaction of the RhB solution (20 mg/L). g-C$_3$N$_4$ nanosheets exhibit a favorable degradation rate, and the rate of that is 1.3 times that of thick g-C$_3$N$_4$.

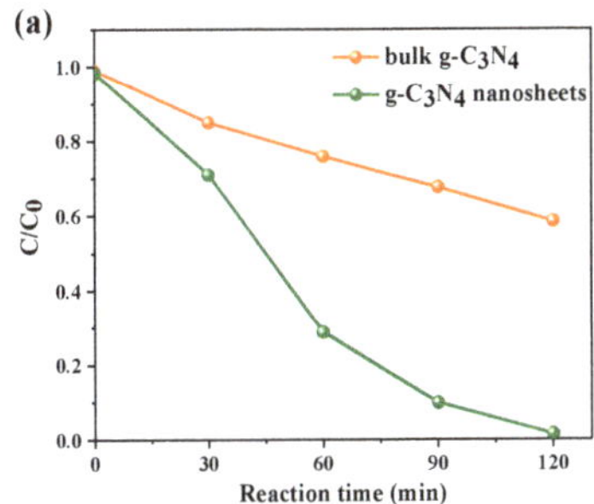
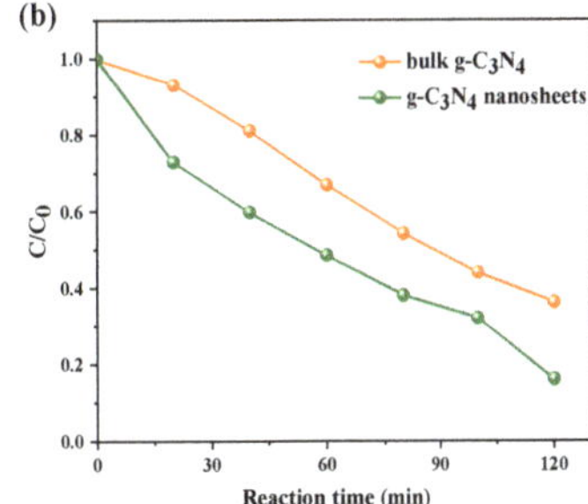

**Figure 9.** Photocatalytic activity of bulk g-C$_3$N$_4$ and as prepared g-C$_3$N$_4$ nanosheets for the degradation of (**a**) MO and (**b**) RhB under visible light irradiation.

Therefore, 2D structure nanosheets obtain by plasma have better photocatalytic performance under visible light irradiation, which was consistent with the previous report on two-dimesional nanosheet materials, further proving that the DBD plasma can be used as a facile method for preparing MoS$_2$ and g-C$_3$N$_4$ nanosheets.

## 4. Conclusions

Plasma as the environmentally benign approach provides a general platform for fabricating ultrathin nanosheets with prospective applications as photocatalysts for H$_2$ evolution and pollutant degradation. MoS$_2$ nanosheets and g-C$_3$N$_4$ nanosheets were successfully prepared utilizing a novel atmospheric plasma method at a moderate temperature. The gas etching and gas expansion process along with the repulsive force between electrons are the main reasons for plasma exfoliate the bulk precursors of layered structure to ultrathin nanosheets. The obtained MoS$_2$ and g-C$_3$N$_4$ nanosheets have several atomic layer thickness. Furthermore, a 2D structure MoS$_2$ and g-C$_3$N$_4$ nanosheets exhibits a larger surface area, and more flexible photophysical properties. The nano structure could ensure an easier electron transfer during the reaction. The plasma method show potential for the fast, solvent-free, low temperature, and large-scale application for the preparation of 2-dimensional nanomaterials.

**Supplementary Materials:** The following are available online at http://www.mdpi.com/2079-4991/9/8/1139/s1. Figure S1: Schematic of DBD plasma system, Figure S2: IR image of the reaction system during DBD treatment, Figure S3: Image of (a) g-C$_3$N$_4$ and (b) MoS$_2$ before DBD treatment and after DBD treatment, Figure S4: XPS survey spectra of (a) MoS$_2$ and (b) g-C$_3$N$_4$, Figure S5: FTIR spectra of bulk g-C$_3$N$_4$ and g-C$_3$N$_4$ nanosheets, Figure S6: SEM of (a) bulk g-C$_3$N$_4$ and (b) g-C$_3$N$_4$ nanosheets, Table S1: The atomic concentration of all elements in MoS$_2$, Table S2: The atomic concentration of all elements in g-C$_3$N$_4$.

**Author Contributions:** Z.W. (Zhao Wang) and Q.X.Y. conceived the experiments. Z.W. (Zhenhai Wang) conducted the preparation experiment. B.Z. conducted the photocatalysis experiment. X.P. and L.Z. set up the experimental device. B.Z. and Z.W. (Zhenhai Wang) wrote the original draft of the manuscript.

*Nanomaterials* **2019**, *9*, 1139

**Funding:** The support from the National Key Research and Development Program of China (No. 2016YFF0102503) and National Natural Science Foundation of China (No. 21878214) are greatly appreciated.

**Conflicts of Interest:** The authors declare no competing interests.

## References


1. Fu, J.W.; Yu, J.G.; Jiang, C.J.; Cheng, B. g-C$_3$N$_4$-Based Heterostructured Photocatalysts. *Adv. Energy Mater.* **2018**, *8*, 1701503. [CrossRef]

2. Hisatomi, T.; Kubota, J.; Domen, K. Recent advances in semiconductors for photocatalytic and photoelectrochemical water splitting. *Chem. Soc. Rev.* **2014**, *43*, 752–7535. [CrossRef] [PubMed]

3. Li, Y.Q.; Wang, W.; Wang, F.; Di, L.B.; Yang, S.C.; Zhu, S.J.; Yao, Y.B.; Ma, C.H.; Dai, B.; Yu, F. Enhanced Photocatalytic Degradation of Organic Dyes via Defect-Rich TiO$_2$ Prepared by Dielectric Barrier Discharge Plasma. *Nanomaterials* **2019**, *9*, 720. [CrossRef] [PubMed]

4. Di, L.B.; Xu, Z.J.; Wang, K.; Zhang, X.L. A facile method for preparing Pt/TiO$_2$ photocatalyst with enhanced activity using dielectric barrier discharge. *Catal. Today* **2013**, *211*, 109–113. [CrossRef]

5. Haque, F.; Daeneke, T.; Kalantar-Zadeh, K.; Ou, J.Z. Two-Dimensional Transition Metal Oxide and Chalcogenide-Based Photocatalysts. *Nano-Micro Lett.* **2018**, *10*, 23. [CrossRef] [PubMed]

6. Choi, W.B.; Choudhary, N.; Han, G.H.; Park, J.H.; Akinwande, D.; Lee, Y.H. Recent development of two-dimensional transition metal dichalcogenides and their applications. *Mater. Today* **2017**, *20*, 116–130. [CrossRef]

7. Tan, C.L.; Cao, X.H.; Wu, X.J.; He, Q.Y.; Yang, J.; Zhang, X.; Chen, J.Z.; Zhao, W.; Han, S.K.; Nam, G.; et al. Recent Advances in Ultrathin Two-Dimensional Nanomaterials. *Chem. Rev.* **2017**, *117*, 6225–6331. [CrossRef]

8. Chen, Y.; Sun, H.; Peng, W. 2D Transition Metal Dichalcogenides and Graphene-Based Ternary Composites for Photocatalytic Hydrogen Evolution and Pollutants Degradation. *Nanomaterials* **2017**, *7*, 62. [CrossRef]

9. Liu, P.B.; Zhang, Y.Q.; Yan, J.; Huang, Y.; Xia, L.; Guang, Z.X. Synthesis of lightweight N-doped graphene foams with open reticular structure for high-efficiency electromagnetic wave absorption. *Chem. Eng. J.* **2019**, *368*, 285–298. [CrossRef]

10. Truong, L.; Jerng, S.; Roy, S.B.; Jeon, J.H.; Kim, K.; Akbar, K.; Yi, Y.; Chun, S. Chrysanthemum-Like CoP Nanostructures on Vertical Graphene Nanohills as Versatile Electrocatalysts for Water Splitting. *ACS Sustain. Chem. Eng.* **2019**, *7*, 4625–4630. [CrossRef]

11. He, Q.G.; Liu, J.; Tian, Y.L.; Wu, Y.Y.; Magesa, F.; Deng, P.H.; Li, G.L. Facile Preparation of Cu$_2$O Nanoparticles and Reduced Graphene Oxide Nanocomposite for Electrochemical Sensing of Rhodamine B. *Nanomaterials* **2019**, *9*, 958. [CrossRef] [PubMed]

12. Nikokavoura, A.; Trapalis, C. Graphene and g-C$_3$N$_4$ based photocatalysts for NOx removal: A review. *Appl. Surf. Sci.* **2018**, *430*, 18–52. [CrossRef]

13. Krasian, T.; Punyodom, W.; Worajittiphon, P. A hybrid of 2D materials (MoS$_2$ and WS$_2$) as an effective performance enhancer for poly (lactic acid) fibrous mats in oil adsorption and oil/water separation. *Chem. Eng. J.* **2019**, *369*, 563–575. [CrossRef]

14. Huang, K.; Li, Z.J.; Lin, J.; Han, G.; Huang, P. Two-dimensional transition metal carbides and nitrides (MXenes) for biomedical applications. *Chem. Soc. Rev.* **2018**, *47*, 5109–5124. [CrossRef] [PubMed]

15. Park, S.Y.; Kim, Y.H.; Lee, S.Y.; Sohn, W.; Lee, J.E.; Kim, D.H.; Shim, Y.; Kwon, K.C.; Choi, K.S.; Yoo, H.J.; et al. Highly selective and sensitive chemoresistive humidity sensors based on rGO/MoS$_2$ van der Waals composites. *J. Mater. Chem. A* **2018**, *6*, 5016–5024. [CrossRef]

16. Xie, J.F.; Zhang, J.J.; Li, S.; Grote, F.B.; Zhang, X.D.; Zhang, H.; Wang, R.X.; Lei, Y.; Pan, B.C.; Xie, Y. Controllable Disorder Engineering in Oxygen-Incorporated MoS$_2$ Ultrathin Nanosheets for Efficient Hydrogen Evolution. *J. Am. Chem. Soc.* **2013**, *135*, 17881–17888. [CrossRef] [PubMed]

17. Cai, Z.Y.; Liu, B.; Zou, X.L.; Cheng, H.M. Chemical Vapor Deposition Growth and Applications of Two-Dimensional Materials and Their Heterostructures. *Chem. Rev.* **2018**, *118*, 6091–6133. [CrossRef] [PubMed]

18. Wang, Z.; Zhang, Y.; Neyts, E.C.; Cao, X.X.; Zhang, X.S.; Jang, B.W.L.; Liu, C.J. Catalyst Preparation with Plasmas: How Does It Work? *ACS Catal.* **2018**, *8*, 2093–2110. [CrossRef]

19. Di, L.B.; Zhang, J.S.; Zhang, X.L. A review on the recent progress, challenges, and perspectives of atmospheric-pressure cold plasma for preparation of supported metal catalysts. *Plasma Process. Polym.* **2018**, *15*, 1700234. [CrossRef]

20. Zhang, J.S.; Di, L.B.; Yu, F.; Duan, D.Z.; Zhang, X.L. Atmospheric-Pressure Cold Plasma Activating Au/P25 for CO Oxidation: Effect of Working Gas. *Nanomaterials* **2018**, *8*, 742. [CrossRef]

21. Sun, Q.D.; Yu, B.; Liu, C.J. Characterization of ZnO Nanotube Fabricated by the Plasma Decomposition of $Zn(OH)_2$ Via Dielectric Barrier Discharge. *Plasma Chem. Plasma Process.* **2012**, *32*, 201–209. [CrossRef]

22. Neyts, E.C.; Ostrikov, K.K.; Sunkara, M.K.; Bogaerts, A. Plasma Catalysis: Synergistic Effects at the Nanoscale. *Chem. Rev.* **2015**, *115*, 13408–13446. [CrossRef] [PubMed]

23. Liu, R.; Wang, Y.Y.; Liu, D.D.; Zou, Y.Q.; Wang, S.Y. Water-Plasma-Enabled Exfoliation of Ultrathin Layered Double Hydroxide Nanosheets with Multivacancies for Water Oxidation. *Adv. Mater.* **2017**, *29*, 1701546. [CrossRef] [PubMed]

24. Wang, Y.Q.; Yu, F.; Zhu, M.Y.; Ma, C.H.; Zhao, D.; Wang, C.; Zhou, A.M.; Dai, B.; Ji, J.Y.; Guo, X.H. N-Doping of plasma exfoliated graphene oxide via dielectric barrier discharge plasma treatment for the oxygen reduction reaction. *J. Mater. Chem. A* **2018**, *6*, 2011–2017. [CrossRef]

25. Peng, X.F.; Wang, Z.H.; Wang, Z.; Gong, J.B.; Hao, H.X. Electron reduction for the preparation of rGO with high electrochemical activity. *Catal. Today* **2019**, *2*, 1–6. [CrossRef]

26. Guo, L.; Yang, Z.; Marcus, K.; Li, Z.; Luo, B.; Zhou, L.; Wang, X.; Du, Y.; Yang, Y. $MoS_2$/$TiO_2$ heterostructures as nonmetal plasmonic photocatalysts for highly efficient hydrogen evolution. *Energy Environ. Sci.* **2018**, *11*, 106–114. [CrossRef]

27. Li, Z.Z.; Meng, X.C.; Zhang, Z.S. Recent development on $MoS_2$-based photocatalysis: A review. *J. Photochem. Photobiol. C Photochem. Rev.* **2018**, *35*, 39–55. [CrossRef]

28. Wang, X.; Maeda, K.; Thomas, A.; Takanabe, K.; Xin, G.; Carlsson, J.M.; Domen, K.; Antonietti, M. A metal-free polymeric photocatalyst for hydrogen production from water under visible light. *Nat. Mater.* **2009**, *8*, 76–80. [CrossRef]

29. Yin, S.M.; Han, J.Y.; Zhou, T.H.; Xu, R. Recent progress in g-$C_3N_4$ based low cost photocatalytic system: Activity enhancement and emerging applications. *Catal. Sci. Technol.* **2015**, *5*, 5048–5061. [CrossRef]

30. Wang, Y.G.; Li, Y.G.; Bai, X.; Cai, Q.; Liu, C.L.; Zuo, Y.H.; Kang, S.F.; Cui, L.F. Facile synthesis of Y-doped graphitic carbon nitride with enhanced photocatalytic performance. *Catal. Commun.* **2016**, *84*, 179–182. [CrossRef]

31. Wang, G.Z.; Zhou, F.; Yuan, B.F.; Xiao, S.Y.; Kuang, A.L.; Zhong, M.M.; Dang, S.H.; Long, X.J.; Zhang, W.L. Strain-Tunable Visible-Light-Responsive Photocatalytic Properties of Two-Dimensional CdS/g-$C_3N_4$: A Hybrid Density Functional Study. *Nanomaterials* **2019**, *9*, 244. [CrossRef] [PubMed]

32. Xie, J.F.; Zhang, H.; Li, S.; Wang, R.X.; Sun, X.; Zhou, M.; Zhou, J.F.; Lou, X.W.D.; Xie, Y. Defect-Rich $MoS_2$ Ultrathin Nanosheets with Additional Active Edge Sites for Enhanced Electrocatalytic Hydrogen Evolution. *Adv. Mater.* **2013**, *25*, 5807–5813. [CrossRef] [PubMed]

33. Niu, P.; Zhang, L.L.; Liu, G.; Cheng, H.M. Graphene-Like Carbon Nitride Nanosheets for Improved Photocatalytic Activities. *Adv. Funct. Mater.* **2012**, *22*, 4763–4770. [CrossRef]

34. Yang, S.; Gong, Y.; Zhang, J.; Zhan, L.; Ma, L.; Fang, Z.; Vajtai, R.; Wang, X.; Ajayan, P.M. Exfoliated Graphitic Carbon Nitride Nanosheets as Efficient Catalysts for Hydrogen Evolution Under Visible Light. *Adv. Mater.* **2013**, *25*, 2452–2456. [CrossRef] [PubMed]

35. Ong, W.J.; Tan, L.L.; Ng, Y.H.; Yong, S.T.; Chai, S.P. Graphitic Carbon Nitride (g-$C_3N_4$)-Based Photocatalysts for Artificial Photosynthesis and Environmental Remediation: Are We a Step Closer To Achieving Sustainability? *Chem. Rev.* **2016**, *116*, 7159–7329. [CrossRef] [PubMed]

36. Jia, T.T.; Li, M.M.J.; Ye, L.; Wise Man, S.; Liu, G.L.; Qu, J.; Nakagawa, K.; Tsang, S.C.E. The remarkable activity and stability of a dye-sensitized single molecular layer $MoS_2$ ensemble for photocatalytic hydrogen production. *Chem. Commun.* **2015**, *51*, 13496–13499. [CrossRef] [PubMed]

37. Choudhary, N.; Islam, M.A.; Kim, J.H.; Ko, T.; Schropp, A.; Hurtado, L.; Weitzman, D.; Zhai, L.; Jung, Y. Two-dimensional transition metal dichalcogenide hybrid materials for energy applications. *Nano Today* **2018**, *19*, 16–40. [CrossRef]

38. Lin, Q.Y.; Li, L.; Liang, S.J.; Liu, M.H.; Bi, J.H.; Wu, L. Efficient synthesis of monolayer carbon nitride 2D nanosheet with tunable concentration and enhanced visible-light photocatalytic activities. *Appl. Catal. B Environ.* **2015**, *163*, 135–142. [CrossRef]

39. Zhang, J.S.; Chen, Y.; Wang, X.C. Two-dimensional covalent carbon nitride nanosheets: Synthesis, functionalization, and applications. *Energy Environ. Sci.* **2015**, *8*, 3092–3108. [CrossRef]

40. Latorre Sánchez, M.; Esteve Adell, I.; Primo, A.; García, H. Innovative preparation of $MoS_2$–graphene heterostructures based on alginate containing (NH4)2MoS4 and their photocatalytic activity for $H_2$ generation. *Carbon* **2015**, *81*, 587–596. [CrossRef]

41. Zheng, S.Z.; Zheng, L.J.; Zhu, Z.Y.; Chen, J.; Kang, J.L.; Huang, Z.L.; Yang, D.C. $MoS_2$ Nanosheet Arrays Rooted on Hollow rGO Spheres as Bifunctional Hydrogen Evolution Catalyst and Supercapacitor Electrode. *Nano-Micro Lett.* **2018**, *10*, 23. [CrossRef] [PubMed]

42. Ma, L.; Wang, G.; Jiang, C.; Bao, H.; Xu, Q. Synthesis of core-shell $TiO_2$ @g-$C_3N_4$ hollow microspheres for efficient photocatalytic degradation of rhodamine B under visible light. *Appl. Surf. Sci.* **2018**, *430*, 263–272. [CrossRef]

43. Li, F.T.; Liu, S.J.; Xue, Y.B.; Wang, X.J.; Hao, Y.J.; Zhao, J.; Liu, R.H.; Zhao, D.S. Structure Modification Function of g-$C_3N_4$ for $Al_2O_3$ in the In Situ Hydrothermal Process for Enhanced Photocatalytic Activity. *Chem. A Eur. J.* **2015**, *21*, 10149–10159. [CrossRef] [PubMed]

44. Xia, P.F.; Zhu, B.C.; Yu, J.G.; Cao, S.W.; Jaroniec, M. Ultra-thin nanosheet assemblies of graphitic carbon nitride for enhanced photocatalytic $CO_2$ reduction. *J. Mater. Chem. A* **2017**, *5*, 3230–3238. [CrossRef]

45. Zhang, S.Q.; Su, C.S.; Ren, H.; Li, M.; Zhu, L.F.; Ge, S.; Wang, M.; Zhang, Z.L.; Li, L.; Cao, X.B. In-Situ Fabrication of g-$C_3N_4$/ZnO Nanocomposites for Photocatalytic Degradation of Methylene Blue: Synthesis Procedure Does Matter. *Nanomaterials* **2019**, *9*, 215. [CrossRef] [PubMed]

46. Li, J.; Liu, E.Z.; Ma, Y.N.; Hu, X.Y.; Wan, J.; Sun, L.; Fan, J. Synthesis of $MoS_2$/g-$C_3N_4$ nanosheets as 2D heterojunction photocatalysts with enhanced visible light activity. *Appl. Surf. Sci.* **2016**, *364*, 694–702. [CrossRef]

47. Zou, L.R.; Huang, G.F.; Li, D.F.; Liu, J.H.; Pan, A.L.; Huang, W.Q. A facile and rapid route for synthesis of g-$C_3N_4$ nanosheets with high adsorption capacity and photocatalytic activity. *RSC Adv.* **2016**, *6*, 86688–86694. [CrossRef]

48. Cai, X.G.; He, J.Y.; Chen, L.; Chen, K.; Li, Y.L.; Zhang, K.S.; Jin, Z.; Liu, J.Y.; Wang, C.M.; Wang, X.G.; et al. A 2D-g-$C_3N_4$ nanosheet as an eco-friendly adsorbent for various environmental pollutants in water. *Chemosphere* **2017**, *171*, 192–201. [CrossRef]

49. Ding, W.; Liu, S.Q.; He, Z. One-step synthesis of graphitic carbon nitride nanosheets for efficient catalysis of phenol removal under visible light. *Chin. J. Catal.* **2017**, *38*, 1711–1718. [CrossRef]

50. Li, Y.F.; Fang, L.; Jin, R.X.; Yang, Y.; Fang, X.; Xing, Y.; Song, S.Y. Preparation and enhanced visible light photocatalytic activity of novel g-$C_3N_4$ nanosheets loaded with $Ag_2CO_3$ nanoparticles. *Nanoscale* **2015**, *7*, 758–764. [CrossRef]

51. Xu, J.; Zhang, L.W.; Shi, R.; Zhu, Y.F. Chemical exfoliation of graphitic carbon nitride for efficient heterogeneous photocatalysis. *J. Mater. Chem. A* **2013**, *1*, 14766–14772. [CrossRef]

52. Miao, H.; Zhang, G.W.; Hu, X.Y.; Mu, J.L.; Han, T.X.; Fan, J.; Zhu, C.J.; Song, L.X.; Bai, J.T.; Hou, X. A novel strategy to prepare 2D g-$C_3N_4$ nanosheets and their photoelectrochemical properties. *J. Alloys Compd.* **2017**, *690*, 669–676. [CrossRef]

53. Xiao, J.W.; Zhang, Y.; Chen, H.J.; Xu, N.S.; Deng, S.Z. Enhanced Performance of a Monolayer $MoS_2$/$WSe_2$ Heterojunction as a Photoelectrochemical Cathode. *Nano-Micro Lett.* **2018**, *10*, 60. [CrossRef] [PubMed]

54. Zhu, M.S.; Zhai, C.Y.; Sun, M.J.; Hu, Y.F.; Yan, B.; Du, Y.K. Ultrathin graphitic $C_3N_4$ nanosheet as a promising visible-light-activated support for boosting photoelectrocatalytic methanol oxidation. *Appl. Catal. B Environ.* **2017**, *203*, 108–115. [CrossRef]

55. Wang, M.; Shen, M.; Zhang, L.X.; Tian, J.J.; Jin, X.X.; Zhou, Y.J.; Shi, J.L. 2D-2D $MnO_2$/g-$C_3N_4$ heterojunction photocatalyst: In-situ synthesis and enhanced $CO_2$ reduction activity. *Carbon* **2017**, *120*, 23–31. [CrossRef]

*Article*

# A Surface Dielectric Barrier Discharge Plasma for Preparing Cotton-Fabric-Supported Silver Nanoparticles

**Zhiyuan Fan, Lanbo Di *, Xiuling Zhang and Hongyang Wang**

College of Physical Science and Technology, Dalian University, Dalian 116622, China
* Correspondence: dilanbo@dlu.edu.cn; Tel.: +86-411-87402712

Received: 5 June 2019; Accepted: 20 June 2019; Published: 1 July 2019

**Abstract:** Cotton-fabric-supported silver nanoparticles (Ag NPs) have aroused great attention due to their remarkable physical and chemical properties and excellent broad-spectrum antibacterial performance.In this work, a surface dielectric barrier discharge (DBD) plasma method is developed and employed to prepare cotton fabric supported Ag NPs (Ag/cotton) for the first time. UV-Vis and X-ray photoelectron spectroscopy (XPS) results confirm the formation of Ag NPs. TEM images show that the size of Ag NPs is in the range 4.8–5.3 nm. Heat-sensitive cotton fabrics are not destroyed by surface DBD plasma according to FTIR and XRDresults. Wash fastness of the Ag/cotton samples is investigated using ultrasonic treatment for 30 min and it is shown that the Ag NPs possess good adhesion to the cotton fabric according to UV-Vis spectra. Antibacterial activity of the Ag/cotton samples shows that obvious bacteriostasis loops are observed around the samples with the appearance of both Gram-negative bacterium *Escherichia coli* (*E. coli*) and Gram-positive bacterium *Bacillus subtilis* (*B. subtilis*). The average diameter of the bacteriostasis loops against both *E. coli* and *B. subtilis* becomes larger with an increasing silver loading amount.This work provides a universal, fast, simple, and environmentally-friendly cold plasma method for synthesizing Ag NPs on heat-sensitive materials at atmospheric pressure.

**Keywords:** surface DBD; atmospheric-pressurecold plasma; cotton fabric; silver; antimicrobial

---

## 1. Introduction

Cotton fabrics are highly popular due to their excellent properties such as comfortability, air permeability, flexibility, bio-degradable ability, and good absorption of water. However, they provide ideal conditions (such as humidity, temperature, and nutrients, etc.) for bacteria and fungi to grow due to their high specific surface areas. In cotton fabrics, therefore, pathogens breed quickly, send out an unpleasant odor, and lead to textile decoloration and value shrinking, which accelerates the cross-spread of infection and aggravates human health. To solve these problems, great efforts have been devoted to modifying cotton fabrics by adding various antibacterial chemicals (including silver, copper, tungsten carbide, graphene derivatives, and tellurium, etc.) to improve their antibacterial activity [1–5]. Among these antibacterial chemicals, silver is widely used due to its low cost, and silver nanoparticles (Ag NPs) generally exhibit remarkable physical and chemical properties and excellent antibacterial performance against a wide range of pathogens [6].

According to the reduction process of silver ions into Ag NPs, the methods for preparing cotton-fabric-supported silver nanocomposites can be categorized into an exsitu reduction method and an insitu reduction method. With regards to theexsitu reduction method, an aqueous colloidal solution of Ag NPs is first prepared and then is supported on cotton fabrics by functional groups [7,8]. For example, Xu et al. [9] prepared antimicrobial cotton fabric by a one-pot modification process using a colloidal solution of Ag NPs via carboxymethyl chitosan (CMC), which was used as a binder and stabilizer. Consequently, the Ag NPs were uniformly distributed on the cotton fabric surface and exhibited remarkable and durable antibacterial activity. In addition, pressurized gyration has been proven to be an efficient exsitu method for mass production of polymeric fibers with antibacterial NPs [10,11]. Details about the method can be found in a feature article written by Edirisinghe et al. [12]. In the case of the insitu reduction method, silver ions arefirst supported on cotton fabric and then reduced by chemical reagents to metallic Ag NPs [13,14]. For example, El-Naggar et al. [15] activated loomstate, scoured, and bleached cotton fabrics by ethanolamine treatment and obtained three fabric-supported Ag NPs with ethanolamine in the absence of another external precursor. The size of the Ag NPs increased with increasing $AgNO_3$ concentration. The loomstate- and scoured-cotton-fabric-supported Ag NPs exhibited excellent durability and antibacterial activity. As mentioned above, both the exsitu method and insitu method have prepared cotton-fabric-supported Ag NPs with high antibacterial activity and durability. However, excess chemical reagents are generally required to be served as reducing agents. Therefore, more attention has been paid to environmentally-friendly preparation methods, such as the bio-synthesis method [16], photo-reduction method [17], and cold plasma method [1].

Cold plasma is characterized by a high electron temperature and low gas temperature (which can be close to room temperature). It has been proven to be an environmentally-friendly and fast method for reducing metal ions to synthesize supported metal catalysts due to its non-thermal equilibrium properties [18,19]. The synthesized supported metal catalysts possess small and high dispersion metal NPs [20,21] and an enhanced metal-support interaction due to the strong electric field in plasma [22–24]. Hence, synthesized metal catalysts generally exhibit high catalytic activity and stability. Li et al. [25] prepared cotton-fabric-supported Ag NPs by room temperature cold plasma reduction at low pressure and the samples exhibited high antibacterial activity against the bacteria *E. coli* and *B. subtilis*. Compared to low-pressure (LP) cold plasma, atmospheric-pressure (AP) cold plasma, which does not require high-cost and sophisticated vacuum equipment, is more attractive. However, since the electron temperature of AP cold plasma is generally low at atmospheric pressure, hydrogen-containing species ($H_2$, $CH_4$, $NH_3$, and ethanol, etc.) should be used as the reducing agents [26–28]. Consequently, the generally adopted AP cold plasma jet, plate-to-plate, and coaxial dielectric barrier discharges(DBD) are not suitable for preparing cotton-fabric-supported Ag NPs due to the open-air circumstances or the direct contact between the heat-sensitive cotton fabrics and the discharge zone.

Surface DBD is one important type of DBD discharge in which the discharges occur between the dielectric and discharge electrodes. It has been widely used in aerodynamic flow control [29], seed treating [30], film deposition [31], ozone production [32], and bacteria sterilization [33], etc. For materials treatment, surface DBD may allow any substrate thickness and is promising for large scale treatment. In this work, a surface DBD plasma was employed to reduce silver ions to prepare cotton-fabric-supported Ag NPs at atmospheric pressure for the first time. The heat-sensitive cotton fabric is not closely contacted with the discharge zone but is placed below it at a 4 mm distance. Hence, the generated active hydrogen species in surface DBD plasma is able to reduce the silver ions supported on cotton fabric into metallic Ag NPs without destroying thecotton fabric structure. The prepared samples exhibit high antibacterial activity against both the Gram-negative bacterium *E. coli* and the Gram-positive bacterium *B. subtilis*.

## 2. Materials and Methods

### 2.1. Preparation of the Ag/Cotton Samples

Prior to preparing the cotton-fabric-supported silver (Ag/cotton) samples, 100% plain-woven cotton fabrics (320 $g \cdot m^{-2}$) were cut into small pieces ($2.5 \times 2.5$ $cm^2$) and cleaned by Triton X-100 at 60 °C for 1 h. Silver nitrate ($AgNO_3$) and polyvinyl pyrrolidone (PVP, molecular weight = 1000–1,300,000) were purchased from Kemiou Chemical Reagent Co., Ltd. (Tianjin, China). Ag/cotton samples were prepared by impregnation, which was followed by cold plasma treatment. Typically, a piece of cotton fabric was put into the $AgNO_3$ aqueous solution and PVP and kept in the dark for 2 h. Then, the wet samples were centrifuged at 6000 rpm for 10 min to remove the excess $AgNO_3$. Afterwards, the samples were flattened and put into asurface DBD reactor.

Figure 1 illustrates a schematic diagram of the surface DBD device used in preparing the Ag/cotton samples, as well as the electrode pattern for the surface DBD. The surface DBD electrode system consisted of a comb-like discharge electrode and an induction electrode separated by an alumina plate with 1 mm thickness. Both the discharge electrode and induction electrode were made of tungsten. The discharge electrode comprised 11 stripes (1 mm wide and 4 mm strip-to-strip) connected all together at the end and covering $9.0 \times 5.0$ cm$^2$ in area. The induction electrode was a $7.6 \times 4.2$ cm$^2$ rectangle. A mixture of high purity Ar and $H_2$ (>99.999%) was adopted as the working gas and the generated active hydrogen species (H, H$^*$, H$_2^*$) were used to reduce the silver ions into Ag NPs [11,12]. The total flow rate of the working gas was 100 mL·min$^{-1}$ and the gas ratio of Ar to $H_2$ was 1:1. A CTP-2000K power supply (Nanjing Suman Electronic Co. Ltd., Nanjing, China) was employed to ignite the surface DBD electrode system to generate AP cold plasma. The discharge voltage was a 5.7 kV peak-to-peak sinusoidal high voltage and the discharge frequency was kept at 10.4 kHz. The Ag/cotton samples were placed below the surface DBD electrode system at a 4 mm distance. The treatment was performed three times, taking 2 min for each treatment. After surface DBD plasma treatment, the as-prepared Ag/cotton samples were dried at 120 °C for 2 h under vacuum in the dark. Then, they were rinsed with deionized water to get rid of the residual water-soluble $AgNO_3$ salt and dried under vacuum for another 2 h. Various $AgNO_3$ concentrations (1, 2, 5, and 10 mM·L$^{-1}$) were adopted, while the PVP concentration was kept constant at 10 mg·ml$^{-1}$. Accordingly, the Ag/cotton samples prepared by surface DBD plasma were marked as Ag/Cotton-1, Ag/Cotton-2, Ag/Cotton-5, and Ag/Cotton-10, respectively.

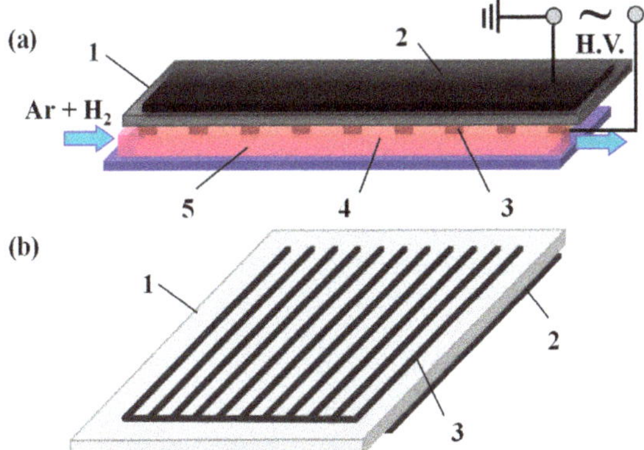


**Figure 1.** (a) Schematic diagram of the surface dielectric barrier discharge (DBD) device for preparing Ag/cotton samples and (b) the electrode pattern for the surface DBD. 1: alumina plate, 2: induction electrode, 3: discharge electrode, 4: Ag/cotton sample, 5: cold plasma.

### 2.2. Characterization

X-ray photoelectron spectroscopy (XPS) experiments were performed on a Thermo Fisher Scientific K-Alpha XPS spectrometer (Waltham, MA, USA) to investigate the valence state of the silver element. The binding energies were calibrated using a C1s peak at 284.6 eV as the reference. Infrared spectra of the samples were recorded on a Thermo Scientific Nicolet 6700 Fourier Transform Infrared Spectrometer (Waltham, MA, USA) using attenuated total reflection in the region of 6000–600 cm$^{-1}$ with a resolution of 4 cm$^{-1}$ by accumulating 32 scans. UV-Vis diffuse reflectance spectra (DRS) of the samples were collected for the spectral range 200–800 nm using a Hitachi U-3900 UV-Vis spectrometer (Tokyo, Japan) with BaSO$_4$ as a reference. XRD data of the samples were examined via a Dandong Haoyuan DX-2700

X-ray diffractometer (Dandong, China) using $CuK_{\alpha1}$ radiation ($\lambda = 1.54178$ Å). The morphology of the samples was examined using a Zeiss Sigma500 SEM (Jena, Germany) at a 5 kV accelerating voltage. Prior to SEM investigation, the cotton fabric samples were coated with a thin film of gold by sputtering. The SEM images were recorded at magnifications of 5000× and 10,000×. TEM and high resolution TEM (HRTEM) experiments were carried out using a FEI Tecnai G2 F20 S-Twin TEM (Hillsboro, OR, USA). Prior to testing, the samples were cut into pieces and placed in ethanol with ultrasonic dispersion for 120 min. Then, a few drops of the suspension were supported on an ultra-thin carbon film and dried in air at room temperature. The sizes and size distribution of the Ag NPs were obtained by calculating sizes for more than 300 Ag NPs.

### 2.3. Wash Fastness

Wash fastness of the Ag/cotton samples was estimated by ultrasonic leaching method. Typically, a piece of Ag/cotton sample ($2.5 \times 2.5$ cm$^2$) was immersed in a Pyrex beaker with 30 mL of deionized water. Then, the beaker was put into a KQ2200DB digital control ultrasonic generator (Kunshan Ultrasonic Instrument Co. Ltd., Kunshan, China) with a frequency of 40 kHz and an ultrasonic input power of 100 W. Ultrasonic leaching was performed for 30 min at a 30 °C sonication bath temperature. Then, the samples were dried under vacuum at 120 °C for 2 h. The wash fastness of the samples was tested by comparing their UV-Vis DRS spectra before and after ultrasonic leaching treatment.

### 2.4. Antibacterial Activity

The antibacterial activity of the Ag/Cotton samples against the Gram-negative bacterium *E. coli* and the Gram-positive bacterium *B. Subtilis* (Shanghai Luwei Technology Co. Ltd., Shanghai, China) was evaluated by the disk diffusion method and dilution method of plate counting. The process was similar to that used in a previous work [25]. Mueller-Hinton (MH) agar medium (6 g·L$^{-1}$ beef powder, 1.5 g·L$^{-1}$ soluble starch, 17.5 g·L$^{-1}$ acid hydrolyzed complex protein, and 17 g·L$^{-1}$ agar), deionized water, and the petridish were autoclaved prior to use. To compare antibacterial activity, blank cotton fabric and the Ag/cotton samples were cut into pieces of equal size ($1.0 \times 1.0$ cm$^2$). The *E. coli* and *B. subtilis* bacteria were cultured with MH agar medium in a constant temperature incubator at 37 °C for 8 h, respectively. Afterwards, the bacteria cultures were diluted (*E. coli*: folds 100, ~10$^7$ CFU·mL$^{-1}$; *B. subtilis*: folds 100, ~10$^7$ CFU·mL$^{-1}$. CFU is cell colony forming unit) with deionized water and MH agar, and the inoculums were evenly spread over the petridish, respectively. Lastly, the samples were planted onto the agar plates and incubated at 37 °C for 24 h to test their antibacterial activity by examining the diameter of the growth-inhibition zone.

## 3. Results and Discussion

Figure 2 shows the photographs of the blank cotton and the surface DBD plasma-prepared Ag/cotton samples with different silver loading amounts. Clearly, the color of the Ag/cotton samples has changed from colorless to yellow following surface DBD plasma treatment; this is generally thought to be the color of metallic Ag NPs. This change reveals that the silver ions in AgNO$_3$ can be reduced to metallic silver by surface DBD plasma. Moreover, by increasing the loading amount of silver, the color of the Ag/cotton samples changes from light yellow to brownish yellow, indicating that more Ag NPs are formed at a high silver loading amount. Similar phenomena were also observed in the preparation of Ag/cotton samples by DC glow discharge cold plasma at low pressure [25].

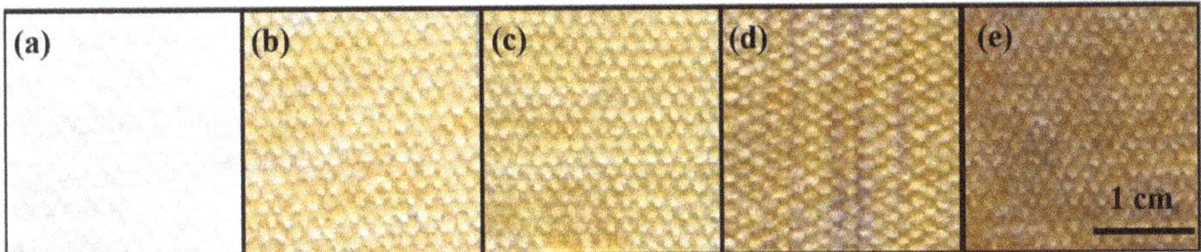


**Figure 2.** Photographs of (**a**) blank cotton, (**b**) Ag/Cotton-1, (**c**) Ag/Cotton-2, (**d**) Ag/Cotton-5, and (**e**) Ag/Cotton-10.

To further confirm the reduction ability of the surface DBD plasma, an XPS spectrum of Ag3d in Ag/Cotton-5 was measured, as is presented in Figure 3. The Ag3d XPS spectrum can be fitted with two peaks which correspond to metallic silver and Ag(I). According to the XPS data, the composition of metallic silver in the Ag/Cotton-5 sample is as high as 83.3%. This further verifies that surface DBD plasma is an efficient method for reducing silver ions supported on cotton fabric. The incomplete reduction of silver ions may be ascribed to the abundant oxygen-containing functional groups on the cotton fabric surface, which can be identified from the FTIR spectra of the samples (Figure 4). Similar phenomena have been found during the reduction of metal ions supported on activated carbon by plate-to-plate DBD plasma [34].

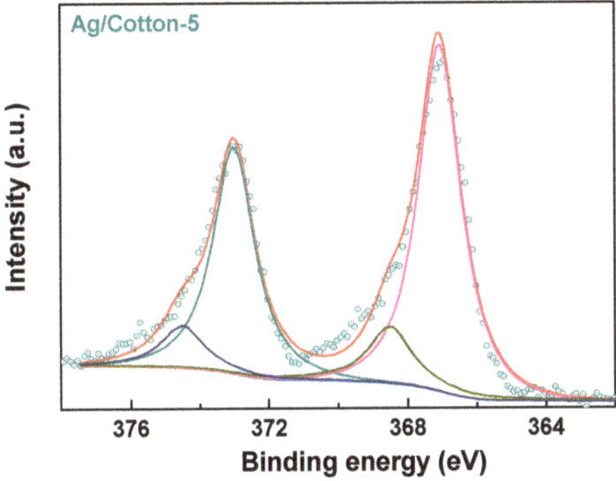


**Figure 3.** X-ray photoelectron spectroscopy (XPS) spectrum of Ag3d for the Ag/Cotton-5 sample.

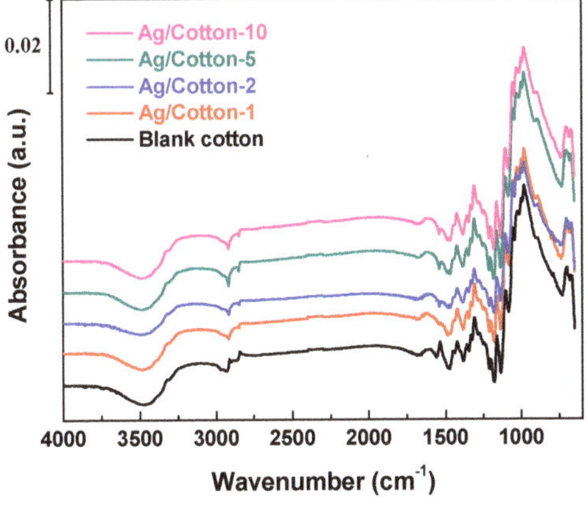


**Figure 4.** FTIR spectra of the Ag/cotton samples along with the blank cotton.

As shown in Figure 4, the functional groups in the Ag/cotton samples and the blank cotton were identified by FTIR to investigate the effect of the surface DBD plasma treatment. For all the samples, characteristic peaks for cellulose were observed in the range 1000–1200 cm$^{-1}$ [35]. The broad and strong peak band observed at around 3322 cm$^{-1}$ for the samples corresponds to the stretching vibration band of the O-H groups. The medium bands at 2800–3000 cm$^{-1}$ may be ascribed to the symmetric and asymmetric stretching vibrations of the C-H groups in the cellulose samples [36]. The band at 1682 cm$^{-1}$ can be assigned to adsorbed water in the samples. The peaks at 1484 cm$^{-1}$ and 1385 cm$^{-1}$ correspond to the wagging and bending of the C-H functional groups in cellulose, respectively. All of the Ag/cotton samples prepared by surface DBD plasma exhibited similar FTIR spectra to the blank cotton and no new peaks were able to be detected from the FTIR spectra of the Ag/cotton samples. This indicates that surface DBD plasma has no obvious influence on the structure of the cotton fabric support. The nondestructive property of surface DBD plasma is attributed to the fact that the heat-sensitive cotton fabric was not closely contacted with the discharge zone but below it at a 4 mm distance. Hence, the generated active hydrogen species in surface DBD plasma can reduce the silver ions into metallic silver without destroying the cotton fabric.

Figure 5 shows the UV-Vis DRS spectra of the Ag/cotton samples prepared by surface DBD plasma as well as the blank cotton. Compared to the blank cotton, surface plasmon resonance (SPR) peaks at around 420 nm can be observed for all of the Ag/cotton samples, which further suggests that surface DBD plasma can reduce silver ions into metallic Ag NPs. This is consistent with the XPS result (Figure 3). From Figure 5, we can also see that the intensity of the SPR peak has been enhanced while increasing the loading amount of silver, revealing that more Ag NPs are formed. In addition, the SPR peak becomes broader as the silver loading amount is increased, and redshifts of the SPR peaks are observed for the Ag/cotton samples. The SPR peaks are 415, 419, 423, and 425 nm, respectively, for Ag/Cotton-1, Ag/Cotton-2, Ag/Cotton-5, and Ag/Cotton-10. This is in line with the color change of the samples from light yellow to brownish yellow as seen in Figure 2, demonstrating the size growth of Ag NPs with an increasing silver loading amount. According to the Mie theory, the size of the Ag NPs is in the range 40–70 nm when the SPR peak is located within 415–425 nm [37].

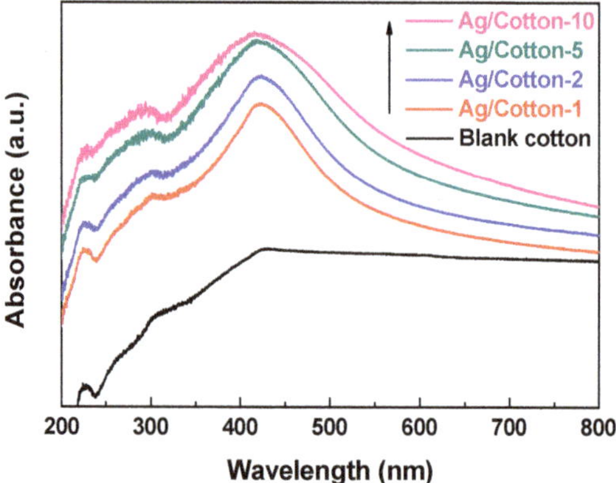


**Figure 5.** UV-Vis diffuse reflectance spectra (DRS) spectra of the Ag/cotton samples in addition to the blank cotton sample.

As illustrated in Figure 6, XRD patterns of the Ag/cotton samples and the blank cotton were obtained to further investigate the structure of the samples. For all of the Ag/cotton samples and the blank cotton sample, diffraction peaks at 14.7° and 16.9° were detected, which can be assigned to the (101) plane of cellulose I crystalline form [38]. In addition, characteristic diffraction peaks of cellulose I crystalline at 22.6° and 34.3° corresponding to the (002) and (040) crystal planes are also observed. Obviously, there is no distinct difference among the samples, which further demonstrates that surface

DBD plasma did not change the phase structure of the cotton fabric. Interestingly, no characteristic diffraction peak corresponding to silver species was detected for the Ag/cotton samples. A similar phenomenon, which was attributed to the small size of Ag NPs, and the low sensitivity of XRD towards metal detection on organic support, has also been observed in a previous work [39]. It should be noted that the aggregation of Ag NPs as opposed to single Ag NPs results in a color change and SPR absorption peaks broadening for the Ag/cotton samples [40]. This is further supported by SEM and TEM analyses.

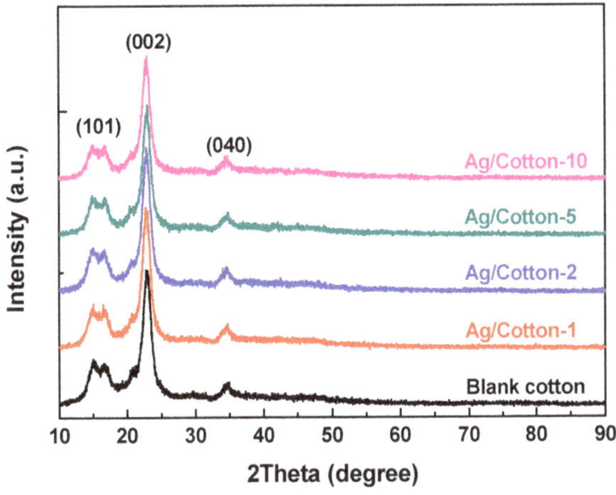


**Figure 6.** XRD patterns of the Ag/cotton samples as well as the blank cotton sample.

To determine the influence of silver deposition on the morphology of the cotton fabric and the distribution of the Ag NPs, typical SEM images of Ag/Cotton-1, Ag/Cotton-10, and the blank cotton sample wereobtained, as shown in Figure 7. The blank cotton without deposition of Ag NPs exhibits a clean and smooth surface. By contrast, the Ag/cotton samples show a rough structure, and aggregated Ag NPs are observed. The size of the Ag NPs is in the tens of nanometers and the size of the Ag NPs tends to grow with an increasing silver loading amount. This is consistent with the UV-Vis DRS results.

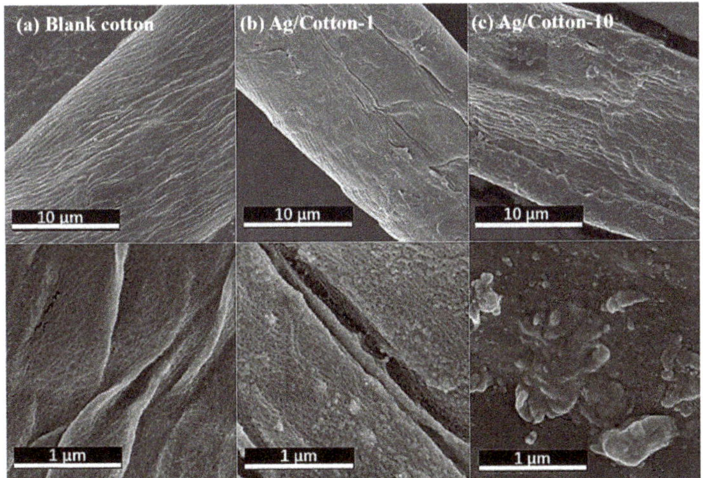


**Figure 7.** Typical SEM images of (**a**) blank cotton, (**b**) Ag/Cotton-1, and (**c**) Ag/Cotton-10.

To clearly disclose the distribution of Ag NPs in the Ag/cotton samples, TEM images of Ag/Cotton-1 and Ag/Cotton-10 were obtained, as shown in Figure 8. The Ag NPs in the Ag/cotton samples exhibit a spherical shape are homogeneously distributed. The average size of the Ag NPs ($D_{Ag}$) was obtained by measuring more than 300 Ag NPs, and the histograms of size distribution of the AgNPs are also illustrated in Figure 8. $D_{Ag}$ in Ag/Cotton-1 and Ag/Cotton-10 were found to be 4.8 ± 1.7 and 5.3 ± 1.9 nm, respectively, demonstrating a slight increase in the size of the Ag NPs at a higher silver loading amount. The size of the Ag NPs detected by TEM was much smaller than that obtained by SEM, indicating the tendency of aggregation of Ag NPs. The Ag NPs were thought to be aggregated after cold plasma reduction due to the hydrophilic property of cellulose [41]. The corresponding HRTEM images of the Ag NPs were also measured, as shown in the insets in Figure 8. Clear lattice fringes with a lattice spacing of 0.236 nm can be observed for the Ag/cotton samples which may be attributed to the face-centered cubic (fcc) Ag (111) planes (*Fm3m*, a = 4.086Å, JCPDS card, file no. 65-2871). This is in line with the XPS and UV-Vis DRS results, revealing that metallic Ag NPs with high crystalline quality are obtained by surface DBD plasma treatment.

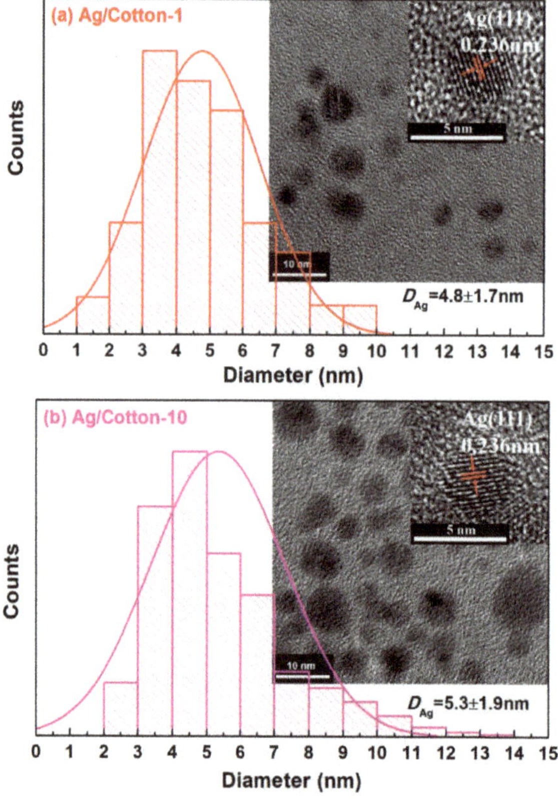


**Figure 8.** Typical TEM images and histograms of silver nanoparticles (Ag NPs) for (**a**) Ag/Cotton-1 and (**b**) Ag/Cotton-10, and the corresponding high resolution TEM (HRTEM) images of the Ag NPs (insets).

Wash fastness of the Ag/cotton samples was estimated using ultrasonic treatment. UV-Vis DRS spectra of the Ag/cotton samples as-synthesized and after 30 min of ultrasonic treatment are presented in Figure 9. Regarding the UV-Vis DRS spectra of Ag/Cotton-1, Ag/Cotton-2, and Ag/Cotton-5 as-synthesized and after 30 min of ultrasonic treatment, there is no distinct difference that can be observed. However, a slight decrease in the SPR peak intensity may be observed for Ag/Cotton-10 after 30 min ultrasonic treatment due to the high loading amount of silver. These results indicate that surface DBD plasma is a fast and environmentally-friendly method for preparing Ag/cotton samples with a relatively low silver loading amount.

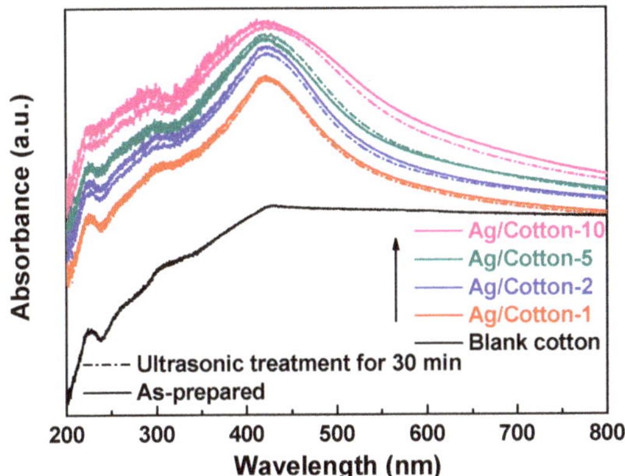


**Figure 9.** UV-Vis DRS spectra of the Ag/cotton samples as-synthesized and after 30 min of ultrasonic treatment, in addition to the blank cotton sample.

The Gram-negative bacterium *E. coli* and Gram-positive bacterium *B. subtilis* with the same concentration were employed to evaluate the antibacterial effect of the Ag/cotton samples, and the results are illustrated in Figure 10. The blank cotton exhibited no inhibition for bacterial growth of *E. coli* and *B. subtilis*. However, obvious bacteriostasis loops can be observed around the Ag/cotton samples with the appearance of *E. coli* or *B. subtilis*. These findings indicate that the surface DBD plasma is efficient for preparing Ag/cotton samples with high antibacterial activity, which may be ascribed to the small size and high distribution of Ag NPs. The average diameters of the bacteriostatic loops for Ag/Cotton-1, Ag/Cotton-2, Ag/Cotton-5, and Ag/Cotton-10 against *E. coli* were found to be 1.38, 1.39, 1.45, and 1.63 cm, respectively. By contrast, the average diameters of the bacteriostatic loops for Ag/Cotton-1, Ag/Cotton-2, Ag/Cotton-5 and Ag/Cotton-10 against *B. subtilis* were 1.85, 1.86, 2.05, and 2.18 cm, respectively. The average diameters of the bacteriostasis loops against both *E. coli* and *B. subtilis* became larger with an increase in the silver loading amount, demonstrating that Ag/cotton samples with a higher silver concentration possess higher antibacterial activity. Moreover, the diameters of the bacteriostasis loops against *B. subtilis* for the same Ag/cotton samples were larger those that against *E. coli*, suggesting that the Ag/cotton samples exhibit higher antibacterial activity against Gram-positive bacterium *B. subtilis* than Gram-negative bacterium *E. coli*, which is in line with the results of Li et al. [25]. This may be due to the structure of the cell wall of the bacteria [42].

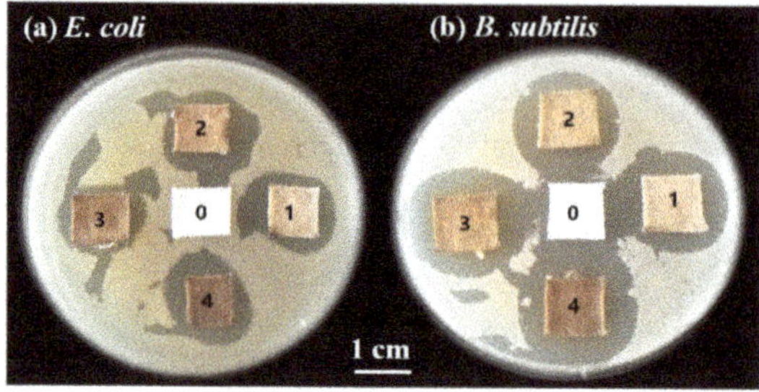


**Figure 10.** Growth inhibition of (**a**) *E. coli* and (**b**) *B. subtilis* for the samples. 0: blank cotton, 1: Ag/Cotton-1, 2: Ag/Cotton-2, 3: Ag/Cotton-5, 4: Ag/Cotton-10.

## 4. Conclusions

Cotton-fabric-supported Ag NPs (Ag/cotton) have been successfully prepared in this work using an environmentally-friendly and fast surface DBD plasma method for the first time. No additional reducing chemicals except the active hydrogen species generated in cold plasma served as the reducing agents. The color of the samples changed from colorless to yellow after surface DBD plasma treatment and the color changed from light yellow to brownish yellow with an increase in the loading amount of silver. Moreover, SPR peaks and the Ag3d XPS results further confirm that the Ag species mainly exist in the metallic state. TEM images show that the Ag NPs are uniformly distributed on the fiber surface and that the size of the Ag NPs is in the range 4.8 to 5.3 nm. FTIR and XRD results demonstrate that heat-sensitive cotton fabrics are not destroyed by surface DBD plasma, which may be due to the fact that the cotton fabric is not in close contact with the discharge zone during the treatment but below it at a 4 mm distance. Wash fastness estimated using ultrasonic treatment revealed that the Ag NPs in the Ag/cotton samples possess high adhesion ability to the cotton fabric. The Ag/cotton samples exhibit high antibacterial activity for both the Gram-negative bacterium *E. coli* and the Gram-positive bacterium *B. subtilis*. This work provides a universal, fast, simple, and environmentally-friendly cold plasma method for synthesizing Ag NPs on heat-sensitive materials at atmospheric pressure.

**Author Contributions:** Conceptualization, L.D. and X.Z.; Data curation, Z.F. and H.W.; Investigation, Z.F. and H.W.; Project administration, L.D.; Supervision, L.D. and X.Z.; Writing—original draft, L.D.; Writing—review & editing, X.Z.

**Funding:** This work was supported by the National Natural Science Foundation of China (grant nos. 21773020, 21673026, and 11505019), the Liaoning Innovative Talents in University (grant no. LR2017025), and the Natural Science Foundation of Liaoning Province (grant no. 20180550085).

**Conflicts of Interest:** The authors declare no conflict of interest.

## References

1.  Nikiforov, A.; Deng, X.; Xiong, Q.; Cvelbar, U.; DeGeyter, N.; Morent, R.; Leys, C. Non-thermal plasma technology for the development of antimicrobial surfaces: A review. *J. Phys. D Appl. Phys.* **2016**, *49*, 204002. [CrossRef]

2.  Cheong, Y.K.; Calvo-Castro, J.; Ciric, L.; Edirisinghe, M.; Cloutman-Green, E.; Illangakoon, U.E.; Kang, Q.; Mahalingam, S.; Matharu, R.K.; Wilson, R.M.; et al. Characterisation of the chemical composition and structural features of novel antimicrobial nanoparticles. *Nanomaterials* **2017**, *7*, 152. [CrossRef] [PubMed]

3.  Altun, E.; Aydogdu, M.O.; Koc, F.; Crabbe-Mann, M.; Brako, F.; Kaur-Matharu, R.; Ozen, G.; Kuruca, S.E.; Edirisinghe, U.; Gunduz, O.; et al. Novel making of bacterial cellulose blended polymeric fiber bandages. *Macromol. Mater. Eng.* **2018**, *303*, 1700607. [CrossRef]

4.  Matharu, R.K.; Porwal, H.; Ciric, L.; Edirisinghe, M. The effect of graphene–poly (methyl methacrylate) fibres on microbial growth. *Interface Focus* **2018**, *8*, 20170058. [CrossRef] [PubMed]

5.  Matharu, R.K.; Charani, Z.; Ciric, L.; Illangakoon, U.E.; Edirisinghe, M. Antimicrobial activity of tellurium-loaded polymeric fiber meshes. *J. Appl. Polym. Sci.* **2018**, *135*, 46368. [CrossRef]

6.  Simončič, B.; Klemenčič, D. Preparation and performance of silver as an antimicrobial agent for textiles: A review. *Text. Res. J.* **2016**, *86*, 210–223. [CrossRef]

7.  Cheng, F.; Betts, J.W.; Kelly, S.M.; Schaller, J.; Heinze, T. Synthesis and antibacterial effects of aqueous colloidal solutions of silver nanoparticles using aminocellulose as a combined reducing and capping reagent. *Green Chem.* **2013**, *15*, 989–998. [CrossRef]

8.  Wu, M.; Ma, B.; Pan, T.; Chen, S.; Sun, J. Silver-nanoparticle-colored cotton fabrics with tunable colors and durable antibacterial and self-healing superhydrophobic properties. *Adv. Funct. Mater.* **2016**, *26*, 569–576. [CrossRef]

9.  Xu, Q.; Zheng, W.; Duan, P.; Chen, J.; Zhang, Y.; Fu, F.; Liu, X. One-pot fabrication of durable antibacterial cotton fabric coated with silver nanoparticles via carboxymethyl chitosan as a binder and stabilizer. *Carbohyd. Polym.* **2019**, *204*, 42–49. [CrossRef]

10. Illangakoon, U.E.; Mahalingam, S.; Wang, K.; Cheong, Y.K.; Canales, E.; Ren, G.G.; Cloutman-Green, E.; Edirisinghe, M.; Ciric, L. Gyrospun antimicrobial nanoparticle loaded fibrous polymeric filters. *Mater. Sci. Eng. C Mater.* **2017**, *74*, 315–324. [CrossRef]

11. Xu, Z.; Mahalingam, S.; Rohn, J.L.; Ren, G.; Edirisinghe, M. Physio-chemical and antibacterial characteristics of pressure spun nylon nanofibres embedded with functional silver nanoparticles. *Mater. Sci. Eng. C Mater.* **2015**, *56*, 195–204. [CrossRef] [PubMed]

12. Heseltine, P.L.; Ahmed, J.; Edirisinghe, M. Developments in pressurized gyration for the mass production of polymeric fibers. *Macromol. Mater. Eng.* **2018**, *303*, 1800218. [CrossRef]

13. Bacciarelli-Ulacha, A.; Rybicki, E.; Matyjas-Zgondek, E.; Pawlaczyk, A.; Szynkowska, M.I. A new method of finishing of cotton fabric by in situ synthesis of silver nanoparticles. *Ind. Eng. Chem. Res.* **2014**, *53*, 4147–4155. [CrossRef]

14. Raza, Z.A.; Rehman, A.; Mohsin, M.; Bajwa, S.Z.; Anwar, F.; Naeem, A.; Ahmad, N. Development of antibacterial cellulosic fabric via clean impregnation of silver nanoparticles. *J. Clean. Prod.* **2015**, *101*, 377–386. [CrossRef]

15. El-Naggar, M.E.; Shaarawy, S.; Hebeish, A.A. Bactericidal finishing of loomstate, scoured and bleached cotton fibres via sustainable in-situ synthesis of silver nanoparticles. *Int. J. Biol. Macromol.* **2018**, *106*, 1192–1202. [CrossRef] [PubMed]

16. El-Rafie, M.H.; Shaheen, T.I.; Mohamed, A.A.; Hebeish, A. Bio-synthesis and applications of silver nanoparticles onto cotton fabrics. *Carbohyd. Polym.* **2012**, *90*, 915–920. [CrossRef]

17. Rehan, M.; Barhoum, A.; Van Assche, G.; Dufresne, A.; Gätjen, L.; Wilken, R. Towards multifunctional cellulosic fabric: UV photo-reduction and in-situ synthesis of silver nanoparticles into cellulose fabrics. *Int. J. Biol. Macromol.* **2017**, *98*, 877–886. [CrossRef]

18. Di, L.; Zhang, J.; Zhang, X. A review on the recent progress, challenges, and perspectives of atmospheric-pressure cold plasma for preparation of supported metal catalysts. *Plasma Process. Polym.* **2018**, *15*, 1700234. [CrossRef]

19. Wang, Z.; Zhang, Y.; Neyts, E.C.; Cao, X.; Zhang, X.; Jang, B.W.L.; Liu, C.J. Catalyst preparation with plasmas: How does it work? *ACS Catal.* **2018**, *8*, 2093–2110. [CrossRef]

20. Di, L.; Li, Z.; Zhang, X.; Wang, H.; Fan, Z. Reduction of supported metal ions by a safe atmospheric pressure alcohol cold plasma method. *Catal. Today* **2019**. [CrossRef]

21. Dao, V.D.; Choi, Y.; Yong, K.; Larina, L.L.; Shevaleevskiy, O.; Choi, H.S. A facile synthesis of bimetallic AuPt nanoparticles as a new transparent counter electrode for quantum-dot-sensitized solar cells. *J. Power. Sources* **2015**, *274*, 831–838. [CrossRef]

22. Jia, X.; Zhang, X.; Rui, N.; Hu, X.; Liu, C.J. Structural effect of Ni/ZrO$_2$ catalyst on CO$_2$ methanation with enhanced activity. *Appl. Catal. B Environ.* **2019**, *244*, 159–169. [CrossRef]

23. Chu, W.; Xu, J.; Hong, J.; Lin, T.; Khodakov, A. Design of efficient Fischer Tropsch cobalt catalysts via plasma enhancement: Reducibility and performance. *Catal. Today* **2015**, *256*, 41–48. [CrossRef]

24. Di, L.; Zhang, J.; Ma, C.; Tu, X.; Zhang, X. Atmospheric-pressure dielectric barrier discharge cold plasma for synthesizing high performance Pd/C formic acid dehydrogenation catalyst. *Catal. Today* **2019**. [CrossRef]

25. Li, Z.; Meng, J.; Wang, W.; Wang, Z.; Li, M.; Chen, T.; Liu, C.J. The room temperature electron reduction for the preparation of silver nanoparticles on cotton with high antimicrobial activity. *Carbohyd. Polym.* **2017**, *161*, 270–276. [CrossRef] [PubMed]
26. Tu, X.; Gallon, H.J.; Twigg, M.V.; Gorry, P.A.; Whitehead, J.C. Dry reforming of methane over a Ni/Al$_2$O$_3$ catalyst in a coaxial dielectric barrier discharge reactor. *J. Phys. D Appl. Phys.* **2011**, *44*, 274007. [CrossRef]
27. Hu, S.; Li, F.; Fan, Z.; Gui, J. Improved photocatalytic hydrogen production property over Ni/NiO/N–TiO$_{2-x}$ heterojunction nanocomposite prepared by NH$_3$ plasma treatment. *J. Power Sources* **2014**, *250*, 30–39. [CrossRef]
28. Di, L.; Li, Z.; Lee, B.; Park, D.W. An alternative atmospheric-pressure cold plasma method for synthesizing Pd/P25 catalysts with the assistance of ethanol. *Int. J. Hydrog. Energy* **2017**, *42*, 11372–11378. [CrossRef]
29. Audier, P.; Fénot, M.; Bénard, N.; Moreau, E. Flow control of an elongated jet in cross-flow: Film cooling effectiveness enhancement using surface dielectric barrier discharge plasma actuator. *Appl. Phys. Lett.* **2016**, *108*, 084103. [CrossRef]
30. Park, Y.; Oh, K.S.; Oh, J.; Seok, D.C.; Kim, S.B.; Yoo, S.J.; Lee, M.J. The biological effects of surface dielectric barrier discharge on seed germination and plant growth with barley. *Plasma Process Polym.* **2018**, *15*, 1600056. [CrossRef]
31. Di, L.B.; Li, X.S.; Shi, C.; Xu, Y.; Zhao, D.Z.; Zhu, A.M. Atmospheric-pressure plasma CVD of TiO$_2$ photocatalytic films using surface dielectric barrier discharge. *J. Phys. D Appl. Phys.* **2009**, *42*, 032001. [CrossRef]
32. Abdelaziz, A.A.; Ishijima, T.; Seto, T.; Osawa, N.; Wedaa, H.; Otani, Y. Characterization of surface dielectric barrier discharge influenced by intermediate frequency for ozone production. *Plasma Sources Sci. Teachnol.* **2016**, *25*, 035012. [CrossRef]
33. Jeon, J.; Rosentreter, T.M.; Li, Y.; Isbary, G.; Thomas, H.M.; Zimmermann, J.L.; Shimizu, T. Bactericidal Agents Produced by Surface Micro-Discharge (SMD) Plasma by Controlling Gas Compositions. *Plasma Process Polym.* **2014**, *11*, 426–436. [CrossRef]
34. Qi, B.; Di, L.; Xu, W.; Zhang, X. Dry plasma reduction to prepare a high performance Pd/C catalyst at atmospheric pressure for CO oxidation. *J. Mater. Chem. A* **2014**, *2*, 11885–11890. [CrossRef]
35. Chung, C.; Lee, M.; Choe, E.K. Characterization of cotton fabric scouring by FT-IR ATR spectroscopy. *Carbohydr. Polym.* **2004**, *58*, 417–420. [CrossRef]
36. Montazer, M.; Keshvari, A.; Kahali, P. Tragacanth gum/nano silver hydrogel on cotton fabric: In-situ synthesis and antibacterial properties. *Carbohyd. Polym.* **2016**, *154*, 257–266. [CrossRef]
37. Wiley, B.J.; Im, S.H.; Li, Z.Y.; McLellan, J.; Siekkinen, A.; Xia, Y. Maneuvering the surface plasmon resonance of silver nanostructures through shape-controlled synthesis. *J. Phys. Chem. B* **2006**, *110*, 15666–15675. [CrossRef]
38. Nam, S.; Condon, B.D. Internally dispersed synthesis of uniform silver nanoparticles via in situ reduction of [Ag(NH$_3$)$_2$]$^+$ along natural microfibrillar substructures of cotton fiber. *Cellulose* **2014**, *21*, 2963–2972. [CrossRef]
39. Zhou, Y.; Xiang, Z.; Cao, D.; Liu, C.J. Covalent organic polymer supported palladium catalysts for CO oxidation. *Chem. Commun.* **2013**, *49*, 5633–5635. [CrossRef]
40. Wang, W.; Yang, M.; Wang, Z.; Yan, J.; Liu, C. Silver nanoparticle aggregates by room temperature electron reduction: Preparation and characterization. *RSC Adv.* **2014**, *4*, 63079–63084. [CrossRef]
41. Cölfen, H.; Antonietti, M. Mesocrystals: Inorganic superstructures made by highly parallel crystallization and controlled alignment. *Angew. Chem. Int. Ed.* **2005**, *44*, 5576–5591. [CrossRef] [PubMed]
42. Shrivastava, S.; Bera, T.; Roy, A.; Singh, G.; Ramachandrarao, P.; Dash, D. Characterization of enhanced antibacterial effects of novel silver nanoparticles. *Nanotechnology* **2007**, *18*, 225103. [CrossRef]

 *nanomaterials*

*Article*

# Aqueous Gold Nanoparticles Generated by AC and Pulse-Power-Driven Plasma Jet


**Pengcheng Xie [1,2], Yi Qi [1], Ruixue Wang [1,*], Jina Wu [3] and Xiaosen Li [3,*]**

[1]  College of Mechanical and Electrical Engineering, Beijing University of Chemical Technology, Beijing 100029, China; xiepc@mail.buct.edu.cn (P.X.); 2018210368@mail.buct.edu.cn (Y.Q.)

[2]  State Key Laboratory of Organic-Inorganic Composites, Beijing University of Chemical Technology, Beijing 100029, China

[3]  State Key Laboratory of NBC Protection for Civilian, Beijing 102205, China; wujina09@163.com

*  Correspondence: wrx@mail.buct.edu.cn (R.W.); momentday@126.com (X.L.)



Received: 31 August 2019; Accepted: 16 October 2019; Published: 18 October 2019



**Abstract:** In this study, we developed a simple-to-use approach based on an atmospheric pressure plasma jet to synthesize aqueous Au nanoparticles (AuNP). Special attention was paid to the different reaction dynamics and AuNP properties under AC and pulse-power-driven plasma jets (A-Jet and P-Jet, respectively). The morphology of the AuNP, optical emissions, and chemical reactions were analyzed. Further, a copper mesh was placed above the reaction cell to evaluate the role of electrons and neutral species reduction. A visible color change was observed after the A-Jet treatment for 30 s, while it took 3 min for the P-Jet. The A-Jet treatment presented a much higher AuNP growth rate and a smaller AuNP diameter compared with the P-Jet treatment. Further analysis revealed an increase in chemical concentrations ($Cl^-$ and $H_2O_2$) and liquid conductivity after plasma treatment, with a higher increased amplitude for the A-Jet case. Moreover, the electrons alone had little effect on AuNP generation, while neutral species showed a clear $Au^+$ reduction effect, and a unique coupling effect between both reactions was observed. The different reaction dynamics between the A-Jet and P-Jet were attributed to their different local heating effects and different discharge power during the reaction.

**Keywords:** aqueous gold nanoparticles; AC-powered plasma jet; pulse-powered plasma jet; gold nanoparticle generation; chemical reactions


---

## 1. Introduction

High-quality gold nanoparticles (AuNPs) play a crucial role in numerous fields, including catalysis, optoelectronics, medical imaging, and sensors [1,2]. Highly purified surfactant-free AuNPs are very important because residual components limit their applications. Numerous methods, including chemical reduction, biosynthesis, and laser ablation, have been reported for AuNP synthesis [3,4]. However, the high cost associated with the synthesis of monodisperse size and shape-controlled nanoparticles and the use of toxic reducing agents makes them undesirable [5].

Plasma–liquid interaction has received much attention as a novel nanoparticle generation approach [6]. The electron energy generated by atmospheric pressure plasma can reach up to a few electron volts; thus, it is able to initiate the direct reduction of metal ions in the gas–liquid interface. Other oxidative species, such as OH· and $H_2O_2$, can have possible reactions with chloroauric acid $[AuCl_x(OH)_{4-x}]^-$ to form AuNP because $AuCl^-_x$ has a strong oxidation ability [7]. Various types of plasma can be applied to produce nanoparticles [8]: (1) remote plasma or plasma jet: gas discharge between an electrode and the electrolyte surface, (2) direct plasma: discharge between the electrodes. Microplasma, as one of the remote plasmas, usually works as a cathode electrode [9]. The high-energy electrons are believed to be responsible for initiating AuNP growth [10]. Normally, the electrolytic cell is driven by a DC power supply with several kV, and a large resistor (100 k$\Omega$) is connected to the circuit to limit the discharge current. Most of the power is consumed on the resistor instead of discharge. In addition, due to the small diameter of microplasma (hundreds of micrometers), the production rate is limited. In direct plasma, two metallic electrodes were immersed in liquids [11]. By applying radio frequency (RF) power, metallic nanoparticles are produced through the erosion of a metallic electrode exposed to plasma. The particle size is determined by the quenching rate of the surrounding water.

Recently, the continuous synthesis of colloidal gold nanoparticles by introducing liquid droplets into a plasma reactor has been proposed [12,13]. The picoliter reactor volume and droplet effect result in a high synthesis rate. However, the ability to scale-up the reactor might make this technology more practical. Ionic liquids have a very low vapor pressure and thus can operate at a low pressure where traditional large-sized processing plasmas can be applied [14]. Although there is no need for additional surfactant to stabilize the synthesized NPs, it is difficult to change their surface function. Most importantly, specific metal salts or ionic liquids are required due to the low solubility of many metal salts in ionic salts [15].

Atmospheric pressure plasma, working at room temperature, has drawn an immense amount of interest due to it being easy to scale up, its low cost, and its compact size [16–18]. A plasma jet, generated in open air rather than in confined chambers, can be used for direct treatment without the limitation of the object's size [19]. However, the treatment area of single plasma is small (less than 1 cm$^2$) due to its inherent structure [20]. One convenient method is to form a jet array by grouping a number of individual plasma jets units together. Although the interaction mechanisms between individual jets require further investigation, a 2D jet array with 45 single jets has been successfully implemented to form a large treatment area [21]. The scale-up ability of plasma jets makes them practical for further industrial application.

In this study, we developed a one-step approach based on an atmospheric pressure plasma jet that directly interacts with liquids to generate aqueous AuNPs. By applying the plasma jet above the liquid surface, the AuNP can be produced by plasma–liquid interaction. Different from the microplasma mentioned above, there is no secondary electrode needed. In addition, the treatment area of the plasma jet can be enlarged by grouping multiply plasma jets together, which provides the possibility for industrial application. The different reaction dynamics and AuNP properties were compared with AC and pulse-power-driven plasma jets (A-Jet and P-Jet, respectively). The plasma-induced chemistry was studied by monitoring different chemical concentrations ($Cl^-$, $H_2O_2$, and $NO_3$), the pH value, and the conductivity of the liquid during reaction processes. The role of electrons and neutral species for $Au^+$ reduction was analyzed by isolating these two factors into different reaction cells. Finally, the chemical reactions in the liquid by A-Jet and P-Jet were analyzed to explain the synthesis mechanisms involved in the A-Jet and the P-Jet.

## 2. Materials and Methods

In order to synthesize aqueous AuNPs, an atmospheric pressure plasma jet was placed 1 cm above an aqueous solution mixed with $HAuCl_4$ and sodium citrate (total volume: 4 mL). The plasma jet was constructed with a typical needle–ring electrode structure, the details of which can be found in reference [22]. A hollow stainless-steel tube was inserted into a quartz tube (inner diameter: 1.5 mm, wall thickness: 0.1 mm, length: 150 mm), working as a high-voltage electrode. Copper tape with a width of 10 mm and a thickness of 180 µm was wrapped around the quartz tube surface and worked as a ground electrode. $HAuCl_4$ and sodium citrate were purchased from Sinopharm and were used without further purification. Once the plasma jet was ignited, high-energy electrons, reactive species, and UV were generated in the gas phase. These species subsequently went through the plasma–liquid interface to the bulk liquid. In the liquid region, the reduction of $Au^+$ into $Au^0$ occurred, and AuNPs started to grow with the existence of the stabilizer (sodium citrate). Due to the high recombination rate of reactive species, only long-lived species (such as $H_2O_2$, OH-, $NO^-_2$, $NO^-_3$) penetrated into the liquid. At the same time, the color of the $HAuCl_4$ solution turned gradually from light yellow to pink, which indicated the formation of AuNP. After several seconds or minutes (depending on power supply) of the plasma jet treatment, the solution containing AuNP was collected.

The plasma jet was powered either by an AC power supply (Nanjing Suman Plasma Technology, CTP-2000K, Nanjing, China) or a pulsed power supply (Xi'an Smart Maple Electronic Technology, HVP-22P, Xi'an, China), and the discharge parameters were verified. Research-grade Ar gas (99.999%) with a gas flow rate of 2 slm (standard liter per minute) was used as the working gas. Once the plasma was ignited, the high-energy electrons that cascaded into the liquid solution were responsible for the initiation of AuNP growth. The schematic diagram of the plasma AuNP synthesis system is shown in Figure 1.

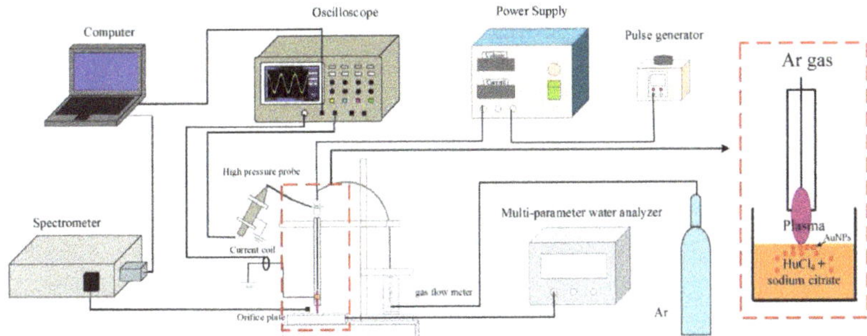


**Figure 1.** A schematic diagram of the plasma jet setup used for AuNP synthesis (inset shows enlarged diagram of plasma jet–liquid system).

The voltage and current characteristics were monitored by a high-voltage probe (Tektronix, P6015A, 1000:1, OR, USA) and a current probe (Pearson, model4100, 1 V/A) through a digital oscilloscope (Lecory WR204Xi, NYC, USA), respectively. The input power was calculated by the voltage–charge Lissajous method. A small capacitor with a capacitance of 1000 μF was connected to the plasma jet ground electrode. The area ($A$) enclosed by the power voltage, and the capacitor charge ($Q$) was calculated. The input power ($P$) was determined by

$$P = fA, \tag{1}$$

where $f$ is the pulse frequency.

A fiber optic cable was placed near the exit nozzle of the plasma jet to guide the light emission to the spectrometer (Ocean optics, Maya 2000 pro, FL, USA). Visible images of the as-synthesized AuNPs were captured by a digital SLR camera (Nikon D3200, Tkyo, Japan) coupled with a zoom lens (Nikkor, S-line, Nikon, Tkyo, Japan) with an exposure time of 0.5 ms. The AuNPs were characterized by an ultraviolet–visible absorption spectrometer (IMPLEN Nanophotometr N60, MUC, Germany) in the wavelength range from 200 to 900 nm. The size and shape distribution of the AuNPs was analyzed by a transmission electron microscope (TEM, TECNAIF30, OR, USA). A selected area electron diffraction (SAED) device coupled with the TEM was used to identify the crystal structure of the AuNPs.

The concentration of Cl⁻ was measured by a water quality analyzer (Leici DZS-708, Shanghai, China). For Cl⁻ measurement, the measuring electrode and reference electrode were immersed in a 0.001 mol/L KCl standard solution for 2 h. KCl solutions with five gradients (100, 10, 1, 0.1, and 0.01 mmol/L) were used for Cl⁻ calibration. After each measurement, the electrodes were rinsed with deionized water to reduce the experimental error. The conductivity and pH value of the samples were measured by a conductivity meter (Leici TM-03 Pen-shaped conductivity meter, Shanghai China) and a pH meter (Leici PHSJ-3F, Shanghai, China), respectively. A hydrogen peroxide kit (LOHAND Test Strips Series 0–25 mg/L, Hangzhou, China) measured the concentration of hydrogen peroxide.

## 3. Results

### 3.1. Characterization of A-Jet and P-Jet

Figure 2 represents the typical characteristics of A-Jet by measuring the voltage-current waveform, the calculated Lissajous figure, and the discharge image. During discharge, the applied voltage and frequency were kept consistently as 6.8 kV and 90 kHz, respectively. The discharge occurred at both voltage rise and fall times during one discharge period, as shown in Figure 2a. There were several small current peaks followed by the main current peak at pulse rise and fall times, which confirmed a filamentary discharge mode in our condition (inset image). The development of a filamentary micro-discharge can be divided into three stages [23]: (1) pre-breakdown stage with negative charge accumulation, (2) ionization wave propagation towards the cathode, and (3) discharge filament bridged the gap, and a bright plasma channel is formed. To calculate the discharge power during discharge, a 100 µF capacitor was series-connected in the circuit. The obtained Lissajous figure is shown in Figure 2b. The average power is equal to the product of the repetition frequency and the energy in a discharge period. So, the discharge power was calculated as 30.1 W. The input power of the AC power supply was 68.1 W; thus, the power efficiency was 45%.

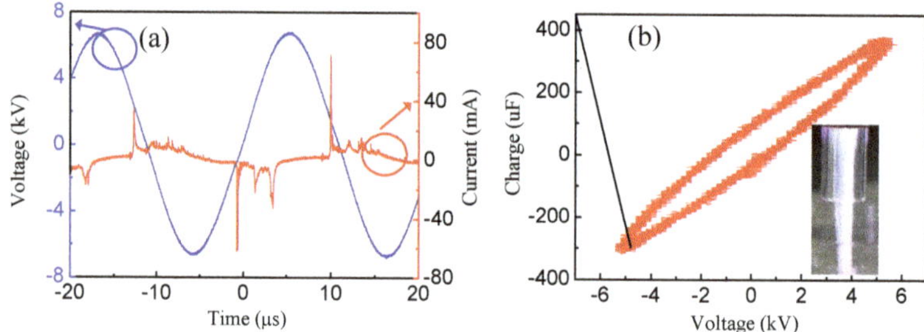

**Figure 2.** (**a**) Voltage and current discharge waveforms and (**b**) the corresponding Lissajous figure (The inset shows discharge image of A-Jet).

Different from the AC power supply, pulsed power supply with rapid pulse rise time provides high reduced electric field intensity (E/N) to accelerate electrons. In a pulsed discharge mode, high energy electrons are believed to ionize the gas to generate secondary electrons instead of depending on space ionization [24], so, the discharge is more spatially uniform, and the gas temperature is much lower in P-Jet compared to A-Jet. Figure 3 shows the typical waveforms of applied voltage, plasma current, discharge image, and consumed energy during one pulse duration. The discharge power was set as follows: a pulse rise and fall time of 50 ns, a pulse duration time of 5 µs, and a pulse frequency of 8 kHz. One can see that the amplitudes of applied voltage and plasma current of the pulsed plasma jet were 8.5 kV and 0.2 A, respectively (Figure 3a), with bipolar plasma current behavior. The instantaneous power consumption was 2.2 kW, and the energy consumption during one pulse duration was 2.34 mJ, respectively (Figure 3b). With an applied frequency of 8 kHz in this case, the energy power consumption was 16.4 W, which was much lower than that of the A-Jet (30.1 W). Compared to A-Jet, the discharge of P-Jet was much gentler and more homogenous, as shown in the inset image.

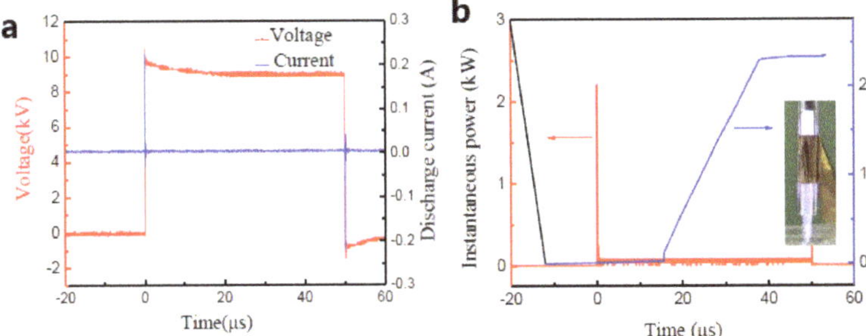

**Figure 3.** (**a**) Voltage and current discharge waveforms and (**b**) instantaneous power and energy waveforms of the P-Jet (the inset shows the discharge image of P-Jet).

### 3.2. Synthesis of AuNP by A-Jet and P-Jet

To generate AuNP, 1 mL $HAuCl_4$ (1.214 mM) and 3 mL sodium citrate (34 mM) solutions were mixed together and treated with the A-Jet for different times. The time dependence of AuNP generation was studied first. As shown in Figure 4a, with a plasma jet treatment of 30 s, the mixture solution changed from a shallow yellow to a dark violet, confirming the generation of AuNPs. As the plasma treatment time increased, the solution gradually became a red color. The shift in color was basically due to light absorption, depending on the particle size. When the size of the nanoparticles decreased, smaller wavelengths would be absorbed, and, so, a red color would be reflected. The absorption peak showed a similar trend: the absorption peak shifted towards lower wavelengths as the process time increased (Figure 4b). The absorption peaks were centered at 584, 566, and 535 nm for different plasma processing times of 30, 60, and 90 s, respectively. As the plasma treatment time increased to 120 s, the absorption peak intensity stopped increasing, and the central absorption peak shifted towards higher wavelengths. This phenomenon was more obvious when the plasma processing time increased to 150 s, at which time the central absorption peak increased to 545 nm, and the peak width became wider. This indicated that after 90 s, the $Au^+$ in the solution was totally consumed and AuNP began to aggregate. After 90 s of reaction, the synthesized AuNPs were isolated from the solution by centrifugation for 20 min at $1 \times 10^4$ rad/min and dried overnight at room temperature. A total of 9 mg of AuNP was collected, and the formation rate was calculated as 0.4 mg/s.

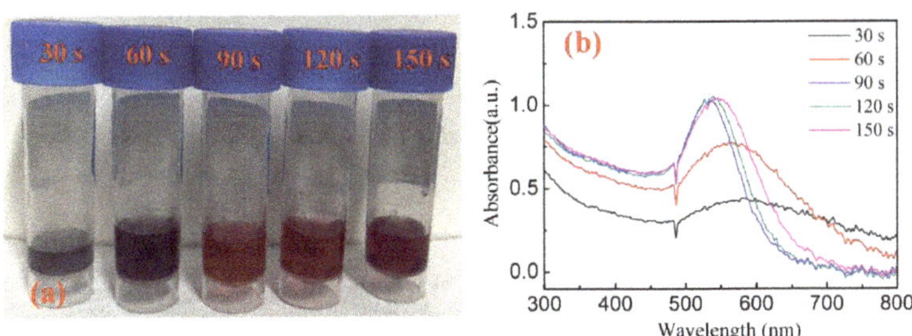


**Figure 4.** (**a**) Photos and (**b**) UV–Vis absorption spectra of AuNP for process times of 30, 60, 90, 120, and 150 s.

With variations in the $HAuCl_4$/sodium citrate ratio, the absorption peaks of AuNP also changed (Figure 5). The central absorption peak position shifted to lower wavelengths with decreasing of the $HAuCl_4$/sodium citrate ratio. This "blue shift" of the peak indicated a decrease in the average nanoparticle size as the sodium citrate volume increased. This was understandable since the sodium citrate acted as a stabilizer as well as a reducing agent. In addition, the absorption peak intensity decreased as the sodium citrate volume increased due to fewer Au seeds provided by $HAuCl_4$. A higher volume of sodium citrate led to more Au seeds and prevented the aggregation of AuNPs [25]. This was further confirmed by a broad distribution of UV–Vis absorption spectra with a very small volume of sodium citrate. The TEM images (Figure 6a–c) showed that the AuNP were generally of a range of shapes (spherical, cylindrical, and hexagon). The average diameters of the synthesized AuNPs were 18.2 ± 9.0, 32.9 ± 14.1, and 180.6 ± 20.5 nm with $HAuCl_4$/sodium citrate ratios of 1:3, 1:1.75, and 1:0.3 (Figure 6a–c), respectively. The size distribution of the synthesized AuNPs measured by TEM was consistent with the UV-Vis spectra results (Figure 5), which confirmed that the higher sodium citrate concentration reduced the AuNP diameters. The dotted rings in the SAED pattern in Figure 6d suggest that these AuNP had a pronounced crystal structure. The *d*-spacing of the rings suggests that the dotted rings represented the Bragg reflection of the [111], [200], [220], and [222] crustal planes, indicating a face-centered cubic (fcc) crystal structure. In addition, the high-resolution TEM (HRTEM) image of one typical single AuNP (Figure 5e) showed a crystal lattice fringe spacing of 0.236 nm, corresponding to the [111] lattice planes of the AuNP.

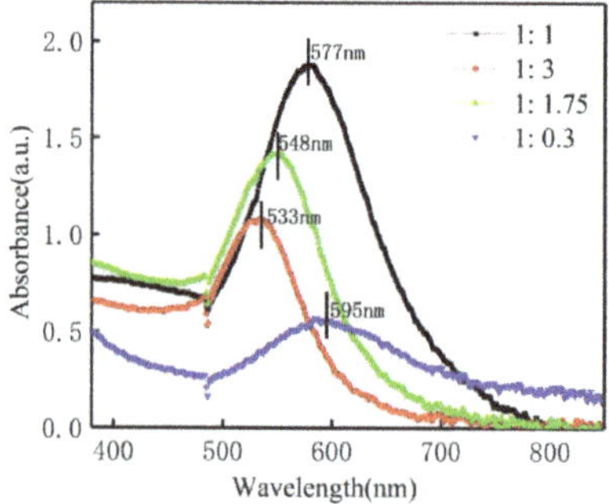


**Figure 5.** UV–Vis absorption spectra of AuNP for different $HAuCl_4$/sodium citrate ratios: (**a**) 1:1, (**b**) 1:3, (**c**) 1:1.75, and (**d**) 1:0.3. The total volume was kept at 40 mL, and the plasma treatment time was 90 s.

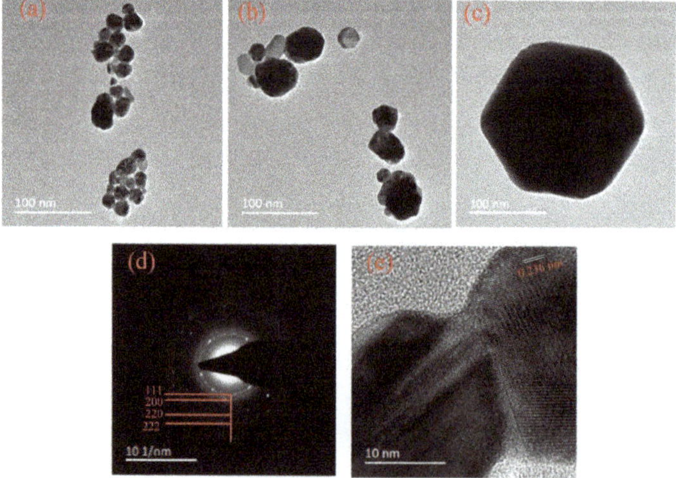


**Figure 6.** TEM images with different $HAuCl_4$/sodium citrate ratios: (**a**) 1:3, (**b**) 1:1.75, and (**c**) 1:0.3. (**d**) Selected area electron diffraction (SAED) pattern of an ensemble of AuNP, and (**e**) image of a typical single crystal of AuNP.

Unlike the A-Jet case, the mixture solution did not change colors until 3 min of pulse plasma treatment ($HAuCl_4$/sodium citrate ratio of 1:3). Further, the solution presented a brick-red color after 3 min of reaction and seemed cloudy (Figure 7a). Accordingly, the UV–Vis absorption peak was centered at 590 nm, which confirmed a large AuNP diameter (Figure 7b). As the pulsed plasma treatment time increased, the UV–Vis spectra shifted to a lower wavelength (582 nm at 7 min). The further increase of reaction time resulted in decreased absorption peak intensity and a broad absorption peak width (indicating a broad size distribution). After 7 min, the $Au^+$ in the solution was totally consumed and AuNP began to aggregate. The generated AuNP were collected after 7 min reaction, and the formation rate was calculated as 9.5 µg/s. Comparing the reactions initiated by the A-Jet and the P-Jet, the following differences were found: (1) the AuNP generation rate by the A-Jet was much faster than that of the P-Jet; (2) the AuNP size distribution was narrower in the A-Jet case; and (3) the AuNP size control range was broader in the A-Jet case. These differences were due to the different power consumption rates and different chemical reaction pathways introduced by the A-Jet and the P-Jet, which is well explained in the next section.

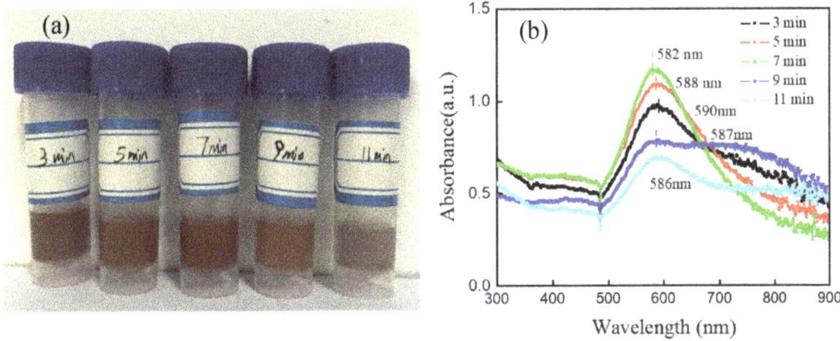

**Figure 7.** (**a**) Photos and (**b**) UV–Vis absorption spectra of AuNP for process times of 3, 5, 7, 9, and 11 min under P-Jet treatment (pulse frequency: 8 kHz, voltage: 8 kV, $HAuCl_4$/sodium citrate ratio of 1:3).

The TEM images of AuNP produced with a $HAuCl_4$/sodium citrate ratio of 1:3 for 7 min are shown in Figure 8. As shown in Figure 8a, the AuNP also presented a broad range of shapes, including spheres and polygons, and had an average diameter of 82.5 ± 21.5 nm. The dotted rings in the SAED pattern in Figure 8b represent the Bragg reflection of the [111], [200], [220], [222], and [420] crystal planes, indicating an fcc crystal structure. The HRTEM image of one typical single AuNP (Figure 8c) showed a crystal lattice fringe spacing of 0.204 nm (corresponding to the [200] lattice planes), which confirmed the crystallinity of the particles.

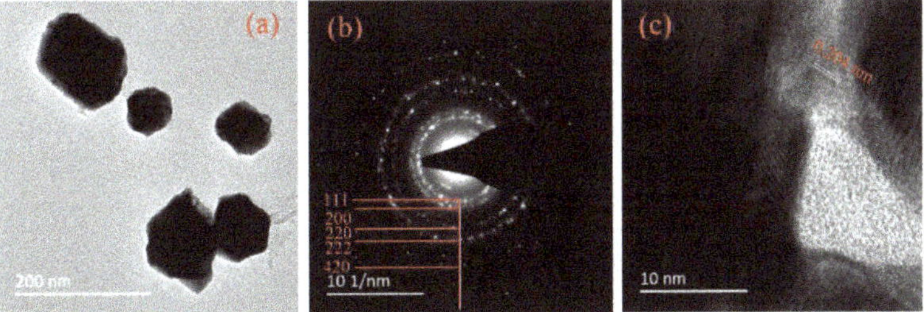

**Figure 8.** (**a**) TEM images of AuNP produced with a $HAuCl_4$/sodium citrate ratio of 1:3 with pulsed plasma treatment for 7 min. (**b**) SAED pattern of an ensemble of AuNP. (**c**) high-resolution TEM (HRTEM) image of a typical single crystal of AuNP.

As the diameter of synthesized AuNP increased with the decrease of sodium citrate concentration, and with $HAuCl_4$/sodium citrate ratio of 1:3, the average diameter of particle size was 82.5 ± 21.5 nm. So, there was no need to change the $HAuCl_4$/sodium citrate ratio as we did in A-Jet. Instead, the pulse repetition frequency was varied. Figure 9 shows that the intensity of the UV–Vis absorption peaks varied with the changes in pulse frequency. With the increase of pulse frequency, the UV–Vis absorption peak presented a blue shift, and the peak intensity increased. The increase of pulse frequency enhanced the input energy, thus corresponding to a higher AuNP generation rate.

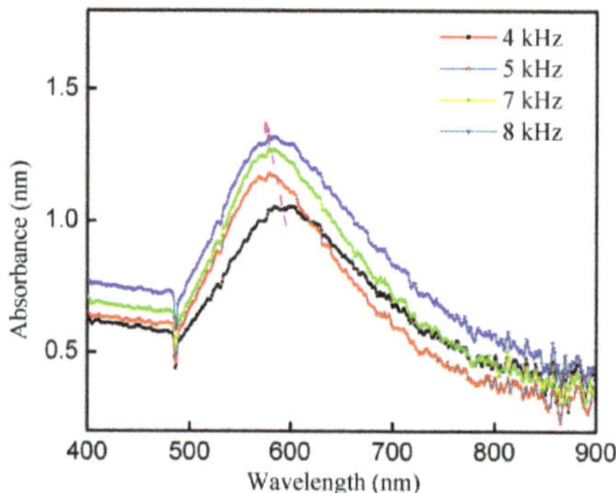


**Figure 9.** UV–Vis absorption spectra of AuNP under different pulse frequencies (applied voltage: 8.4 kV, treatment time: 7 min, $HAuCl_4$/sodium citrate ratio of 1:3).

Tables 1 and 2 summarized the influence of variation of experimental parameters on AuNP properties by A-Jet and P-Jet, respectively. The particle sizes and shapes were obtained by TEM images of the synthesized AuNP. For the particle size distribution, around 100 particles were measured, and their average particle size was calculated. The AuNP with spherical, cylindrical, and hexagon shapes were found in all cases, confirming that these parameters had a little influence on particle shape. In other words, it was very difficult to control particle shape in our system. However, the particle diameters changed. For time-dependence, A-Jet and P-Jet showed a similar tendency. The UV–Vis absorption peak first decreased then increased with a plasma treatment time increase. In the first stage, a loose self-assembly of small Au nanoclusters formed coarse particles and disassembled into small nanoparticles at a low pH value. With the decrease of the $AuHCl_4$/sodium citrate ratio, the UV–Vis absorption peak showed "red shift", and particle size increased in A-Jet. It was interesting to see that the diameter of AuNP in A-Jet and P-Jet was 18.2 ± 9.0 nm and 82.5 ± 21.5 nm at $AuHCl_4$/sodium citrate ratio of 1:3, indicating smaller AuNP were produced with A-Jet. A further decrease in sodium citrate ratio concentration would induce a larger particle size. So, it was hard to generate AuNP with a smaller size with P-Jet. That was why the variation of the $AuHCl_4$/sodium citrate ratio was not conducted in the P-Jet case. Instead, we changed the pulse repetition frequency and found that although the UV-Vis spectra peak moved to a lower wavelength, the particle diameter did not change too much if we compared it to the A-Jet case.

**Table 1.** Summarize of AuNP properties treated by A-Jet.

| Parameters | | A-Jet | | | |
|---|---|---|---|---|---|
| | | Wavelength (nm) | Absorbance (a.u.) | Average Size (nm) | Particles Type |
| Time (s) | 30 | 584 | 0.437 | - | - |
| | 60 | 566 | 0.772 | - | - |
| | 90 | 535 | 1.028 | 20.3 ± 12.2 | Spherical, cylindrical, hexagon |
| | 120 | 541 | 1.053 | 27.4 ± 10.4 | Spherical, cylindrical, hexagon |
| | 150 | 545 | 1.046 | - | - |
| AuHCl$_4$/Sodium citrate ratio | 1:1 | 577 | 1.877 | - | |
| | 1:3 | 533 | 1.072 | 18.2 ± 9.0 | Spherical, cylindrical, hexagon |
| | 1:1.75 | 548 | 1.410 | 32.9 ± 14.1 | Spherical, cylindrical, hexagon |
| | 1:0.3 | 595 | 0.550 | 180.6 ± 20.5 | Spherical, cylindrical, hexagon |

**Table 2.** Summarize of AuNP properties treated by P-Jet.

| Parameters | | P-Jet | | | |
|---|---|---|---|---|---|
| | | Wavelength (nm) | Absorbance (a.u.) | Average Size (nm) | Particles Type |
| Time (min) | 3 | 590 | 0.977 | - | - |
| | 5 | 588 | 1.087 | - | - |
| | 7 | 582 | 1.167 | 81.2 ± 19.2 | Spherical, cylindrical, hexagon |
| | 9 | 587 | 0.779 | 100.2 ± 20.3 | Spherical, cylindrical, hexagon |
| | 11 | 586 | 0.694 | 95.3 ± 29.8 | Spherical, cylindrical, hexagon |
| AuHCl$_4$/Sodium citrate ratio | 1:1 | 594 | 1.923 | - | - |
| | 1:3 | 582 | 1.167 | 82.5 ± 21.5 | Spherical, cylindrical, hexagon |
| | 1:1.75 | 589 | 1.285 | - | - |
| | 1:0.3 | 603 | 0.716 | - | - |

### 3.3. Plasma-Induced Chemistry Involved in A-Jet and P-Jet

To explain the mechanism involved in A-Jet and P-Jet AuNP synthesis, optical emission spectroscopy was used to investigate the active species generated in the plasma jet. As shown in Figure 10, OH· emission at 306–309 nm and $N_2$ second positive system ($C^3\Pi_u$-$B^3\Pi_g$) at 337, 353, 380, and 405 nm were clearly observable in both cases; however, the peak intensity in the A-Jet case was much higher than in the P-Jet case. Similar Ar emissions from 656 to 850 nm presented in both cases and the intensity of these emissions were close. Near the gas–liquid interface, the dissociative electron attachment of $H_2O$ formed OH· radicals and subsequently combined into long-lived species ($H_2O_2$), which played an important role in the AuNPs synthesis:

$$e^-_{gas} + H_2O \rightarrow H^- + OH \tag{2}$$

$$OH\cdot + OH\cdot \rightarrow H_2O_2. \tag{3}$$

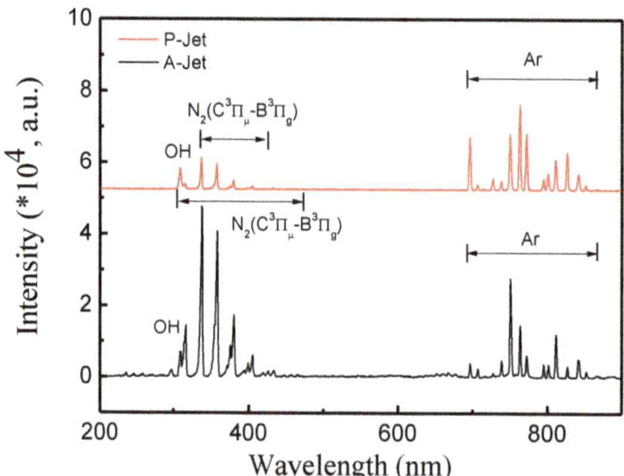

**Figure 10.** Optical emission spectra of the A-Jet and P-Jet near the liquid surface.

However, the H radical was not detected in both cases, which confirmed that water decomposition into the atomic H and OH· radical pathway was not dominant in our experiments. The presence of the $N_2$ second positive system was caused by excitation, quenching processes, associative excitation, pooling reactions, transfer of energy between collisional partners, Penning excitation [26], etc. The $N_2$ species were then combined into $NO_3^-$ and $NO_2^-$ long-lived species in the liquid. Although the energy of metastable-state Ar is lower than the threshold excitation energy of $O_2$ (13.6 eV), metastable-state Ar can dissociate oxygen molecules and excite the state by Penning excitation. However, the O emission lines were not detected due to their low intensity.

Furthermore, the liquid chemistry was studied by measuring the conductivity, pH value, and $Cl^-$ concertation. The applied voltage for the A-Jet was kept at 6.4 kV, and it was 8 kV for the P-Jet in these experiments, unless otherwise specified. As shown in Figure 11a, the pH value decreased almost linearly with the increase of the plasma treatment time, while the conductivity showed the opposite tendency. After AC plasma treatment for 150 s, the pH value decreased from 6.32 to 5.36, while the conductivity increased from 396 to 500 μs/cm. With the pulsed power plasma treatment, the pH value and conductivity showed a tendency similar to the AC plasma jet, but with a smaller change. For example, after 11 min of treatment, the pH decreased from 6.32 to 5.78, and the conductivity increased from 396 to 462 μs/cm. The slower changes of pH and conductivity in the P-Jet compared with the A-Jet were consistent with its slower AuNP generation rate. A similar evolution of pH and conductivity was observed in discharges generated above the water surface [27]. The changes in pH and conductivity may be attributed to the water hydrolysis initiated by electrons or reactive species generated by the plasma jet [28]. Water hydrolysis produced $H^+$ and led to a decreased pH value during the reaction:

$$2H_2O - 4e^- \rightarrow O_2 \uparrow + 4H^+. \tag{4}$$

In addition, the reduction of $Au^{3+}$ by $H_2O_2$ consumed $OH^-$:

$$3H_2O_2 + 3OH^- + Au^{3+} \rightarrow Au^0 + 3HO_2 + 3H_2O. \tag{5}$$

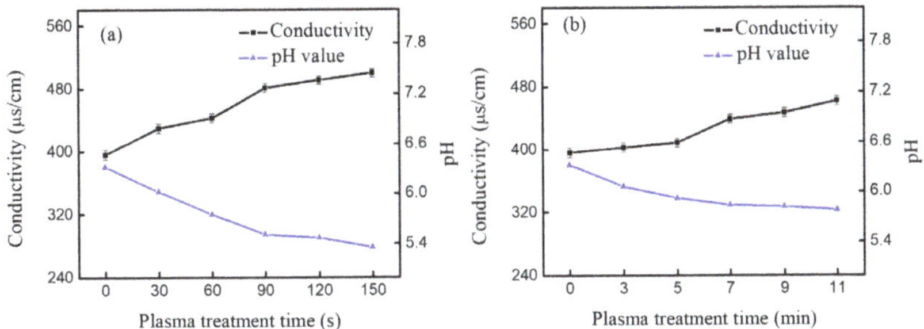

**Figure 11.** The conductivity and pH value of the mixture solution after plasma reaction: (**a**) A-Jet for 90 s (voltage: 6.4 kV, frequency: 90 kHz) and (**b**) P-Jet for 7 min (voltage: 8 kV, frequency: 8 kHz).

The pH variation in the solution was essential for AuNP diameter control. In the first state, the reduction of $AuCl^-$ by plasma formed a loose self-assembly (100 nm or more) of small Au nanoclusters (less than 1 nm) [29]. With the pH drop in the solution, the self-assembly of Au nanoclusters disassembled into smaller sizes than the previous assembly. Because at low pH values, the self-assembly of Au nanoclusters are easy to disassemble. This explains why AuNP presented a much larger size when generated by P-Jet compared with the A-Jet. As confirmed by the optical emission spectra, newly generated $NO_3^-$, $NO_2^-$, and $Cl^-$ long-lived species cascaded by $N_2$ species might have been responsible for the conductivity increase in the liquid.

Figure 12 shows the $Cl^-$ and $H_2O_2$ concentrations as treatment time increased for the A-Jet (a) and P-Jet (b). After plasma treatment, the $H_2O_2$ concentration was 0.56 mM for the A-Jet (reaction time: 150 s) and 0.59 mM for the P-Jet (reaction time: 11 min). Although the $H_2O_2$ concentration for the A-Jet and P-Jet processes reached almost the same value, clearly, the $H_2O_2$ production rate was much faster in the A-Jet process. The growth of $Cl^-$ concentration was even faster in the A-Jet process. After a reaction time of 150 s, the $Cl^-$ concentration reached 1.1 mM, while the $Cl^-$ concentration was 0.4 mM in the P-Jet process after 11 min of reaction. These results proved that the strong chemical reactions involved in the A-Jet were responsible for its faster growth rate of AuNP.

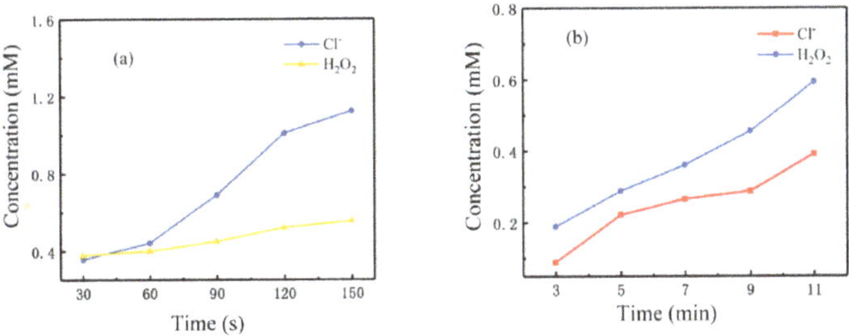

**Figure 12.** $Cl^-$ and $H_2O_2$ generation in the (**a**) A-Jet and (**b**) P-Jet cases.

## 3.4. The Role of Electrons and Neutral Species

The generation of AuNP initiated the reduction of metal ions ($Au^+$) in the solution either by electrons or neutral species. In the liquid cathode case [22], the Ar ions driven by the voltage fall attached to the liquid surface, creating some secondary effects, such as the generation of secondary electrons. The dissolved secondary electrons had a strong reducing ability for AuNP synthesis [30]. However, in our experiments, without a noble anode in the liquid, the voltage drop could be neglected, and the role of electron reduction could have been suppressed. To verify our assumption, a square copper grid with a diameter of 4 × 4 cm was placed above reaction cell A, and the copper grid was connected to a copper wire in cell B. So, only the electrons or ions could flux into cell B, while other reactive species were transported into cell A, and the effect of neutral species was studied. The schematic diagram of the experimental setup is shown in Figure 13. The validation of this experimental setup was proved in [31]. Cell C, which did not have a copper grid and wire, was studied as the control group.

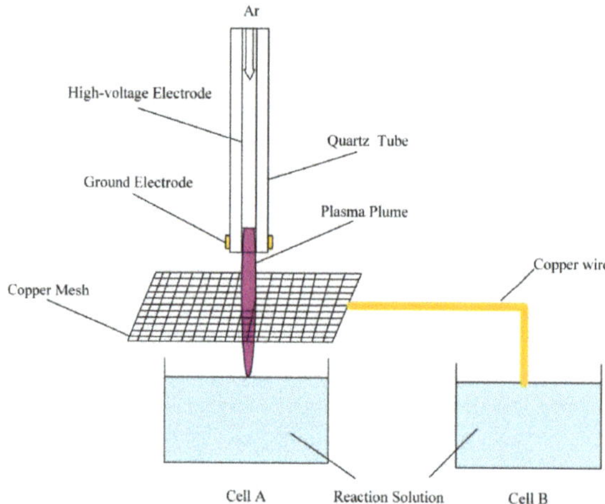


**Figure 13.** Schematic diagram of the experimental setup.

Figure 14 shows the UV–Vis spectra in these three cells after plasma treatment for different times (90 s for A-Jet and 7 min for P-Jet). Compared with cell C, cell B showed a red-shift absorption peak and a lower peak intensity. There were no absorption peaks detected in cell A, which confirmed that electron reduction alone was not dominant in our experiments. Visual observation indicated the same trend: the color of the AuNP changed from red-purple to dark-purple in the A-Jet case and changed from dark-purple to shallow pink. Cell A showed no color changes in both cases. These results also confirmed that the electrons and neutral species had a notable synergetic effect on AuNP synthesis, causing a higher absorption intensity in cell C.

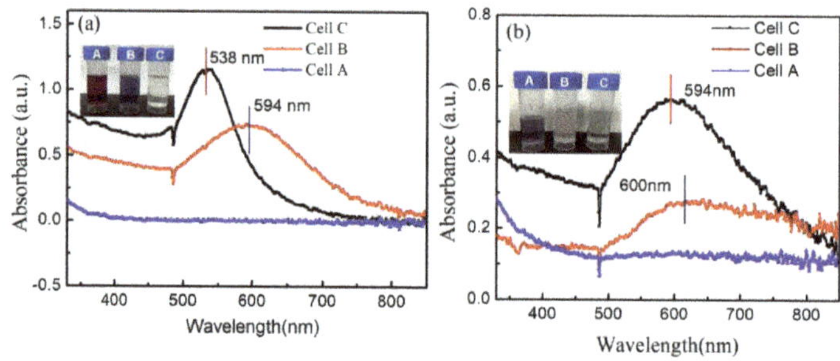


**Figure 14.** UV–Vis absorption spectra of AuNP for cells A–C with the interaction of (**a**) the A-Jet and (**b**) the P-Jet (A, B, and C in the vials stands for Cell A, Cell B and Cell C).

As shown in Figure 15a, the variation of the pH value in Cell A, B, and C was 6.15, 6.27, and 5.51 for the A-Jet and 6.07, 6.31, and 5.84 for the P-Jet, respectively. This confirmed that the electrons transferred to the liquid had little effect on pH value changes. In cell A, the pH value reduced to 6.15 and 6.07 for the A-Jet and the P-Jet, respectively. The combination of natural species and electrons resulted in a maximum pH value reduction of 5.51 and 5.84 for the A-Jet and the P-Jet, respectively.

Similar phenomena were observed for the conductivity measured in the three cells. An increase in conductivity was only observed in cells A and C. These results were consistent with the UV–Vis absorption spectra.

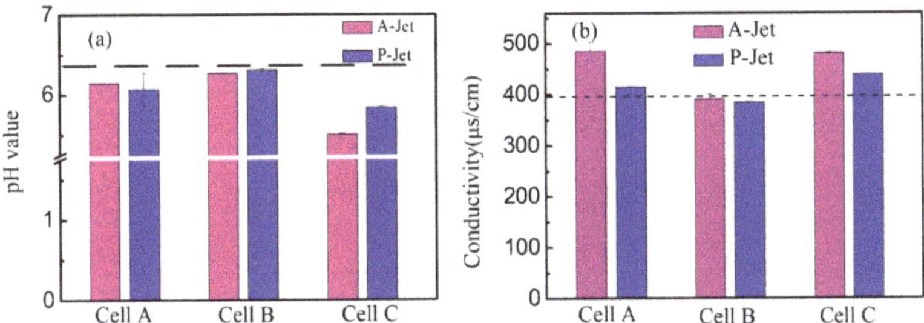

**Figure 15.** The conductivity and pH value of the different cells at: (**a**) A-Jet for 90 s (voltage: 6.4 kV, frequency: 90 kHz) and (**b**) P-Jet for 7 min (voltage: 8 kV, frequency: 8 kHz). The dashed line represents the initial pH value or conductivity of the mixture without plasma treatment.

## 4. Discussions

### 4.1. Chemical Reactions for AuNP Generation

In the plasma–liquid interaction, chemical reactions typically occurred in the gas phase, the gas–liquid interface, and the liquid phase simultaneously. In the gas phase, plasma ignition generated high-energy electrons, reactive species, and UV. These species subsequently dissolved in the water and had a major role in influencing the final liquid chemistry [32]. As confirmed by optical emission spectroscopy (Figure 10), excited Ar, OH·, and $N_2$ ($C^3\Pi_u$-$B^3\Pi_g$) were directly observed. Under a certain amount of pressure provided by Ar gas, there was a thin layer of water steam above the bulk liquid. The transportation of ions, electrons, and neutral species went through the plasma–liquid interface to the bulk liquid. In this region, the recombination, absorption, or desorption of reactive species and solvation of electrons or ions occurred [33], for example, the formation of hydrogen peroxide by hydroxyl radical recombination ($2OH_{(int)} \rightarrow H_2O_{2(int)}$), followed by incorporation of the hydrogen peroxide into the liquid ($H_2O_{2(int)} \rightarrow H_2O_{2(aq)}$). For ions or low-energy electrons, they were mostly immediately solvated when striking the liquid. The simulations showed that even 100 eV $O^+$ ions do not penetrate beyond the liquid surface by more than 3 nm [34]. As for high-energy electrons, the excitation, dissociation, or ionization of water molecules was expected. Solvated electrons (or hydrated electrons) can hydrolyze water by the following reactions:

$$2e^-_{(aq)} + 2H_2O \rightarrow H_{2(g)} + 2OH^-, \tag{6}$$

$$2e^-_{(aq)} + 2H^+ \rightarrow H + 2OH^-. \tag{7}$$

Electrolytic reactions between plasma electrons and aqueous ions yield an excess of hydroxide ions ($OH^-$), making the solution more basic, while reactions between reactive neutral species formed in the plasma phase and the solution lead to nitrous acid ($HNO_2$), nitric acid ($HNO_3$), and hydrogen peroxide ($H_2O_2$), making the solution more acidic [35]. According to our results, the pH value decreased rather than increased, confirming that the latter process dominated.

In the bulk liquid, the reduction of $AuCl^-$ by a reductant agent ($H_2O_2$, hydrated electrons) occurred. In general, reduction by both long-lived species (e.g., $H_2O_2$) and electrons was considered very important, but it is unclear if one was dominant. By putting a copper mesh over the reaction cell, the role of electrons and neutral species was studied. With electrons, there was no AuNP generation, while neutral species induced visible color changes with plasma treatment, confirming that the neutral species played a more important role for AuNP synthesis. It is interesting to note that there was a unique coupling effect between both reactions. This is because the generation of AuNP also requires a moderate pH value and conductivity. The complex processes of plasma–liquid chemistry for AuNP synthesis is shown in Figure 16.

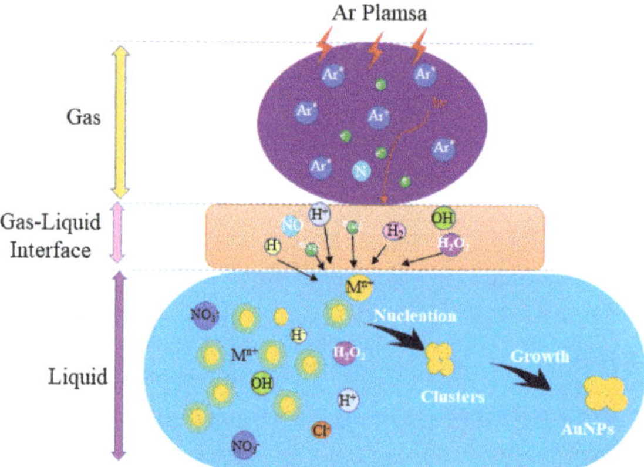


**Figure 16.** Schematic diagram of the synthesis mechanisms for plasma–liquid interaction.

*4.2. The Effect of A-Jet and P-Jet*

The breakdown mechanisms for the A-Jet and the P-Jet are different. For the A-Jet, filament discharge was observed, as confirmed by the waveform of the discharge current shown in Figure 2. Since the excitation voltage is a continuous sine AC wave voltage, the directional movement of ions cannot be neglected, which means the working gas can be heated by positive ions and neutral particles [36]. For the P-Jet, with the narrow pulse duration and short pulse rising time, the electrical energy was consumed to generate energetic electrons during discharge instead of heating plasma gas. The electron energy in P-Jet was higher than that of A-Jet. During one pulse discharge period, a short discharge duration time (about 200 ns, including positive and negative discharges) and a long discharge period (about $1.25 \times 10^8$ ns) guaranteed that there was enough time for the plasma to cool down sufficiently. Thus, a prime difference between A-Jet and P-Jet was their different electron energy and gas temperature. As proved by many studies, the gas temperature of an A-Jet is always higher than that of a P-Jet. The temperature of a P-Jet can remain at room temperature, while an A-Jet can reach 100 °C in similar conditions [34]. During the reaction process, plasma directly touches with the liquid surface, and the reaction occurs mainly around this region. As is known, the reaction temperature is essential for AuNP synthesis, and a higher reaction temperature is beneficial for a higher reaction rate. So, the AuNP growth rate observed here was much higher in the A-Jet (0.4 mg/s) than in the P-Jet (9.5 µg/s). Due to the small diameter of the quartz tubing (the plasma was focused at a very small area of about $1.3 \times 10^{-2}$ cm$^{-2}$) and the short reaction time (especially for the A-Jet), the bulk liquid temperature has little changes. Although P-Jet favors a higher electron energy, as we confirmed in Section 3.4, the electron alone has a minor role in Au$^+$ reduction.

Another consideration is the discharge power. With a short pulse rise time (50 ns), plasma is excited at a high overvoltage. In principle, more highly energetic electrons are produced in P-plasma. However, here, due to the short pulse duration time and small pulse frequency, the input power and energy of the P-Jet was only half that of the A-Jet. As we explained in Section 4.1, the electrons alone had little effect on Au$^+$ reduction. Instead, the concentration of Cl$^-$ and H$_2$O$_2$ were much higher in the A-Jet, and H$_2$O$_2$ can be the main reducing agent for Au$^+$. The higher concentration of Cl$^-$ was beneficial for the pH drop, which thus enhanced the dissociation of Au nanoclusters, resulting in a smaller AuNP diameter.

## 5. Conclusions

In this study, aqueous AuNP were successfully generated by A-Jet and P-Jet, respectively. Different AuNP synthesis processes were observed: (1) faster AuNP growth rate for the A-Jet (more than 40 times faster than the P-Jet) and (2) narrower AuNP size distribution and a broader size control range in the A-Jet case compared to P-Jet case. Further analysis revealed an increase in chemical concentrations ($Cl^-$ and $H_2O_2$) and conductivity after plasma treatment and a higher increased amplitude in the A-Jet case. In addition, the pH value decreased during the reaction. The differences between the A-Jet and the P-Jet were due to their different discharge mechanisms: the local heating and higher discharge power in the A-Jet were very important for AuNP generation. Finally, by putting a copper mesh over the reaction cell, the roles of electrons and neutral species were studied. With electrons, there was no AuNP generation, while neutral species (e.g., OH·) induced visible color changes with plasma treatment. The long-lived species (e.g., $H_2O_2$, combined by OH·) was responsible for $Au^{3+}$ reducing and played a more important role in AuNP synthesis.

**Author Contributions:** Conceptualization, R.W. and X.L.; methodology, P.X. and Y.Q.; resources, J.W.; writing—original draft, R.W. and X.L.

**Funding:** This research was funded by [National Natural Science Foundation of China] grant number [51,877,205] And by [State Key Laboratory of NBC Protection for Civilians] grant number [JH05-2019-01].

**Conflicts of Interest:** The authors declare no conflict of interest.

## References

1. Ma, Y.; Huang, Z.; Li, S.; Zhao, C. Surface-Enhanced Raman Spectroscopy on Self-Assembled Au Nanoparticles Arrays for Pesticides Residues Multiplex Detection under Complex Environment. *Nanomaterials* **2019**, *9*, 426. [CrossRef] [PubMed]

2. Mahan, M.M.; Doiron, A.L. Gold Nanoparticles as X-Ray, CT, and multimodal imaging contrast agents: Formulation, targeting, and methodology. *Nanomaterials* **2018**, *15*, 5837276. [CrossRef]

3. Roya, B.; Hassan, H.; Hossein, F.; Farid, H.H.; Fatemeh, A.; Seyedeh, Z.M. In situ generation of the gold nanoparticles–bovine serum albumin (AuNPs–BSA) bioconjugated system using pulsed-laser ablation (PLA). *Mater. Chem. Phys.* **2016**, *177*, 360–370.

4. Guo, Z.X.; Zhang, M.; Zhao, L.B.; Guo, S.S.; Zhao, X.Z. Generation of alginate gel particles with AuNPs layers by polydimethylsiloxan template. *Biomicrofluidics* **2011**, *5*, 026502. [CrossRef] [PubMed]

5. Peng, C.F.; Duan, X.H.; Xie, Z.J.; Liu, C.L. Shape-controlled generation of gold nanoparticles assisted by dual-molecules: The development of hydrogen peroxide and oxidase-based biosensors. *J. Nanomater.* **2014**, *7*, 576082. [CrossRef]

6. Chen, Q.; Li, J.; Li, Y. A review of plasma–liquid interactions for nanomaterial synthesis. *J. Phys. D* **2015**, *48*, 424005. [CrossRef]

7. Wang, S.; Qian, K.; Bi, X.Z.; Huang, W. Influence of Speciation of Aqueous $HAuCl_4$ on the synthesis, structure, and property of Au colloids. *J. Phys. Chem. C* **2009**, *113*, 6505–6510. [CrossRef]

8. Genki, S.; Tomohiro, A. Nanomaterial synthesis using plasma generation in liquid. *J. Nanomater.* **2015**, *2015*, 123696.

9. Wang, R.; Zuo, S.; Zhu, W.; Zhang, J.; Fang, J. Rapid synthesis of aqueous-phase magnetite nanoparticles by atmospheric pressure non-thermal microplasma and their application in magnetic resonance imaging. *Plasma Process. Polym.* **2014**, *11*, 448–454. [CrossRef]

10. Patel, J.; Nemcova, L.; Maguire, P.; Graham, W.G.; Mariotti, D. Synthesis of surfactant-free electrostatically stabilized gold nanoparticles by plasma-induced liquid chemistry. *Nanotechnology* **2013**, *24*, 245604. [CrossRef]

11. Hattori, Y.; Nomura, S.; Mukasa, S.; Toyota, H.; Inoue, T.; Usui, T. Synthesis of tungsten oxide, silver, and gold nanoparticles by radio frequency plasma in water. *J. Alloy. Compd.* **2013**, *578*, 148–152. [CrossRef]

12. Jiang, N.; Hu, J.; Li, J.; Shang, K.F.; Lu, N.; Wu, Y. Plasma-catalytic degradation of benzene over Ag–Ce bimetallic oxide catalysts using hybrid surface/packed-bed discharge plasmas. *Appl. Catal. B* **2016**, *184*, 355–3635. [CrossRef]

13. Maguire, P.; Rutherford, D.; Montero, M.M.; Mahony, C.; Kelsey, C.; Weedie, M.T.; Martin, F.P.; Mcquid, H.; Diver, D.; Mariotti, D. Continuous In-Flight Synthesis for On-Demand Delivery of Ligand-Free Colloidal Gold Nanoparticles. *Nano Lett.* **2017**, *17*, 1336–1343. [CrossRef] [PubMed]

14. Meiss, S.A.; Rohnke, M.; Kienle, L.; Zein El Abedin, S.; Endres, F.; Janek, J. Employing plasmas as gaseous electrodes at the free surface of ionic liquids: Deposition of nanocrystalline silver particles. *J. Chem. Phys. Chem.* **2007**, *8*, 50–53. [CrossRef]

15. Kaneko, T.; Baba, K.; Hatakeyama, R. Gas–liquid interfacial plasmas: Basic properties and applications to nanomaterial synthesis. *Plasma Phys. Controll. Fusion* **2009**, *51*, 124011. [CrossRef]

16. Wu, S.; Wu, F.; Liu, C.; Liu, X.; Chen, Y.; Shao, T.; Zhang, C. The effects of the tube diameter on the discharge ignition and the plasma properties of atmospheric-pressure micro plasma confined inside capillary. *Plasma Process. Polym.* **2019**, *16*, 1800176. [CrossRef]

17. Wang, Y.Q.; Yu, F.; Zhu, M.Y.; Ma, C.H.; Zhao, D.; Wang, C.; Zhou, A.M.; Dai, B.; Ji, J.Y.; Guo, X.H. N-Doping of plasma exfoliated graphene oxide via dielectric barrier discharge plasma treatment for the oxygen reduction reaction. *J. Mater. Chem. A* **2018**, *6*, 2011–2017. [CrossRef]

18. Di, L.B.; Zhan, Z.B.; Zhang, X.L.; Qi, B.; Xu, W.J. Atmospheric-Pressure DBD Cold Plasma for Preparation of High Active Au/P25 Catalysts for Low-Temperature CO Oxidation. *Plasma Sci. Technol.* **2016**, *18*, 544–548. [CrossRef]

19. Wang, R.; Zhang, C.; Shen, Y.; Zhu, W.; Yan, P.; Shao, T.; Babaeva, N.Y.; Naidis, G.V. Temporal and spatial profiles of emission intensities in atmospheric pressure helium plasma jet driven by microsecond pulse: Experiment and simulation. *J. Appl. Phys.* **2015**, *118*, 123303.

20. Wang, R.; Zhang, K.; Shen, Y.; Zhang, C.; Zhu, W.; Shao, T. Effect of pulse polarity on the temporal and spatial emission of an atmospheric pressure helium plasma jet. *Plasma Sources Sci. Technol.* **2016**, *25*, 015020. [CrossRef]

21. Konesky, G. Dwell time considerations for large area cold plasma decontamination. *Proc. SPIE* **2009**, *7304*, 73040N–73041N.

22. Wang, R.; Gao, Y.; Zhang, C.; Yan, P.; Shao, T. Dynamics of plasma bullets in a microsecond-pulse driven atmospheric pressure He plasma jet. *IEEE Trans. Plasma Sci.* **2016**, *4*, 393–397. [CrossRef]

23. Patrick, V.; Anton, N.; Annemie, B.; Christophe, L. Study of an AC dielectric barrier single micro-discharge filament over a water film. *Sci. Rep.* **2018**, *8*, 10919.

24. Shao, T.; Wang, R.; Zhang, C.; Yan, P. Atmospheric-pressure pulsed discharges and plasmas: Mechanism, characteristics and application. *High Volt.* **2018**, *3*, 14–20. [CrossRef]

25. Wang, R.; Zuo, S.; Wu, D.; Zhang, J.; Zhu, W.; Becker, K.H.; Fang, J. Micro plasma-assisted synthesis of colloidal gold nanoparticles and their use in the detection of cardiac troponin I (cTn-I). *Plasma Process. Polym.* **2015**, *12*, 380–391. [CrossRef]

26. Wang, R.; Sun, H.; Zhu, W.; Zhang, C.; Zhang, S.; Shao, T. Uniformity optimization and dynamic studies of plasma jet array interaction in argon. *Phys. Plasmas* **2017**, *24*, 093507. [CrossRef]

27. Liu, Y.; Zhang, H.; Sun, J.; Liu, J.; Shen, X.; Zhan, J.; Zhang, A.; Ogniger, S.; Cavadias, S.; Li, P. Degradation of aniline in aqueous solution using non-thermal plasma generated in microbubbles. *Chem. Eng. J.* **2018**, *345*, 679–687. [CrossRef]

28. Dang, T.H.; Denat, A.; Lesaint, O.; Teissedre, G. Degradation of organic molecules by streamer discharges in water: Coupled electrical and chemical measurements. *Plasma Sources Sci. Technol.* **2008**, *17*, 024013. [CrossRef]

29. Saito, N.; Hieda, J.; Takai, O. Synthesis process of gold nanoparticles in solution plasma. *Thin Solid Film.* **2009**, *518*, 912–917. [CrossRef]

30. Hart, E.J.; Anbar, M. *The Hydrated Electron*; Wiley: New York, NY, USA, 1970; Volume 276, p. 5126764.

31. Chen, Z.; Liu, D.; Xu, H.; Xia, W.; Liu, Z.; Xu, D.; Rong, M.; Kong, G. Decoupling analysis of the production mechanism of aqueous reactive species induced by a helium plasma jet. *Plasma Sources Sci. Technol.* **2019**, *28*, 025001. [CrossRef]

32. Tani, A.; Ono, Y.; Fukui, S.; Ikawa, S.; Kitano, K. Free radicals induced in aqueous solution by non-contact atmospheric-pressure cold plasma. *Appl. Phys. Lett.* **2012**, *100*, 254103. [CrossRef]

33. Bruggeman, P.J.; Kushner, M.J.; Locke, B.R.; Gardeniers, J.G.; Graham, W.G.; Graves, D.B.; Hofman-Caris, R.C.; Maric, D.; Reid, J.P.; Ceriani, E.; et al. Plasma-liquid interactions: A review and roadmap. *Plasma Sources Sci. Technol.* **2016**, *25*, 053002. [CrossRef]

34. Minagawa, Y.; Shirai, N.; Uchida, S.; Tochikubo, F. Analysis of effect of ion irradiation to liquid surface on water molecule kinetics by classical molecular dynamics simulation. *Jpn. J. Appl. Phys.* **2014**, *53*, 010210. [CrossRef]
35. Rumbach, P.; Witzke, M.; Sankaran, R.M.; Go, D.B. Decoupling interfacial reactions between plasmas and liquids: Charge transfer vs plasma neutral reactions. *J. Am. Chem. Soc.* **2013**, *135*, 16264–16267. [CrossRef] [PubMed]
36. Zhang, L.; Yang, D.; Wang, W.; Wang, S.; Yuan, H.; Zhao, Z.; Sang, C.; Jia, L. Needle-array to plate DBD plasma using sine AC and nanosecond pulse excitations for purpose of improving indoor air quality. *Sci. Rep.* **2016**, *6*, 25242. [CrossRef]

 *nanomaterials*

*Article*

# DBD Plasma Combined with Different Foam Metal Electrodes for $CO_2$ Decomposition: Experimental Results and DFT Validations

Ju Li [1,†], Xingwu Zhai [1,2,†], Cunhua Ma [1,*,†], Shengjie Zhu [1], Feng Yu [1], Bin Dai [1], Guixian Ge [2,*,†] and Dezheng Yang [2,3,*]

[1] Key Laboratory for Green Processing of Chemical Engineering of Xinjiang Bingtuan, School of Chemistry and Chemical Engineering, Shihezi University, Shihezi 832003, China; leej222@163.com (J.L.); zxw1725910806@163.com (X.Z.); zsj497262724@gmail.com (S.Z.); yufeng05@mail.ipc.ac.cn (F.Y.); db_tea@shzu.edu.cn (B.D.)
[2] Key Laboratory of Ecophysics, College of Sciences, Shihezi University, Shihezi 832003, China
[3] Laboratory of Plasma Physical Chemistry, School of Physics, Dalian University of Technology, Dalian 116024, China
* Correspondence: mchua@shzu.edu.cn (C.M.); geguixian@shzu.edu.cn (G.G.); yangdz@dlut.edu.cn (D.Y.); Tel.: +86-0993-205-8775 (C.M. & G.G. & D.Y.)
† These authors contributed equally to this work.

Received: 28 August 2019; Accepted: 22 October 2019; Published: 11 November 2019

**Abstract:** In the last few years, due to the large amount of greenhouse gas emissions causing environmental issue like global warming, methods for the full consumption and utilization of greenhouse gases such as carbon dioxide ($CO_2$) have attracted great attention. In this study, a packed-bed dielectric barrier discharge (DBD) coaxial reactor has been developed and applied to split $CO_2$ into industrial fuel carbon monoxide (CO). Different packing materials (foam Fe, Al, and Ti) were placed into the discharge gap of the DBD reactor, and then $CO_2$ conversion was investigated. The effects of power, flow velocity, and other discharge characteristics of $CO_2$ conversion were studied to understand the influence of the filling catalysts on $CO_2$ splitting. Experimental results showed that the filling of foam metals in the reactor caused changes in discharge characteristics and discharge patterns, from the original filamentary discharge to the current filamentary discharge as well as surface discharge. Compared with the maximum $CO_2$ conversion of 21.15% and energy efficiency of 3.92% in the reaction tube without the foam metal materials, a maximum $CO_2$ decomposition rate of 44.84%, 44.02%, and 46.61% and energy efficiency of 6.86%, 6.19%, and 8.85% were obtained in the reaction tubes packed with foam Fe, Al, and Ti, respectively. The $CO_2$ conversion rate for reaction tubes filled with the foam metal materials was clearly enhanced compared to the non-packed tubes. It could be seen that the foam Ti had the best $CO_2$ decomposition rate among the three foam metals. Furthermore, we used density functional theory to further verify the experimental results. The results indicated that $CO_2$ adsorption had a lower activation energy barrier on the foam Ti surface. The theoretical calculation was consistent with the experimental results, which better explain the mechanism of $CO_2$ decomposition.

**Keywords:** dielectric barrier discharge plasma; foam metal electrodes; $CO_2$ decomposition; density functional theory

---

## 1. Introduction

Recently, the environmental impact of carbon dioxide ($CO_2$) has been extensively studied. The contribution of $CO_2$ to the greenhouse effect has been confirmed, so solutions to the problem of excessive $CO_2$ need to be found. Nowadays, the greenhouse effect originated from $CO_2$ emission is one of the main challenges for human beings. Meanwhile, $CO_2$ is a ubiquitous and universally available C1 (a substance containing one carbon atom) resource in the world. In this regard, three major strategies have been proposed: carbon capture, storage, and utilization [1–5]. As one of the carbon utilization pathways, $CO_2$ direct decomposition has aroused special interest, because it can convert the greenhouse gas into value-added carbon monoxide (CO), which can be used not only as a fuel, but also as a widely used chemical raw material [6].

$CO_2$ thermal decomposition requires temperature excess of 2000 K to form CO and $O_2$ [7]. Hence, this is a high consumption energy process. In recent years, some alternative processes for $CO_2$ decomposition have been developed, among them non-thermal plasma (NTP) is an emerging alternative to $CO_2$ decomposition that has many merits, such as exciting high-energy electrons for reaction under environmental conditions. Meanwhile, the gas temperature of NTP can be very low, which ensures a low energy cost due to reduced heat loss [8]. Therefore, various NTP configurations have been tested, such as corona discharge [9], glow discharge [10], gliding arc discharge [11], microwave discharge [12], and radio frequency discharge [13]. In the near future, dielectric barrier discharge (DBD) plasma will be used in $CO_2$ conversion [14].

DBD is one of the NTPs, which can replace the catalytic chemical process under high-temperature operating conditions. There are numerous microdischarges in the DBD reaction tube. Microdischarges are composed of a number of small current filaments, which are generally short-lived, only a few nanoseconds, and evenly distributed around the high-voltage electrode [15]. In a DBD, the average temperature of energetic electrons is very high, over the range of 10,000–100,000 K, but the actual gas temperature is close to the environment temperature [16]. Except for the generation of high-energy particles, DBD also produces ultraviolet-visible light, ozone species, excited species, radicals, ions, etc. These reactive species are responsible for the efficient initiation and propagation of reactions [17,18].

In general, the DBD reactor performance depends on reactor configuration, flow velocity, power, ambient gas, and catalyst/packing material [19,20]. Hueso et al. [21] conducted a DBD technique to reform methane and directly decompose methanol under normal and low temperature conditions. Depending on the applied voltage, feed ratio, reactants' residence time, or reactor configuration, the conversion can reach 20–80% in the case of methane and 7–45% for $CO_2$. Under similar experimental conditions, methanol direct decomposition is up to 60–100%. Tu et al. [22] observed shifts in $CO_2$ conversion and process energy efficiency by changing DBD plasma processing parameters. Conclusion can be made that increasing the discharge power or lowering the gas flow velocity can improve $CO_2$ conversion with other parameters unchanged, but with lower energy efficiency. Snoeckx et al. [23] conducted extensive and in-depth studies on DBD plasma decomposition of $CO_2$. It can be found that the presence of $N_2$ in 50% of $N_2$ gas mixture has little effect on $CO_2$ conversion and energy efficiency. However, a higher $N_2$ fraction results in a decrease in $CO_2$ conversion and energy efficiency. Michielsen et al. [24] reported $CO_2$ decomposition in a packed-bed DBD reactor packed with glass wool, quartz wool, and $SiO_2$, $ZrO_2$, $Al_2O_3$, and $BaTiO_3$ spherical beads of different sizes. Among the many filler materials studied, $BaTiO_3$ has a maximum conversion of up to 25% and an energy efficiency of 4.5%. Uytdenhouwen et al. [25] used a DBD microplasma reactor to investigate the effect of gap size reduction combined with packing material on the conversion and efficiency of $CO_2$ dissociation. The results were compared to a conventional size reactor as a reference. Even though the energy efficiency is low, decreasing the discharge gap can greatly enhance the $CO_2$ conversion rate.

At present, in the literatures we have seen, most of the catalysts used for $CO_2$ decomposition by plasma are metal oxides. However, a few literatures have reported researches on $CO_2$ conversion by using foam metals as catalysts. In this work, we studied DBD plasma junctions and different packed foam metal materials (Fe, Al, and Ti) for $CO_2$ conversion. Compared to metal oxide catalysts, the foam metals can not only serve as a carrier for energy transformation, but the foam metals distributed in the discharge gap also consume a part of $O_2$ and O radicals, so that the reaction proceeds in the positive direction, promoting the decomposition of $CO_2$. The effects of power, flow velocity, and other discharge characteristics are studied to better understand the influence of the filling catalysts on $CO_2$

splitting. Furthermore, we use density functional theory (DFT) to verify the experimental results. The results show that $CO_2$ adsorption has a lower activation energy barrier on the foam Ti surface.

## 2. Experimental

### 2.1. Dielectric Materials

In this paper, foam metals (Fe, Al, and Ti) are used as filler catalysts along the discharge region. All of these catalysts are commercially available and untreated.

### 2.2. Experimental Setup

The schematic diagram of an experimental device for $CO_2$ splitting by DBD plasma is shown in Figure 1. The experimental device includes the following components: plasma generator, DBD reactor (Beijing Synthware glass), and a gas chromatograph (GC) for detecting the product. The coaxial cylindrical reactor has two electrodes, namely a high-voltage (HV) electrode and a low-voltage (LV) electrode, and they produce a discharge between the two electrodes. The HV electrode is a stainless steel rod, 2 mm in diameter, which is connected to an alternating current (AC) power source and fixed on the central shaft of the reaction tube. Condensed water at 20 °C is added to the reactor housing and the LV electrode is grounded with a wire. The discharge parameters were measured using a voltage probe, a current monitor, and an oscilloscope. The composition of the off-gas from the plasma reactor was determined by GC. The calculation of the discharge power is based on the area of the Lissajous figure displayed on the oscilloscope.

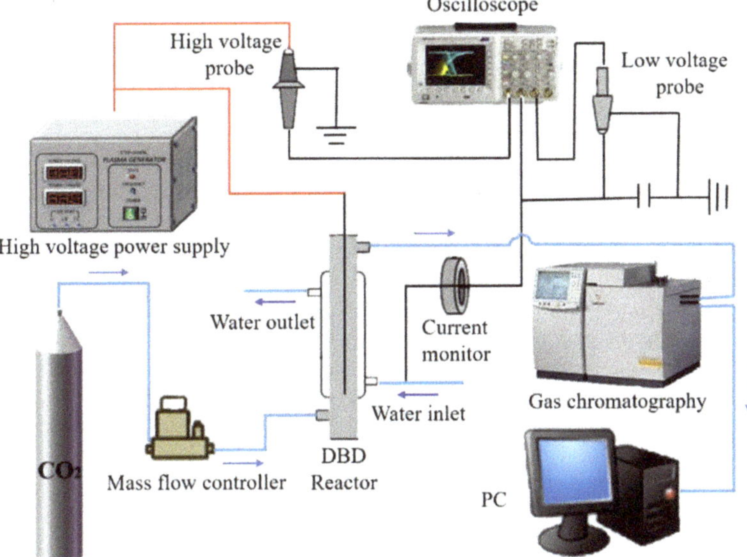


**Figure 1.** Schematic diagram of the experimental device for $CO_2$ splitting.

### 2.3. Gas Analysis and Parameter Definition

Pure $CO_2$ (99.995%) is used as the reaction gas and the inlet feed gas flow velocity was varied from 20 to 100 mL/min through the mass flow controller. After the GC had stabilized, the composition of the gas products was recorded by GC every 15 min. Helium (He) was used as its carrier gas.

CO$_2$ conversion, CO yield, and selectivity are important parameters for evaluating the plasma process performance, as defined below:

$$CO_2 \text{ conversion } (\%) = \frac{CO_2 \text{ converted}}{CO_2 \text{ input}} \times 100,$$ (1)

$$CO \text{ yield } (\%) = \frac{CO \text{ formed}}{CO_2 \text{ input}} \times 100,$$ (2)

$$CO \text{ selectivity } (\%) = \frac{CO \text{ formed}}{CO_2 \text{ converted}} \times 100.$$ (3)

Furthermore, specific input energy (SIE) is another important evaluation of system performance, defined as follows:

$$SIE = \frac{\text{discharge power}}{CO_2 \text{ flow rate}} \times 60.$$ (4)

The energy efficiency (EE) of the plasma splitting CO$_2$ process is defined as the molar number of CO$_2$ decomposed per unit plasma power and is formulated by the following equation:

$$EE \ (\%) = \frac{\Delta H_{298K} \times CO_2 \text{ converted}}{SIE \times 24.5} \times 100,$$ (5)

where $\Delta H$ represents the reaction enthalpy of one mole CO$_2$ splitting.

## 3. Results and Discussion

### 3.1. Effect of Packing Materials on Discharge Characteristics

Figure 2 presents the variation of the Lissajous figure with discharge voltage. From the picture, we can observe that the area of the Lissajous figure increases with increasing voltage. The impact of different filled foam metals on the discharge characteristics at the same input power of 70 W is shown in Figure 3. Compared with non-packing materials, there are much less filamentary discharges in the reactor packed with foam Fe, Al, and Ti. It is apparent that the filling of foam metals in the reactor causes changes in discharge characteristics and discharge patterns, from the original filamentary discharge to both the current filamentary discharge and surface discharge. The results demonstrate that the introduction of the packed foam metals into the DBD reactor can promote the discharge characteristics. We have discussed similar phenomena in our previous studies. For example, when ZrO$_2$ or foam Ni catalyst is introduced into the reactor, the discharge type changes and the filamentation discharge decreases dramatically [26–28]. Moreover, at the same input power, the Lissajous pattern of the electrode made of foam Ti has a larger area than the other two, which means that the foam Ti can further enhance the discharge power. Furthermore, from the slopes of the graph, it can be concluded that the foam Ti electrode transfers more charges than the other two foam metals. Filling the materials into the tube can generate a strong electric field. Near the particle contact points, the field intensity and the number of high-energy electrons increase.

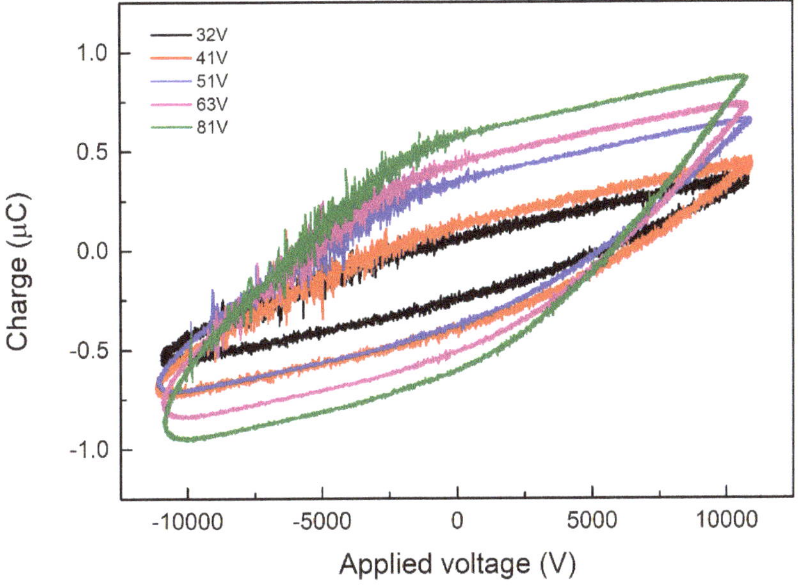

**Figure 2.** Lissajous diagram of a dielectric barrier discharge (DBD) reactor at different voltages.

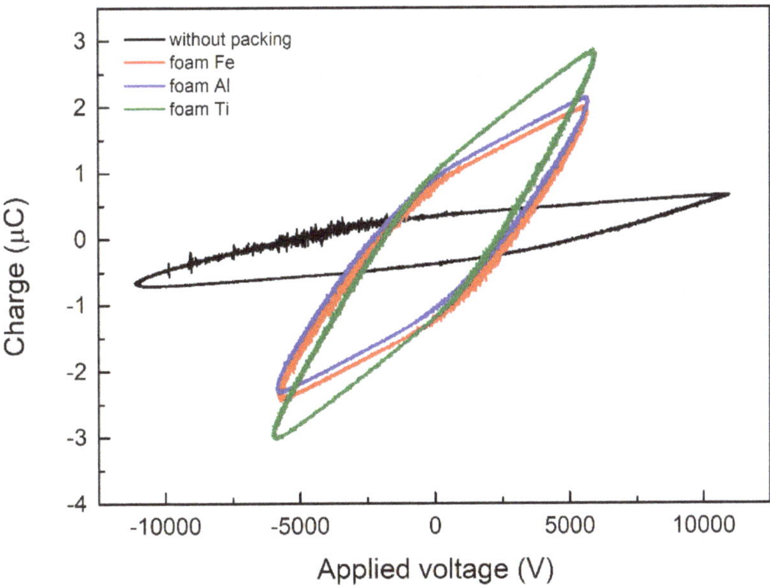

**Figure 3.** Lissajous figure of $CO_2$ decomposition in the DBD reactor with or without foam metals.

From the voltage and current waveforms of the $CO_2$ discharge, changes of the discharge characteristics in the reactor with and without the packing materials can be seen. A typical filamentous discharge can be seen in the empty tube, which can be verified by a plurality of burr peaks in the current signal of Figure 4a. On the contrary, as is exhibited in Figure 4b–d, filling the foam metals into the discharge region produces a typical packed-bed effect and results in the discharge characteristics to change from a filamentary discharge to a combination of surface discharge and filamentary discharge, because the spike discharge signal is reduced as demonstrated. In the packed bed DBD reactor, the filamentary discharge can only be emerged in a small gap between the particle-particle and the particle-tube walls, while the surface discharge can be generated on the particle surfaces near the particles.

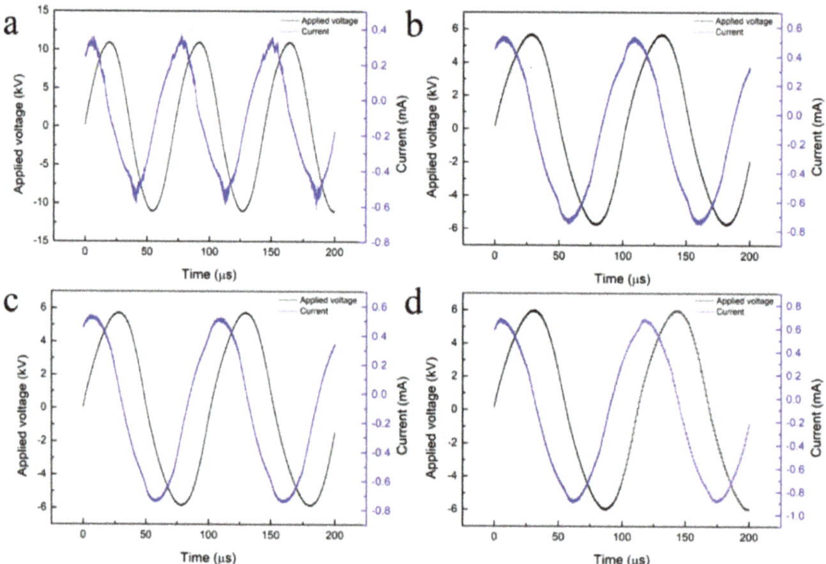

**Figure 4.** Discharge characteristics of $CO_2$ in the DBD reactor without or with filling foam metals: (**a**) Without packing; (**b**) Foam Fe; (**c**) Foam Al; (**d**) Foam Ti.

### 3.2. Effect of Packing Materials on $CO_2$ Conversion and CO Yield

The $CO_2$ conversions and CO yield when different filling foam metals were placed in the discharge gap are displayed in Figure 5. Clearly, packed bed DBD that was filled with three foam metals shows higher $CO_2$ conversion and CO yield than DBD without packing. Notably, $CO_2$ conversion and CO yield are greatly increased in the foam Ti-packed reactor, which are approximately three times more than that of the $CO_2$ conversion and CO yield in the non-packed tube. In the foam Fe-packed tube, $CO_2$ conversion and CO yield are lower than the foam Ti-filled tube, but higher than the foam Al-packed tube.

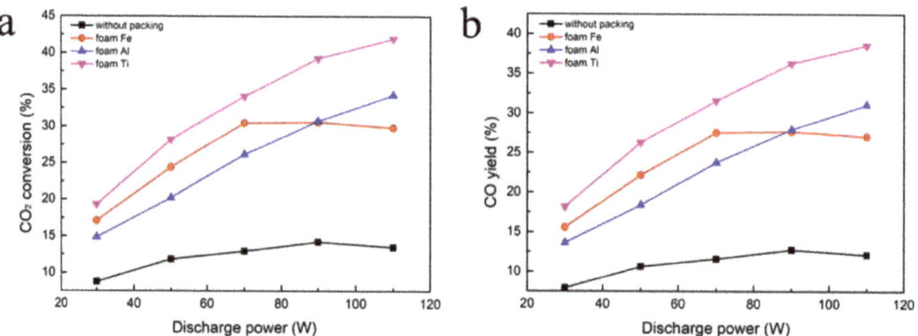

**Figure 5.** Effect of packing materials on: (**a**) $CO_2$ conversion; (**b**) CO yield.

### 3.3. Effect of Discharge Power and Gas Flow Rate on $CO_2$ Conversion

The influence of discharge power and gas flow velocity on $CO_2$ decomposition is depicted in Figure 6. One of the critical factors affecting the $CO_2$ conversion by DBD plasma is the discharge power. The discharge can change the electric field of the plasma reactor, thereby changing the amount of active electrons and active radicals [29]. For an empty tube, it can be noted that increasing the discharge power from 30 W to 90 W results in an improvement in $CO_2$ conversion. However, it drops as the discharge power is further increased to 110 W. When the flow rate is fixed at 20 mL/min, the empty tube reaches a maximum conversion of 21.15% at a power of 90 W. The $CO_2$ conversion tends to be saturated and slightly decreased if the discharge power is continuously increased over 90 W, and 90 W may be the optimum power under this condition, so from the viewpoint of energy-saving, an appropriate discharge power range is required. Similarly, for foam Fe, when the flow rate is 20 mL/min, the maximum conversion rate of 44.84% is achieved at a power of about 70 W. However, further increases in discharge power have no obvious impact on $CO_2$ conversion, because the discharge filaments cover the electrode surface and the quantity of microdischarges is not obviously increased [30].

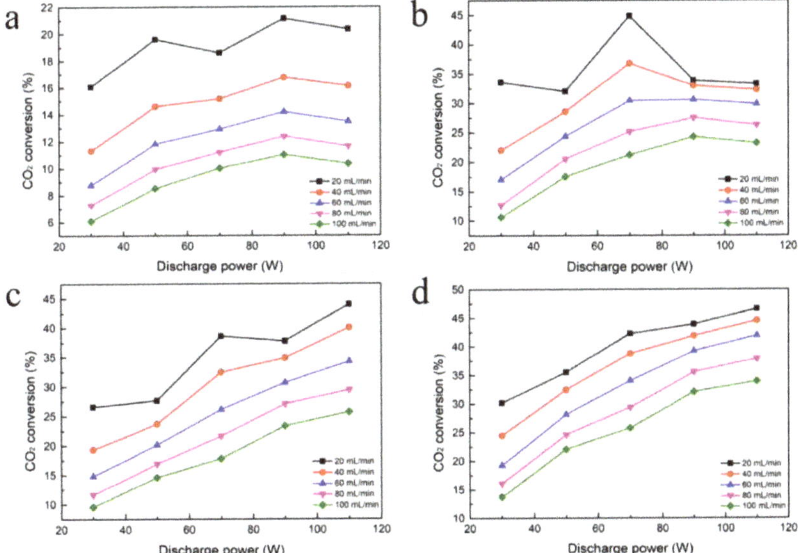

**Figure 6.** Effect of discharge power and gas flow velocity on $CO_2$ conversion: (**a**) Without packing; (**b**) Foam Fe; (**c**) Foam Al; (**d**) Foam Ti.

However, for foam Al and Ti, increasing the discharge power increases $CO_2$ conversion. As the plasma power changes from 30 to 110 W, $CO_2$ conversion increases from 26.58% and 30.16% to 44.02% and 46.61%, respectively. Increasing the plasma discharge power will increase the amplitude and number of current pulses, as depicted in Figure 7. A higher input power means more energy is put into the reaction system, which leads to a higher current density, producing more cation $CO^{2+}$ and anion $O^{2-}$ in DBD plasma [31–33]. This phenomenon indicates that, as the discharge power increases, the quantity of microdischarges are effectively increased, which suggests that more chemical reaction channels and chemically active species are generated for $CO_2$ processing [22,34]. Sufficient energy will activate the electrons and reactant molecules to increase the chances of mutual collision frequency between the active materials, so more chemical bonds will be destroyed and more active species will be produced, resulting in a higher $CO_2$ conversion [32,35].

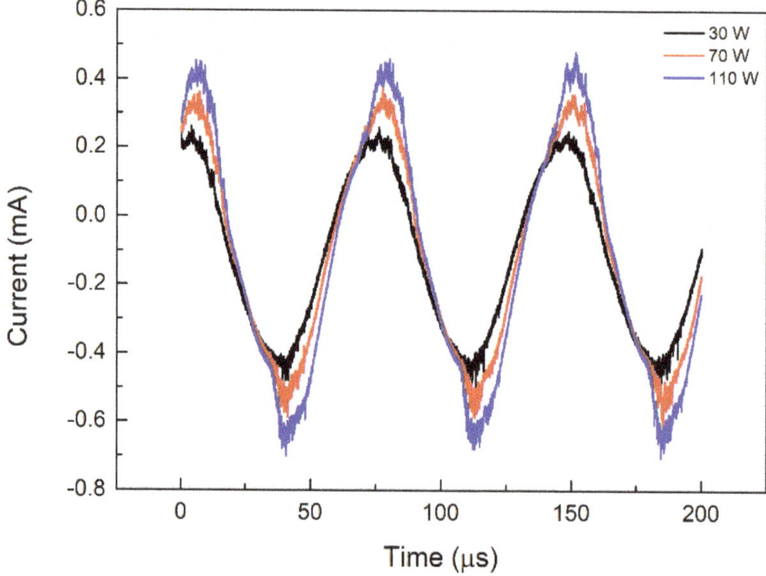

**Figure 7.** Effect of discharge power on the current signals for empty tube.

It has been proved that the feed flow velocity is also one of the important factors affecting the $CO_2$ conversion. It can be noticed from Figure 7 that the $CO_2$ conversion rate is the largest at the minimum $CO_2$ flow rate of 20 mL/min, whether it is an empty tube or a filling foam metal tube. Similar phenomena have been reported in a non-packed DBD reactor [22]. When the other parameters are not changed, enhancing the $CO_2$ flow rate decreases the residence time in the discharge area, which reduces the possibility of $CO_2$ conversion caused by collision with high-energy electrons and chemically reactive species, and thus reduces the $CO_2$ conversion rate. For example, when the $CO_2$ flow velocity varies within a range of 20 to 100 mL/min, $CO_2$ conversion is reduced from 46.61% to 33.98% at 110 W for foam Ti.

### 3.4. Effect of Discharge Power and Gas Flow Rate on Energy Efficiency

Energy efficiency is a big concern in plasma-assisted $CO_2$ decomposition processes. Among various plasmas, microwave plasma decomposition can achieve higher energy efficiency, owing to the selective excitation of the vibration level of $CO_2$ [36]. However, the energy efficiency of $CO_2$ splitting is generally around 10% in DBD plasma, because the active particle impact of $CO_2$ is a vital procedure in the plasma [12,37].

Figure 8 denotes the influencing factors of energy efficiency, namely, discharge power and gas flow velocity. The energy efficiency in filled foam metal tubes is higher than that of empty tubes over the input power range of 30 to 110 W that was studied. From the figure, we can conclude that energy efficiency drops with increasing discharge power. For foam Ti, the higher the power, the higher the $CO_2$ conversion; the energy efficiency decreased from 8.85% to 5.96%, while the plasma discharge power is varied from 30 to 110 W.

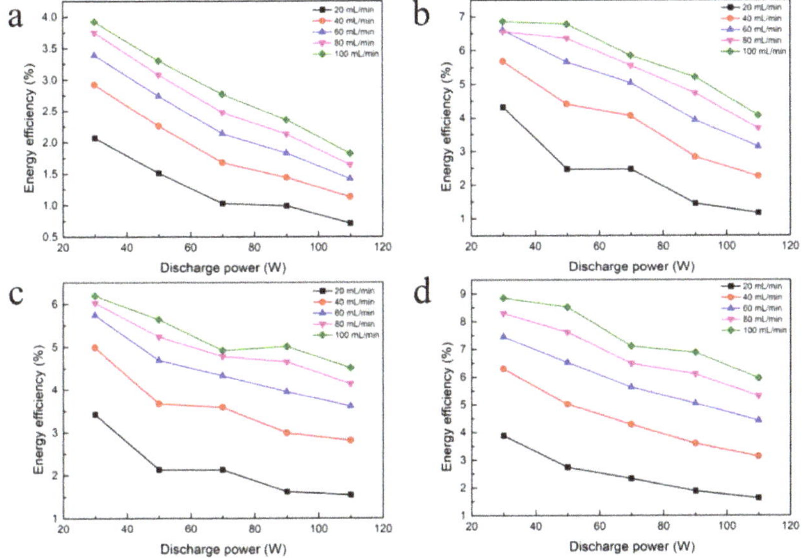

**Figure 8.** Effect of discharge power and gas flow velocity on energy efficiency: (**a**) Without packing; (**b**) Foam Fe; (**c**) Foam Al; (**d**) Foam Ti.

The influence of feed flow velocity and discharge power on the energy efficiency of $CO_2$ splitting by plasma exhibits the opposite behavior. A higher feed flow rate results in lower $CO_2$ conversion, but more energy efficient. For example, by increasing the velocity of feed flow from 20 to 100 mL/min, the energy efficiency increases from 3.88% to 8.85%. Moreover, the reason why energy efficiency decreases with increasing discharge power may be attributed to energy loss, which is released in the form of heat, and it can be verified from an increase in condensate temperature. Under the same plasma operating conditions, the $CO_2$ conversion and energy efficiency cannot simultaneously be taken into account during discharge process. Thus, there is a need to balance the relationship between $CO_2$ decomposition and energy efficiency during the experiment.

### 3.5. Effect of SIE on $CO_2$ Conversion and Energy Efficiency

According to the definition of SIE (Equation (4)), the variation of SIE is determined by the feed flow velocity and discharge power. Generally speaking, SIE is regarded as one of the important individual factors for $CO_2$ transformation process [6,38,39]. Obviously, $CO_2$ conversion and energy efficiency have opposite trends with changes in SIE. As a consequence, there is a demand to investigate the impact of the optimal SIE calculated by different combinations on the $CO_2$ transformation process [40].

Figure 9a shows the impact of SIE on $CO_2$ conversion. The introduction of foam metal catalysts into the DBD reactor has a significant effect on the $CO_2$ decomposition rate, which increases with increasing SIE. Compared with the results of without packing, the foam Fe, Al, and Ti prominently raise $CO_2$ conversion by 2.14, 2.42, and 2.63 times, respectively. Among the three catalysts, foam Ti catalyst has the best catalytic effect.

The effect of SIE on energy efficiency is shown in Figure 9b. The energy efficiency of all the reactors decreases with increasing SIE [41]. For instance, for foam Ti, when the SIE is increased from 42 to 210 kJ/L, the energy efficiency is decreased from 7.11% to 2.33%. Therefore, the interaction effects of power and $CO_2$ flow velocity should be taken into account while using a suitable SIE to achieve high $CO_2$ conversion and energy efficiency simultaneously.

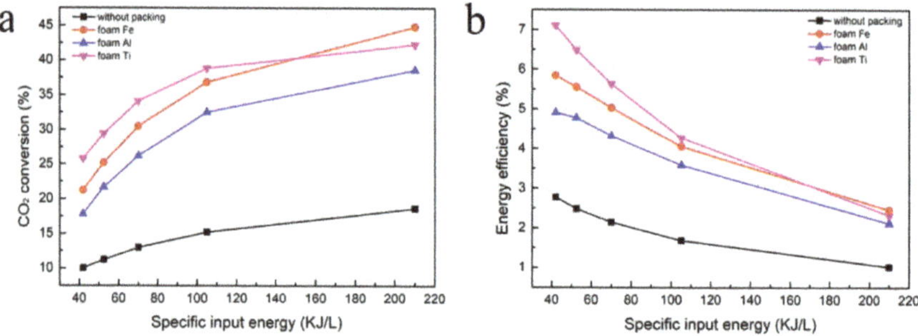

**Figure 9.** Effect of specific input energy of different foam metals on: (**a**) $CO_2$ conversion; (**b**) energy efficiency.

### 3.6. Effect of Different Packing Materials on CO Selectivity

The influence of different packed foam metals on CO selectivity at different discharge powers is exhibited in Figure 10. For foam Fe and Al, the CO selectivity during the whole reaction process is maintained at about 88–96%. Compared with the empty tube discharge, the difference in CO selectivity between the two at different discharge powers is not significant under our experimental conditions. It can be inferred that, for foam Fe and Al, discharge power plays a secondary role in CO selectivity. However, for foam Ti, the CO selectivity varies between 91% and 97%, which is much higher than the other two foam metals, indicating that the foam Ti has a good catalytic effect. The CO selectivity based on carbon atoms (Equation (3)) is close to 100%, which means that the stoichiometric conversion of $CO_2$ to CO has been achieved in this study, and the generation of CO is mainly due to $CO_2$ dissociation. The electron collision dissociation of $CO_2$ is likely to lead to ground state CO and both the ground state and metastable O atoms, whereas studies have revealed that CO bands are observed in experiments, indicating that CO can also be formed in the excited state [42].

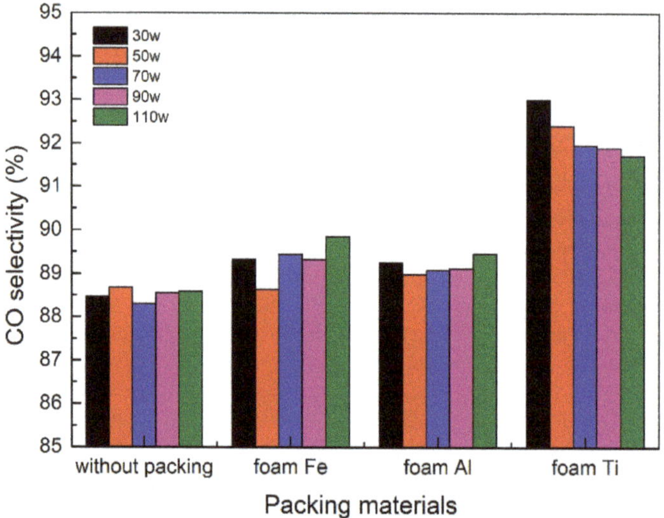

**Figure 10.** CO selectivity of different foam metals at different discharge powers.

*3.7. DFT Validations*

A prerequisite for highly efficient catalysts is the adsorption of $CO_2$ on the catalysts' surface. To gain a clear insight into the adsorption performance of catalysts, an in-depth investigation of the $CO_2$ adsorption was carried out. According to X-ray diffraction (XRD) analysis, the (101) plane of Ti, (111) plane of Al, and (110) plane of Fe are investigated carefully. Based on the above results, we performed DFT calculations to investigate the adsorption of $CO_2$ on catalysts' surface as shown in Figure 11. The results indicate that the adsorption energy of $CO_2$ on the (101) plane of Ti is −2.49 eV (Figure 11a), which is lower than that observed for the (111) plane of Al (Figure 11b) and (110) plane of Fe (Figure 11c). This indicates that $CO_2$ would have more favorable adsorption on the (101) plane of Ti rather than the (111) plane of Al and (110) plane of Fe. These findings explain the origin of the high catalytic activity of the (101) plane of Ti. Accordingly, the (101) plane of Ti is superior to the (111) plane of Al and (110) plane of Fe for the $CO_2$ conversion process.

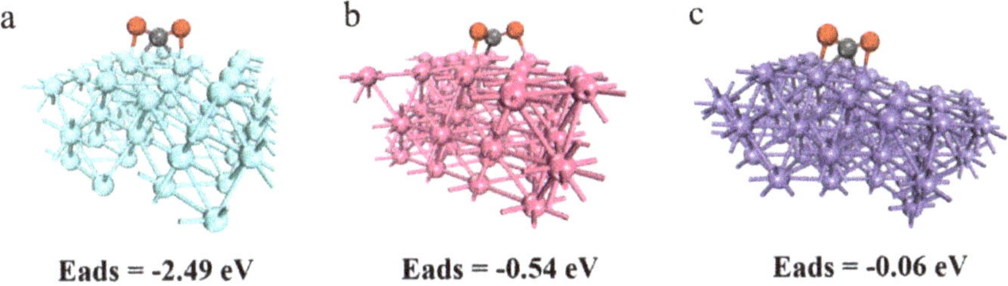


**Figure 11.** Side views of the optimized geometric structures of $CO_2$ adsorption on: (**a**) (101) plane of Ti; (**b**) (111) plane of Al; (**c**) (110) plane of Fe. Gray and red balls stand for the C and O atoms, respectively.

## 4. Conclusions

In this study, we investigated a coaxial DBD plasma junctions and different packed foam metal materials (foam Fe, Al, and Ti) placed into the discharge gap for $CO_2$ conversion. The effects of power, flow velocity, and other discharge characteristics were studied to better understand the influence of the filling catalysts on $CO_2$ decomposition. The filling of foam metals in the reactor caused changes in discharge characteristics and discharge patterns, from the original filamentary discharge to both the current filamentary discharge and surface discharge. Compared with the maximum $CO_2$ conversion of 21.15% and energy efficiency of 3.92% in the reaction tube without the foam metal electrode, a maximum $CO_2$ decomposition rate of 44.84%, 44.02%, and 46.61% and energy efficiency of 6.86%, 6.19%, and 8.85% were obtained in the reaction tubes packed with foam Fe, Al, and Ti, respectively. The results showed that the $CO_2$ conversion rate for the filled foam metal electrode was obviously enhanced when compared with the non-packed tube. It was seen that the foam Ti had the best $CO_2$ decomposition rate among the three foam metals.

Furthermore, we used density functional theory (DFT) to further verify the experimental results. According to X-ray diffraction (XRD) analysis, the (101) plane of Ti, (111) plane of Al, and (110) plane of Fe were carefully investigated. Based on the above results, we performed DFT calculations to investigate the adsorption of $CO_2$ on catalysts surface. The results indicated that the adsorption energy of $CO_2$ on the (101) plane of Ti was −2.49 eV, which was lower than the (111) plane of Al and (110) plane of Fe. The results showed that $CO_2$ adsorption had a lower activation energy barrier on the foam Ti surface. The theoretical calculation is consistent with the experimental results, which better explains the mechanism of $CO_2$ decomposition.

**Author Contributions:** Conceptualization, C.M., G.G., D.Y., B.D. and F.Y.; methodology, C.M., G.G., D.Y., and F.Y.; formal analysis, C.M., G.G., D.Y., F.Y., J.L., X.Z. and S.Z.; investigation, C.M., G.G., D.Y., F.Y., J.L., and X.Z.; data curation, C.M., G.G., D.Y., and F.Y.; writing—original draft preparation, C.M., G.G., D.Y., and F.Y.; writing—review and editing, J.L., X.Z., C.M., G.G., D.Y., and F.Y.; supervision, C.M., G.G., D.Y., and F.Y.

**Funding:** This research was funded by the National Natural Science Foundation of China (No.21663022) and Science and Technology Innovation Talents Program of Bingtuan (No.2019CB025).

**Conflicts of Interest:** The authors declare no conflict of interest.

## References


1. Havran, V.; Dudukovic, M.P.; Lo, C.S. Conversion of methane and carbon dioxide to higher value products. *Ind. Eng. Chem. Res.* **2011**, *50*, 7089–7100. [CrossRef]

2. Aresta, M.; Dibenedetto, A.; Angelini, A. The changing paradigm in $CO_2$ utilization. *J. CO2 Util.* **2013**, *3*, 65–73. [CrossRef]

3. Markewitz, P.; Kuckshinrichs, W.; Leitner, W.; Linssen, J.; Zapp, P.; Bongartz, R.; Schreiber, A.; Müller, T.E. Worldwide innovations in the development of carbon capture technologies and the utilization of $CO_2$. *Energy Environ. Sci.* **2012**, *5*, 7281–7305. [CrossRef]

4. Saikia, P.; Saikia, J.; Sarmah, S.; Goswamee, R. Mesoporous oxidic holey nanosheets from Zn-Cr LDH synthesized by soft chemical etching of $Cr^{3+}$ and its application as $CO_2$ hydrogenation catalyst. *J. CO2 Util.* **2017**, *21*, 40–51. [CrossRef]

5. Abdelaziz, O.Y.; Hosny, W.M.; Gadalla, M.A.; Ashour, F.H.; Ashour, I.A.; Hulteberg, C.P. Novel process technologies for conversion of carbon dioxide from industrial flue gas streams into methanol. *J. CO2 Util.* **2017**, *21*, 52–63. [CrossRef]

6. Nunnally, T.; Gutsol, K.; Rabinovich, A.; Fridman, A.; Gutsol, A.; Kemoun, A. Dissociation of $CO_2$ in a low current gliding arc plasmatron. *J. Phys. D Appl. Phys.* **2011**, *44*, 274009. [CrossRef]

7. Mei, D.; Zhu, X.; He, Y.L.; Yan, J.D.; Tu, X. Plasma-assisted conversion of $CO_2$ in a dielectric barrier discharge reactor: understanding the effect of packing materials. *Plasma Sources Sci. Technol.* **2014**, *24*, 015011. [CrossRef]

8. Zhang, H.; Li, L.; Li, X.; Wang, W.; Yan, J.; Tu, X. Warm plasma activation of $CO_2$ in a rotating gliding arc discharge reactor. *J. CO2 Util.* **2018**, *27*, 472–479. [CrossRef]

9. Horvath, G.; Skalný, J.; Mason, N. FTIR study of decomposition of carbon dioxide in dc corona discharges. *J. Phys. D Appl. Phys.* **2008**, *41*, 225207. [CrossRef]

10. Wang, J.Y.; Xia, G.G.; Huang, A.; Suib, S.L.; Hayashi, Y.; Matsumoto, H. $CO_2$ decomposition using glow discharge plasmas. *J. Catal.* **1999**, *185*, 152–159. [CrossRef]

11. Indarto, A.; Yang, D.R.; Choi, J.W.; Lee, H.; Song, H.K. Gliding arc plasma processing of $CO_2$ conversion. *J. Hazard. Mater.* **2007**, *146*, 309–315. [CrossRef] [PubMed]

12. Kozák, T.; Bogaerts, A. Splitting of $CO_2$ by vibrational excitation in non-equilibrium plasmas: a reaction kinetics model. *Plasma Sources Sci. Technol.* **2014**, *23*, 045004. [CrossRef]

13. Savinov, S.Y.; Lee, H.; Song, H.K.; Na, B.K. Decomposition of methane and carbon dioxide in a radio-frequency discharge. *Ind. Eng. Chem. Res.* **1999**, *38*, 2540–2547. [CrossRef]

14. Lu, N.; Zhang, C.; Shang, K.; Jiang, N.; Li, J.; Wu, Y. Dielectric barrier discharge plasma assisted $CO_2$ conversion: understanding the effects of reactor design and operating parameters. *J. Phys. D Appl. Phys.* **2019**, *52*, 224003. [CrossRef]

15. Li, R.; Tang, Q.; Yin, S.; Sato, T. Plasma catalysis for $CO_2$ decomposition by using different dielectric materials. *Fuel Process. Technol.* **2006**, *87*, 617–622. [CrossRef]

16. Zhang, K.; Zhang, G.; Liu, X.; Phan, A.N.; Luo, K. A study on $CO_2$ decomposition to CO and $O_2$ by the combination of catalysis and dielectric-barrier discharges at low temperatures and ambient pressure. *Ind. Eng. Chem. Res.* **2017**, *56*, 3204–3216. [CrossRef]

17. Zhang, H.; Zhu, F.; Li, X.; Cen, K.; Du, C.; Tu, X. Enhanced hydrogen production by methanol decomposition using a novel rotating gliding arc discharge plasma. *RSC Adv.* **2016**, *6*, 12770–12781. [CrossRef]

18. Zhang, H.; Wang, W.; Li, X.; Han, L.; Yan, M.; Zhong, Y.; Tu, X. Plasma activation of methane for hydrogen production in a $N_2$ rotating gliding arc warm plasma: a chemical kinetics study. *Chem. Eng. J.* **2018**, *345*, 67–78. [CrossRef]

19. Tu, X.; Gallon, H.J.; Twigg, M.V.; Gorry, P.A.; Whitehead, J.C. Dry reforming of methane over a $Ni/Al_2O_3$ catalyst in a coaxial dielectric barrier discharge reactor. *J. Phys. D Appl. Phys.* **2011**, *44*, 274007. [CrossRef]
20. Paulussen, S.; Verheyde, B.; Tu, X.; Bie, C.D.; Martens, T.; Petrovic, D.; Bogaerts, A.; Sels, B. Conversion of carbon dioxide to value-added chemicals in atmospheric pressure dielectric barrier discharges. *Plasma Sources Sci. Technol.* **2010**, *19*, 034015. [CrossRef]
21. Rico, V.J.; Hueso, J.L.; Cotrino, J.; González-Elipe, A.R. Evaluation of different dielectric barrier discharge plasma configurations as an alternative technology for green C1 chemistry in the carbon dioxide reforming of methane and the direct decomposition of methanol. *J. Phys. Chem. A* **2010**, *114*, 4009–4016. [CrossRef] [PubMed]
22. Mei, D.; Tu, X. Conversion of $CO_2$ in a cylindrical dielectric barrier discharge reactor: effects of plasma processing parameters and reactor design. *J. CO2 Util.* **2017**, *19*, 68–78. [CrossRef]
23. Snoeckx, R.; Heijkers, S.; Wesenbeeck, K.V.; Lenaerts, S.; Bogaerts, A. $CO_2$ conversion in a dielectric barrier discharge plasma: $N_2$ in the mix as a helping hand or problematic impurity? *Energy Environ. Sci.* **2016**, *9*, 999–1011. [CrossRef]
24. Michielsen, I.; Uytdenhouwen, Y.; Pype, J.; Michielsen, B.; Mertens, J.; Reniers, F.; Meynen, V.; Bogaerts, A. $CO_2$ dissociation in a packed bed DBD reactor: first steps towards a better understanding of plasma catalysis. *Chem. Eng. J.* **2017**, *326*, 477–488. [CrossRef]
25. Uytdenhouwen, Y.; Alphen, S.V.; Michielsen, I.; Meynen, V.; Cool, P.; Bogaerts, A. A packed-bed DBD micro plasma reactor for $CO_2$ dissociation: does size matter? *Chem. Eng. J.* **2018**, *348*, 557–568. [CrossRef]
26. Zhou, A.M.; Chen, D.; Ma, C.H.; Yu, F.; Dai, B. DBD plasma-$ZrO_2$ catalytic decomposition of $CO_2$ at low temperatures. *Catalysts* **2018**, *8*, 256. [CrossRef]
27. Zhu, S.J.; Zhou, A.M.; Yu, F.; Dai, B.; Ma, C.H. Enhanced $CO_2$ decomposition via metallic foamed electrode packed in self-cooling DBD plasma device. *Plasma Sci. Technol.* **2019**, *21*, 085504. [CrossRef]
28. Zhou, A.M.; Chen, D.; Dai, B.; Ma, C.H.; Li, P.P.; Yu, F. Direct decomposition of $CO_2$ using self-cooling dielectric barrier discharge plasma. *Greenh. Gases Sci. Technol.* **2017**, *7*, 721–730. [CrossRef]
29. Uhm, H.S.; Jang, D.G.; Kim, M.S.; Suk, H. Density dependence of capillary plasma on the pressure and applied voltage. *Phys. Plasmas* **2012**, *19*, 024501. [CrossRef]
30. Dong, L.F.; Li, X.C.; Yin, Z.Q.; Qian, S.F.; Ouyang, J.T.; Wang, L. Self-organized filaments in dielectric barrier discharge in air at atmospheric pressure. *Chin. Phys. Lett.* **2001**, *18*, 1380.
31. Lu, N.; Bao, X.; Jiang, N.; Shang, K.; Li, J.; Wu, Y. Non-thermal plasma-assisted catalytic dry reforming of methane and carbon dioxide over $G$-$C_3N_4$-based catalyst. *Top. Catal.* **2017**, *60*, 855–868. [CrossRef]
32. Shrestha, R.; Tyata, R.; Subedi, D. Effect of applied voltage in electron density of homogeneous dielectric barrier discharge at atmospherice pressure. *Himal. Phys.* **2013**, *4*, 10–13. [CrossRef]
33. Duan, X.; Hu, Z.; Li, Y.; Wang, B. Effect of dielectric packing materials on the decomposition of carbon dioxide using DBD microplasma reactor. *AIChE J.* **2015**, *61*, 898–903. [CrossRef]
34. Mei, D.; He, Y.L.; Liu, S.; Yan, J.; Tu, X. Optimization of $CO_2$ conversion in a cylindrical dielectric barrier discharge reactor using design of experiments. *Plasma Process. Polym.* **2016**, *13*, 544–556. [CrossRef]
35. Wu, A.J.; Zhang, H.; Li, X.D.; Lu, S.Y.; Du, C.M.; Yan, J.H. Determination of spectroscopic temperatures and electron density in rotating gliding arc discharge. *IEEE Trans. Plasma Sci.* **2015**, *43*, 836–845.
36. Spencer, L.; Gallimore, A. $CO_2$ dissociation in an atmospheric pressure plasma/catalyst system: a study of efficiency. *Plasma Sources Sci. Technol.* **2012**, *22*, 015019. [CrossRef]
37. Aerts, R.; Martens, T.; Bogaerts, A. Influence of vibrational states on $CO_2$ splitting by dielectric barrier discharges. *J. Phys. Chem. C* **2012**, *116*, 23257–23273. [CrossRef]
38. Wang, Q.; Yan, B.H.; Jin, Y.; Cheng, Y. Investigation of dry reforming of methane in a dielectric barrier discharge reactor. *Plasma Chem. Plasma Process.* **2009**, *29*, 217–228. [CrossRef]
39. Khassin, A.A.; Pietruszka, B.L.; Heintze, M.; Parmon, V.N. The impact of a dielectric barrier discharge on the catalytic oxidation of methane over Ni-containing catalyst. *React. Kinet. Catal. Lett.* **2004**, *82*, 131–137. [CrossRef]
40. Niu, G.; Qin, Y.; Li, W.; Duan, Y. Investigation of $CO_2$ splitting process under atmospheric pressure using multi-electrode cylindrical DBD plasma reactor. *Plasma Chem. Plasma Process.* **2019**, *39*, 1–16. [CrossRef]

41. Snoeckx, R.; Zeng, Y.; Tu, X.; Bogaerts, A. Plasma-based dry reforming: Improving the conversion and energy efficiency in a dielectric barrier discharge. *RSC Adv.* **2015**, *5*, 29799–29808. [CrossRef]
42. Tu, X.; Whitehead, J. Plasma-catalytic dry reforming of methane in an atmospheric dielectric barrier discharge: understanding the synergistic effect at low temperature. *Appl. Catal. B Environ.* **2012**, *125*, 439–448. [CrossRef]

*Article*

# High-Resolution SEM and EDX Characterization of Deposits Formed by CH₄+Ar DBD Plasma Processing in a Packed Bed Reactor

**Mohammadreza Taheraslani [1,2,*] and Han Gardeniers [1]**

[1]  Mesoscale Chemical Systems, MESA+ Institute for Nanotechnology, University of Twente, P.O. Box 217, 7500 AE Enschede, The Netherlands; j.g.e.gardeniers@utwente.nl
[2]  Faculty of Science and Technology, University of Twente, P.O. Box 217, 7500 AE Enschede, The Netherlands
*  Correspondence: m.taheraslani@utwente.nl; Tel.: +31-53-489-3412

Received: 29 March 2019; Accepted: 8 April 2019; Published: 10 April 2019

**Abstract:** The deposits formed during the DBD plasma conversion of $CH_4$ were characterized by high-resolution scanning electron microscopy (HRSEM) and energy dispersive X-ray elemental analysis (EDX) for both cases of a non-packed reactor and a packed reactor. For the non-packed plasma reactor, a layer of deposits was formed on the dielectric surface. HRSEM images in combination with EDX and CHN elemental analysis of this layer revealed that the deposits are made of a polymer-like layer with a high content of hydrogen (60 at%), possessing an amorphous structure. For the packed reactor, $\gamma$-alumina, Pd/$\gamma$-alumina, $BaTiO_3$, silica-SBA-15, MgO/$Al_2O_3$, and $\alpha$-alumina were used as the packing materials inside the DBD discharges. Carbon-rich agglomerates were formed on the $\gamma$-alumina after exposure to plasma. The EDX mapping furthermore indicated the carbon-rich areas in the structure. In contrast, the formation of agglomerates was not observed for Pd-loaded $\gamma$-alumina. This was ascribed to the presence of Pd, which enhances the hydrogenation of deposit precursors, and leads to a significantly lower amount of deposits. It was further found that the structure of all other plasma-processed materials, including MgO/$Al_2O_3$, silica-SBA-15, $BaTiO_3$, and $\alpha$-alumina, undergoes morphological changes. These alterations appeared in the forms of the generation of new pores (voids) in the structure, as well as the moderation of the surface roughness towards a smoother surface after the plasma treatment.

**Keywords:** plasma catalysis-methane conversion; dielectric barrier discharge; deposits; materials characterization

---

## 1. Introduction

SEM/EDX studies have been extensively utilized to analyse the surface morphology of catalyst samples. The surface morphology of a catalyst can play a key role in determining the activity, selectivity, and stability of the catalyst. In recent years, low-temperature non-thermal plasma systems have been exploited for the preparation and treatment (e.g., reduction) of packing materials and metal-supported catalysts, advantageously preferred to conventional thermal approaches, which demand a quite high operating temperature for the preparation and treatment (e.g., calcination, reduction) of catalysts [1–3].

Plasma treatment of the catalyst can result in effective changes in the morphology and catalytic properties of the surface, depending on the amount of applied power, the exposure time, and the type of gas, which are used to create plasma discharges. Tu et al. [4] reported the treatment of NiO/$Al_2O_3$ using $H_2$/Ar DBD plasma at atmospheric pressure and low temperature (<300 °C). It was found that plasma is capable of reducing NiO to Ni, increasing the surface conductivity, and modifying the discharge characteristics of the plasma. Liu et al. [5] studied the influence of non-thermal plasma on the catalytic properties of various metal-supported catalysts, such as Pt, Pd, and Ni on alumina and HZSM-5 as the supports. These authors reported that the plasma treatment could remarkably enhance the dispersion of the metal on the surface of the support, increasing the activity of the catalyst at low temperatures by generating and redistributing the acidic and basic sites, as well as improving the stability of the catalyst. It was further found that the plasma-treated catalysts show a higher resistance to carbon deposition compared to those prepared by conventional thermal synthesis methods. In addition to the modification of the surface acidity, this was attributed to an improved interaction between the metal and the support after being treated by plasma discharges, as well as to a higher dispersion of metal active sites. Karuppiah et al. [6] investigated the plasma reduction of $Ni/Al_2O_3$ and $CeO_2$-$Ni/Al_2O_3$. Their results indicated that a better dispersion for metal nanoparticles can be obtained due to the treatment of the surface with DBD plasma.

The integration of plasma and catalyst aims at improving the productivity of the process towards desired products, while reducing the generation of unwanted products [7–10]. Despite the synergy of the plasma and catalyst, the formation of undesired products can still take place, which can therefore influence the overall performance of a plasma-catalyst system. As a typical instance, the plasma-driven conversion of hydrocarbons (e.g., $CH_4$) usually produces carbon-containing deposits, which is considered a drawback in such processes, due to its influence on the stability of the plasma, as well as on the catalytic activity of the catalyst. Khoja et al. [11] studied dry reforming of methane in a DBD plasma reactor packed with γ-alumina. SEM images showed the formation of carbonaceous species on alumina. In addition, EDX elemental analysis detected the presence of carbon in the composition of alumina, comparably higher than the fresh γ-alumina powder. Chiremba et al. [12] performed SEM characterization of $BaTiO_3$ after 80 hr exposure to $CH_4$ plasma. $BaTiO_3$ spheres were used as packing for the conversion of methane in a packed-bed DBD reactor. The grain structure of $BaTiO_3$ was influenced after the exposure to DBD discharges, where the grains tend to agglomerate to form larger ones, resulting in the formation of a layered structure.

The conversion of $CH_4$ with DBD plasma reactors is accompanied by the formation of deposits, originating from the decomposition of $CH_4$ to $CH_x$ fragments [13,14]. These fragments and their interaction generate other potential precursors (e.g., $C_2H$, $C_2H_3$) for the formation of deposits. The integration of catalyst surfaces with the DBD plasma discharges is one of the approaches that can reduce the formation of deposits; however, some carbonaceous species are still formed on the catalyst surface. The presence of deposits can influence the stability of the operation for packed-bed DBD reactors. The characterization of the formed deposits can identify the nature of the deposits and their morphology, which gives more insight into the mechanism of the carbonaceous species formation, as well as their effect on the performance of the process.

Therefore, in this article, the deposits formed during the plasma conversion of $CH_4$ for both the non-packed reactor and the reactor packed with different packing materials are characterized using high-resolution scanning electron microscopy (HR-SEM) to study the deposits' layer, as well as the structural changes of the catalyst samples after being exposed to $CH_4$+Ar DBD plasma discharges. Energy dispersive X-ray analysis (EDX) is utilized to analyse the elemental composition of the deposits to supplement the results obtained by HR-SEM.

## 2. Materials and Methods

### 2.1. Packing Materials

The following materials were utilized as the packing: γ-alumina (Alfa Aesar, 234 $m^2/gr$), Pd (1 wt%)/γ-alumina (Alfa Aesar, surface area: 160 $m^2/gr$), α-alumina (Alfa Aesar, 0.82 $m^2/gr$), silica-SBA-15 (Sigma Aldrich, 673 $m^2/gr$), $MgO/Al_2O_3$ (Sasol Company, 440 $m^2/gr$), and $BaTiO_3$ (Alfa Aesar, 18 $m^2/gr$). The surface area of the samples, as given in the brackets above, was measured with nitrogen physisorption at 77 K with a Surface area and Porosity Analyzer (Micrometrics, Norcross, GA, USA), TriStar, Micrometrics.

*2.2. Plasma Processing*

The mixture of $CH_4$ and Ar was introduced to the inlet of the DBD reactor, which was made of a quartz tube, acting as the dielectric with an i.d. of 4 mm (inner diameter) and an o.d. of 6 mm (outer diameter). A stainless steel rod with a diameter of 1.6 mm, acting as the high voltage electrode, was fixed at the centre of the quartz tube. The quartz tube was covered with a rigid stainless steel tube with a length of 10 cm, which acted as the ground electrode. In the case of the packed reactor, the powder was sieved within the particle size range of 100–300 μm and packed inside the discharge gap of the DBD plasma reactor, covering the discharge gap (1.2 mm) between the high voltage electrode and the dielectric quartz tube for the 10 cm length of the plasma discharge zone. A total flow rate of 50 mL/min of the mixture of methane and argon was applied, containing 5 vol% of $CH_4$ as the reactant.

The reaction products were analysed with an online Varian 450 Gas Chromatograph (Varian Inc, Middelburg, The Netherlands) equipped with TCD and FID detectors (Varian Inc, Middelburg, The Netherlands). The products were separated by Hayesep T&Q, Molsieve 13x, and PoraBOND Q columns (Agilent, Santa Clara, CA, USA) to analyse gas-phase products. A high-voltage probe (TESTEC TT-HVP15 HF, Frankfurt, Germany), a probe for connecting the ground electrode (TESTEC TT-HV 250, Frankfurt, Germany), a 3.9 nF capacitor, and an oscilloscope (Pico Scope 2000 series, Pico Tech, Cambridgeshire, UK) were used to measure the discharge power. The discharge power was calculated from Q-V Lissajous figures. The DBD plasma was generated between the high voltage electrode and the ground electrode by applying a high voltage of 7–8 kV with a frequency of 23 kHz. The discharge power was maintained in the range of 7–8 Watt during the 2.5 h of the plasma processing. All experiments were performed at ambient conditions. The outlet temperature of the reactor was measured with a thermocouple, which was close to the ambient temperature, and therefore no heating effect was observed at the outlet of the reactor.

*2.3. Scanning Electron Microscopy (SEM)/Energy Dispersive X-ray Analysis (EDX) Characterization*

Scanning electron microscopy (SEM) with a high resolution was performed using a Zeiss MERLIN HR-SEM instrument. This instrument was also equipped with energy dispersive X-ray analysis from OXFORD Instruments for obtaining the elemental composition of the samples. The quartz tube covered with deposits on its inner surface was cut and mounted inside the chamber of the SEM instrument. For the catalyst samples, the typical SEM sample holder was used. A small amount of the catalyst powder was attached to the sample holder using carbon adhesive tape and the holders were then mounted inside the chamber. For SEM images, a low acceleration voltage of around 1 kV was applied, in order to avoid the charging effects that may occur due to the low conductivity of the deposits formed on the quartz tube. The surface charging became extreme when an acceleration voltage higher than 3 kV was applied, where it was not possible to see a clear image of the sample and, where a large white (bright) area on the image was observed instead. For energy dispersive X-rays (EDX) analysis, a higher voltage of 10 kV was applied in order to produce the emission of X-rays from the sample. It should be mentioned that CHN elemental analysis (Organic Elemental Analyzer, Flash 2000, Thermo Fisher Scientific Inc.) was performed to identify the hydrogen and carbon content of the formed deposits, constituted on the inner surface of the dielectric quartz tube.

## 3. Results and Discussion

*3.1. SEM/EDX Characterization of Deposits Formed on the Dielectric Quartz Tube*

A layer of yellowish deposits was formed during the plasma reaction for the non-packed reactor on the inner surface of the dielectric quartz tube, as depicted in Figure 1.

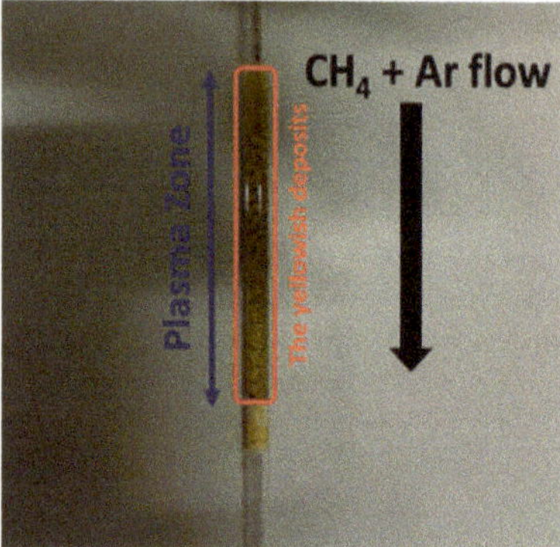



**Figure 1.** The yellowish deposits formed on the dielectric quartz tube.

The SEM image of the layer (Figure 2b) indicates that the deposits possess an amorphous structure with irregular pores and cracks on the surface, showing a combination of both smooth areas and rough areas. For comparison, the SEM image of the clean dielectric quartz tube, before the exposure to plasma, is depicted in Figure 2a, which shows a flat surface.

**Figure 2.** The SEM image of deposits formed on the inner surface of the dielectric quartz tube. (**a**) Before exposure to plasma (clean quartz dielectric tube) and (**b**) after exposure to plasma.

The composition of the deposits was evaluated by EDX analysis, as depicted in Figure 3. It should be noted that it is not possible to detect hydrogen with EDX analysis and therefore, the chemical composition was analysed excluding the content of hydrogen.

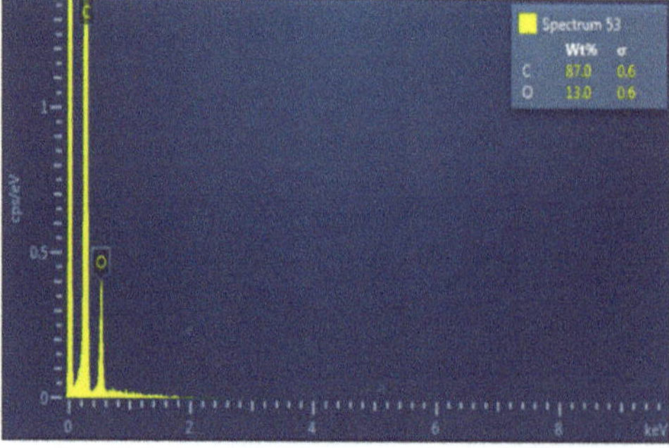



**Figure 3.** Spectra of deposits formed on the inner surface of the quartz tube.

The elemental analysis shows a high content of carbon in the structure of deposits, with an average of 88.6 wt% (91.2 at%) and the rest is oxygen, which either originates from the quartz tube structure ($SiO_2$) or from the moisture, which usually exists in argon flow and which is difficult to avoid, even in a vacuum system. The impurity of oxygen has been reported in previous studies during the growth of plasma polymerized films using the mixture of $CH_4$ and Ar in vacuum systems [15,16].

In order to include the content of hydrogen, the composition of the deposits' layer was analysed using a CHN elemental analyser, which is capable of determining the amount of hydrogen and carbon in the composition of the deposits. It was revealed that the H/C atomic ratio for the deposits is equal to $1.7 \pm 0.1$. From this, it was concluded that the deposits possess a polymer-like structure, considering a higher atomic content of H (~60 at%) [17]. This could further indicate that the layer of deposits is not a highly conductive material, due to a low atomic content of carbon.

### 3.2. SEM/EDX Characterization of γ-Alumina and Pd/γ-Alumina After Exposure to DBD Plasma

The catalyst samples, which were used as packing inside the discharge gap of the implemented DBD plasma reactor, were characterized by high-resolution scanning electron microscopy. The results indicate that the structure of the catalyst samples can be influenced after being exposed to $CH_4$+Ar DBD plasma. The changes on the structure of the catalyst originate from the formation of solid products on the catalyst, resulting from $CH_4$ decomposition, as well as from plasma treatment of the catalyst, while being exposed to plasma-induced species (ions, electrons, radicals) upon the generation of discharges.

Figure 4 shows the images of γ-alumina before and after exposure to plasma. The surface of γ-alumina is covered with amorphous carbon-containing agglomerates after exposure, and has a different structure than the fresh γ-alumina sample, indicating morphological changes during the plasma reactions. The presence of carbonaceous species on the γ-alumina structure was further confirmed by EDX spectra, as depicted in Figure 5.

**Figure 4.** SEM images of γ-alumina (**a,b**) before exposure to DBD plasma; (**c,d**) after exposure to DBD plasma.

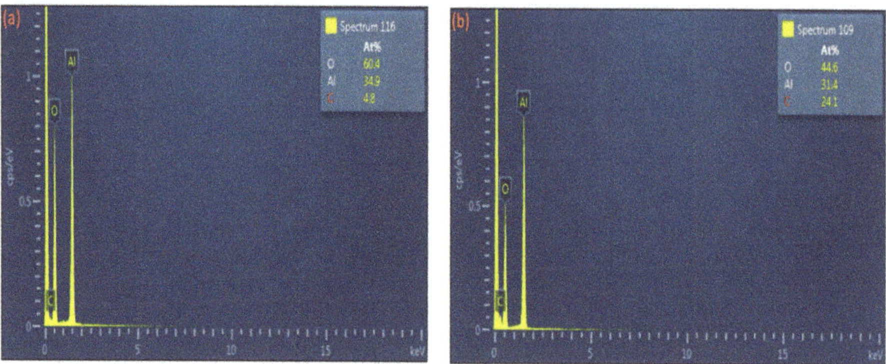

**Figure 5.** EDX spectra of γ-alumina (**a**) before exposure to DBD plasma; (**b**) after exposure to DBD plasma.

These results indicate that γ-alumina undergoes both structural and chemical changes during the exposure to $CH_4$+Ar plasma. The carbon-containing deposits were randomly formed on the γ-alumina structure, as observed by SEM/EDX mapping analysis of γ-alumina after exposure to plasma (Figure 6d). This could further indicate that deposits did not cover the structure uniformly.

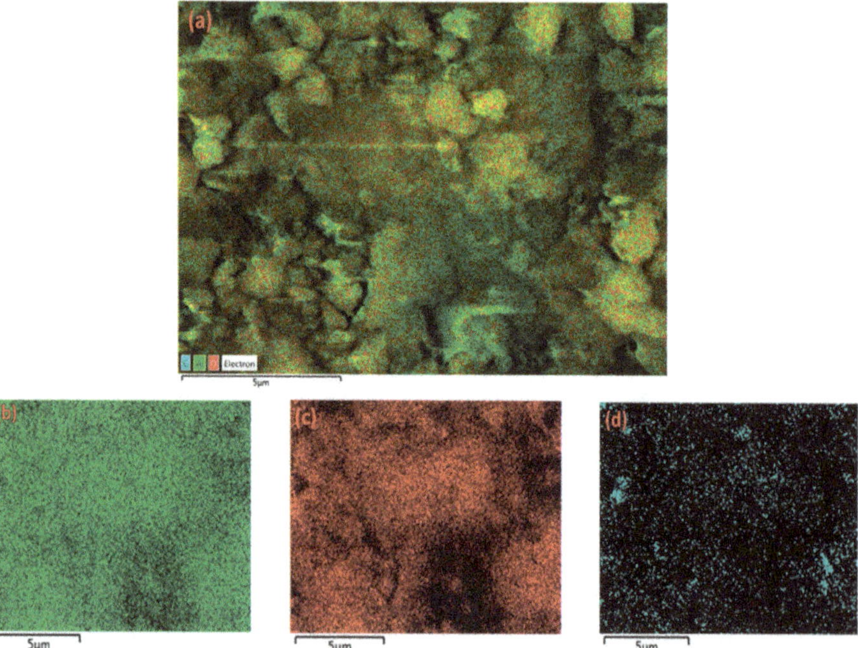

**Figure 6.** EDX mapping of γ-alumina after exposure to $CH_4$+Ar DBD plasma. (**a**) Layered image; (**b**) Al map; (**c**) O map; (**d**) C map.

For some areas of the fresh samples (i.e., before the exposure to plasma), a small amount of carbon was detected in the EDX analysis. This may be ascribed to reagents that were used in the preparation of the catalyst, e.g., carbonate $(CO_3)^2$, which could remain as impurities in the sample. It may also originate from the carbon adhesive tape, used for sampling, which is in direct contact with catalyst particles. Regardless of the source of carbon detected in the fresh samples, the atomic percentage of carbon in the fresh samples was always significantly lower than that detected in the plasma-processed samples.

The formation of agglomerates was not observed in SEM images for Pd/$\gamma$-alumina after exposure to plasma, as shown in Figure 7. This demonstrates that the Pd-containing surface becomes resistant to the deposition of solid species, due to the presence of Pd active sites on $\gamma$-alumina in such a way that the presence of Pd particles promotes the hydrogenation of deposit precursors (e.g., $CH_x$ and $C_2H_y$) on the catalyst surface, resulting in a lower amount of deposits being formed. The results of the conversion of methane and the selectivity of the deposits can be in found in Table S1 of the Supplementary Information. These results indicate that the selectivity of deposits decreases from 61.2% (i.e., for $\gamma$-alumina) to 24.2% (i.e., for Pd/$\gamma$-alumina). This therefore confirms the interaction of deposit precursors (e.g., $CH_x$ and $C_2H_y$) with Pd particles, promoting their hydrogenation towards gas-phase products faster than their deposition on the catalyst surface. Despite the effect of Pd in the reduction of the deposits, the surface morphology of Pd/$\gamma$-alumina has still been influenced after exposure to plasma, where the surface has become smoother, as can be seen in Figure 7c,d, compared to a rough surface observed for the fresh sample before exposure to plasma (Figure 7a,b). The result of EDX mapping for Pd/$\gamma$-alumina (Figure 8) exhibits the presence of carbon-containing areas; however, the amount of carbon is notably lower and more scattered, which therefore could not lead to the formation of carbon-containing agglomerates.

**Figure 7.** SEM images of Pd/$\gamma$-alumina (**a,b**) before exposure to DBD plasma; (**c,d**) after exposure to DBD plasma.

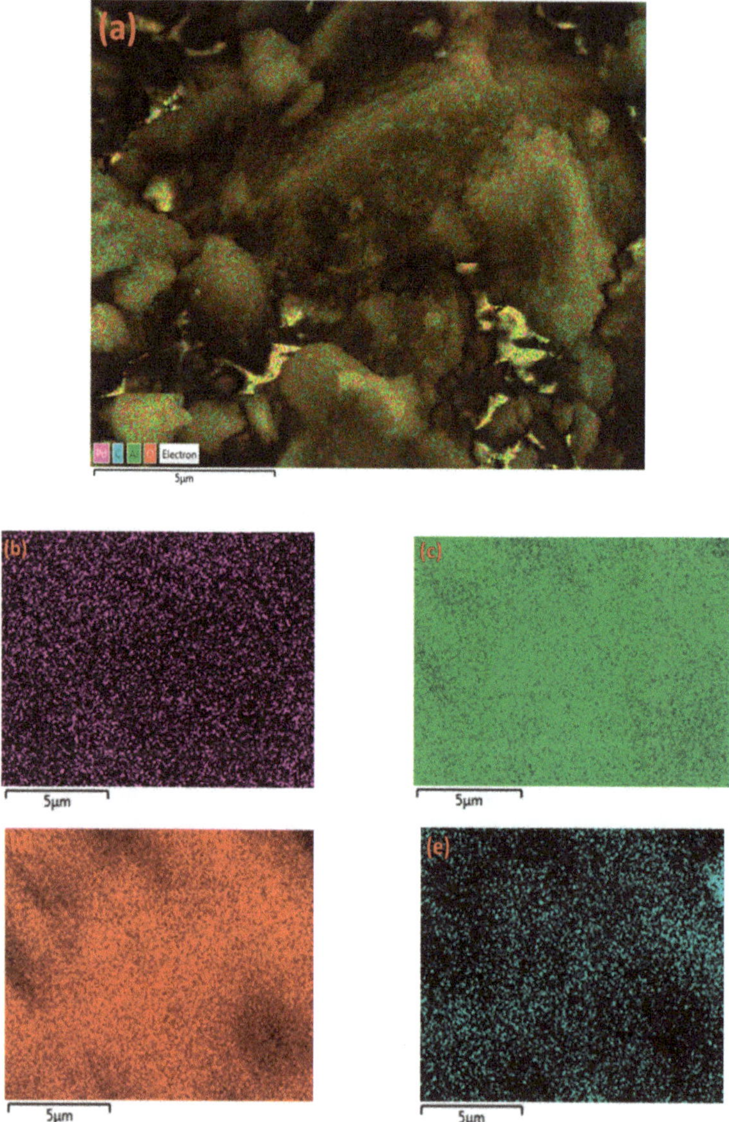

**Figure 8.** EDX mapping of Pd/γ-alumina after exposure to CH$_4$+Ar DBD plasma. (**a**) Layered image; (**b**) Pd map; (**c**) Al map; (**d**) O map; (**e**) C map.

*3.3. SEM/EDX Characterization of High Dielectric BaTiO$_3$ After Exposure to DBD Plasma*

The SEM images of BaTiO$_3$ before and after exposure to plasma are shown in Figure 9a. It can be observed that the tested BaTiO$_3$ has a porous structure. The high porosity of the packing increases the possibility for charges (i.e., those charges transferred between the high voltage electrode and the ground electrode) to be trapped inside the pores. In combination with the high dielectric constant of BaTiO$_3$, this porous structure consequently strengthens the occurrence of partial discharging [18]. Consequently, this causes a weaker electric field at contact points of high dielectric BaTiO$_3$, which leads to a lower conversion of methane.

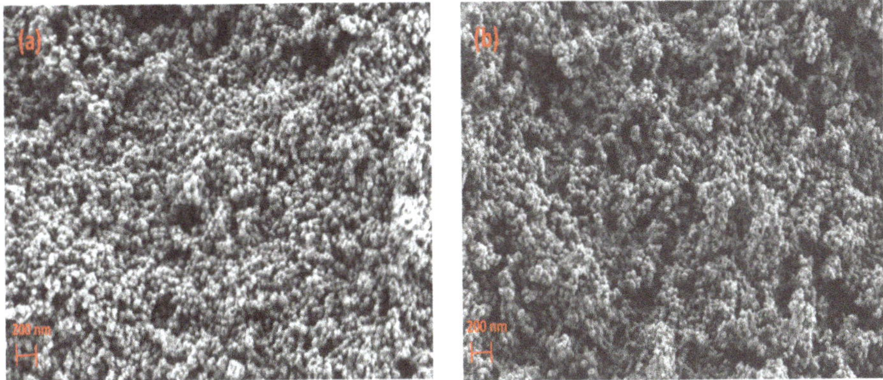

**Figure 9.** SEM images of $BaTiO_3$ (**a**) before exposure to DBD plasma; (**b**) after exposure to DBD plasma.

On the other hand, a lower conversion of methane yields a lower amount of deposits. This is the main reason that the formation of agglomerates is not observed for $BaTiO_3$ after exposure to $CH_4$+Ar plasma, in contrast to $\gamma$-alumina. In agreement with this, the conversion of $CH_4$ for the $BaTiO_3$ packed reactor was 9.3%, which is remarkably lower than the 47.7% obtained for the $\gamma$-alumina packed reactor during the 2.5 h of plasma processing. In this case, a lower amount of solid products became deposited on the surface, which then lowered the probability of the formation of agglomerates during the plasma reaction. However, the exposure of $BaTiO_3$ to DBD discharges could have influenced the morphology in such a way that new voids (pores) were created in the structure of $BaTiO_3$. The formation of pores (voids) inside the structure of plasma-treated materials (e.g., silica) has been previously observed after the treatment of the surface by Ar plasma [19,20].

In order to see the presence of carbon in the elemental composition of $BaTiO_3$, SEM/EDX mapping was performed, as shown in Figure 10. It can be seen that carbon-rich deposits do not form a dense layer on $BaTiO_3$, and their presence is scattered.

**Figure 10.** *Cont.*

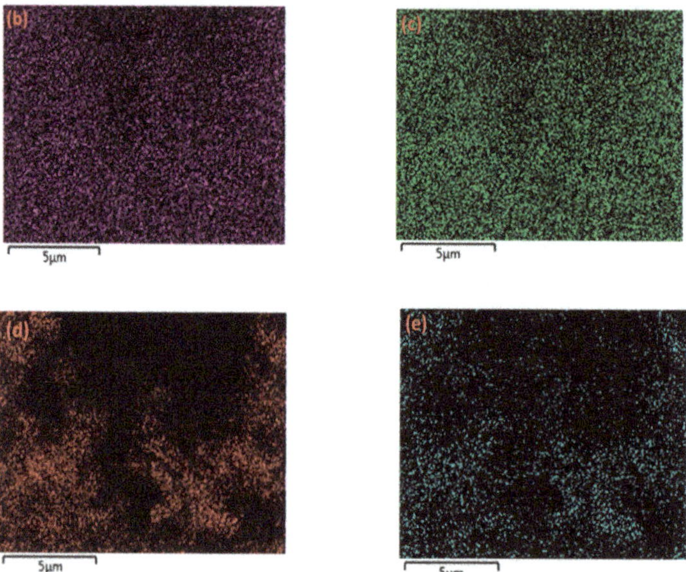

**Figure 10.** EDX mapping of BaTiO$_3$ after exposure to CH$_4$+Ar DBD plasma. (**a**) Layered image; (**b**) Ba map; (**c**) Ti map; (**d**) O map; (**e**) C map.

*3.4. SEM/EDX Characterization of MgO/Al$_2$O$_3$, Silica-SBA-15 and α-Alumina After Exposure to DBD Plasma*

MgO/Al$_2$O$_3$, silica-SBA-15, and α-alumina were processed by exposing them to CH$_4$+Ar DBD discharges. All these materials were morphologically influenced after being treated by plasma. SEM images of MgO/Al$_2$O$_3$ show that its structure was highly affected, where the formation of agglomerates can be clearly seen in Figure 11c,d.

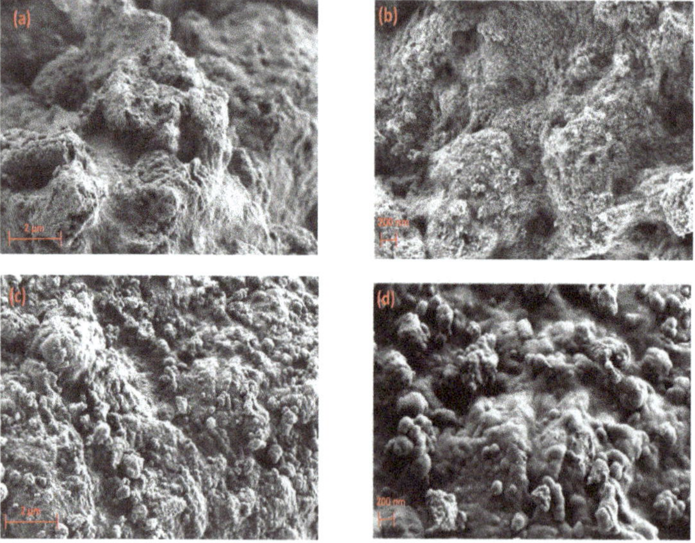

**Figure 11.** SEM images of MgO/Al$_2$O$_3$ (**a**,**b**) before exposure to DBD plasma; (**c**,**d**) after exposure to DBD plasma.

The structure of MgO/Al$_2$O$_3$ is influenced by the presence of carbon-rich deposits randomly, whereas the formation of agglomerates occurs regionally. This can be further seen by SEM/EDX mapping, as shown in Figure 12.

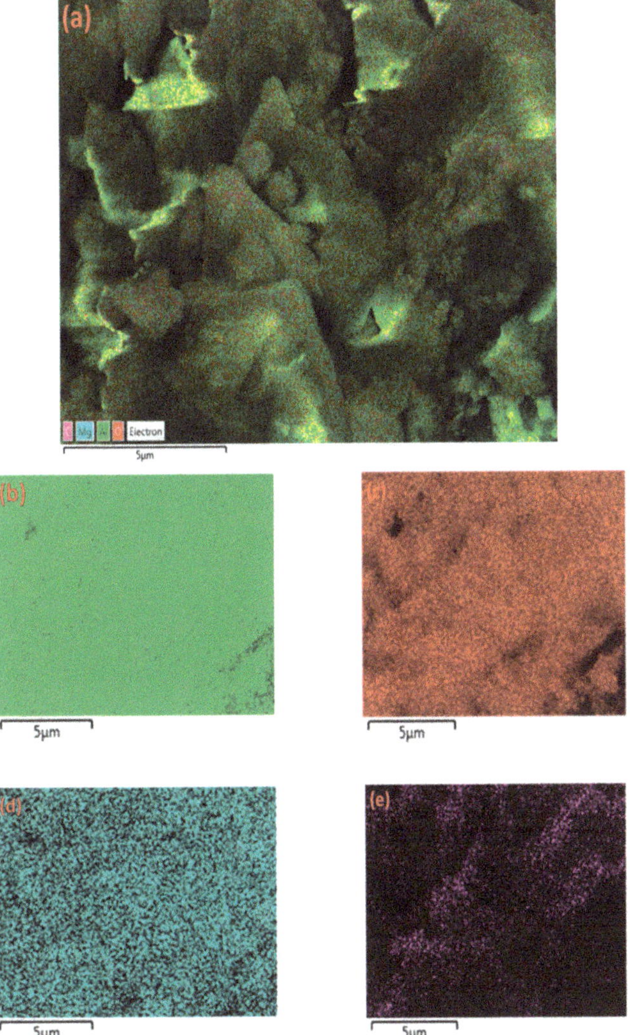


**Figure 12.** EDX mapping of MgO/Al2O3 after exposure to CH$_4$+Ar DBD plasma. (**a**) Layered image; (**b**) Al map; (**c**) O map; (**d**) Mg map; (**e**) C map.

Figure 13 depicts the morphology of silica-SBA-15 for the fresh sample and the one processed by DBD plasma. The tested silica-SBA-15 constitutes a meso-porous rope-like structure, according to Figure 13b.

In addition to the deposition of solid products, silica-SBA-15 undergoes structural changes, which can be seen in the form of the generation of new voids (pores) after exposure to plasma, as shown in Figure 13c. Furthermore, the scattered presence of carbon in the structure of silica-SBA-15 was observed by SEM/EDX mapping, as shown in Figure 14.

**Figure 13.** SEM images of silica-SBA-15 (**a**,**b**) before exposure to DBD plasma; (**c**,**d**) after exposure to DBD plasma.

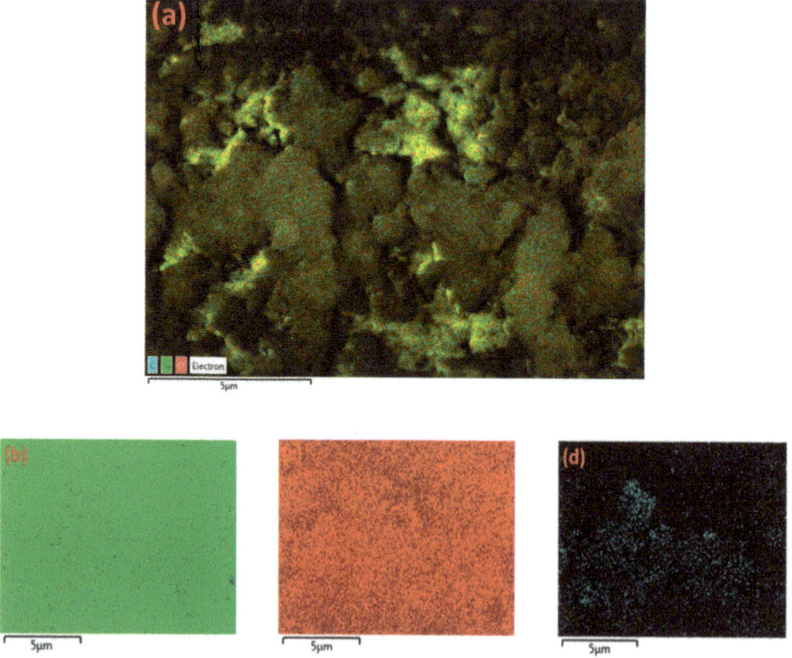

**Figure 14.** EDX mapping of silica-SBA-15 after exposure to $CH_4$+Ar DBD plasma. (**a**) Layered image; (**b**) Si map; (**c**) O map; (**d**) C map.

Similarly, the formation of agglomerates was observed for α-alumina after exposure to $CH_4$ plasma, as shown in Figure 15b. The fresh α-alumina possesses a crystalline structure, according to Figure 15a. This crystalline structure is partially influenced by exposure to the plasma, where the formation of agglomerates can be observed in the structure of α-alumina. Moreover, EDX results indicate that carbon-rich areas are regionally dispersed in the structure of α-alumina, as depicted in Figure 16, for two different regions of the analysed sample. The atomic percentage of carbon changes from 3.6 to 70.6 at%, respectively, for region A and region B of the analysed sample.

**Figure 15.** SEM images of α-alumina (**a**) before exposure to DBD plasma; (**b**) after exposure to DBD plasma.

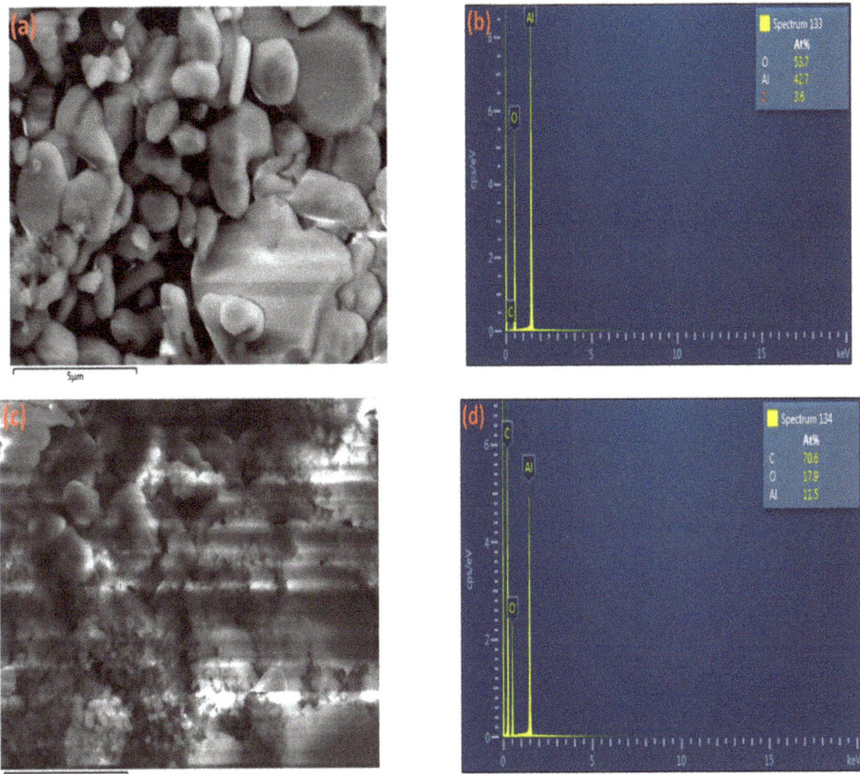


**Figure 16.** EDX electron image and elemental analysis of α-alumina sample after exposure to $CH_4$+Ar DBD plasma. (**a,b**) Region A; (**c,d**) Region B.

As shown in Figure 16c, region B of the sample has been covered with carbon-containing deposits. On the contrary, region A of the sample (Figure 16a) is almost clean, similar to the structure of the fresh sample, as depicted in Figure 15a. EDX mapping of $\alpha$-alumina for region B (Figure 17) also shows the coverage of $\alpha$-alumina with carbonaceous species in the form of agglomerates.

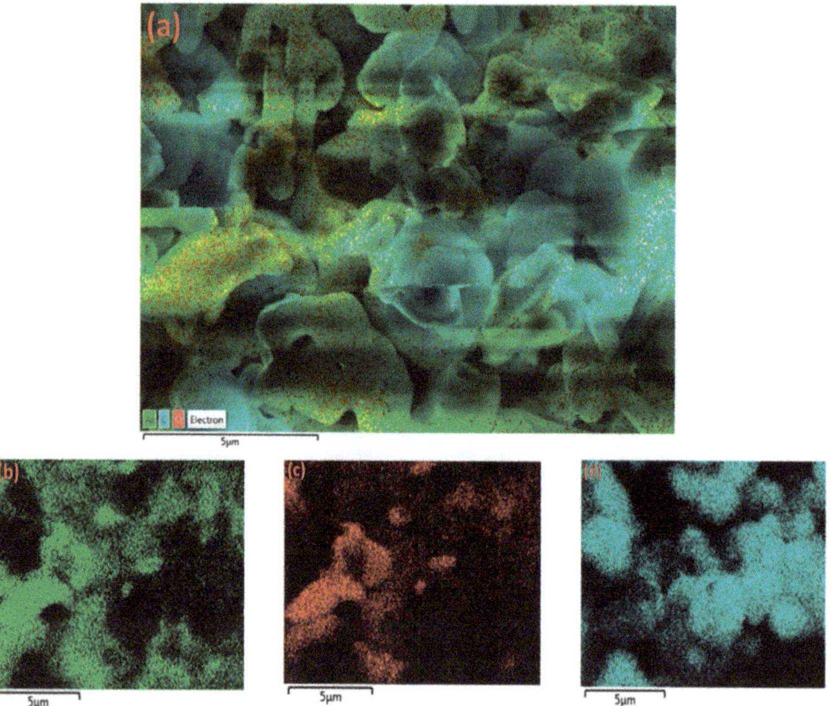


**Figure 17.** EDX mapping of $\alpha$-alumina after exposure to $CH_4$+Ar DBD plasma. (**a**) Layered image; (**b**) Al map; (**c**) O map; (**d**) C map.

These results indicate that all catalyst supports undergo morphological changes. This includes materials such as $MgO/Al_2O_3$ (440 $m^2$/gr) and silica-SBA-15 (673 $m^2$/gr), which both possess high surface areas, as well as $\gamma$-alumina, which possesses a relatively lower surface area (234 $m^2$/gr). Even the structure of a low surface area material (i.e., $\alpha$-alumina) is influenced by the formation of deposits. The results of the conversion of $CH_4$ and the selectivity of the deposits (Table S1) show a similar range of deposits' selectivity for catalyst supports with various surface areas. Therefore, according to these results, no specific correlation between the surface area and the deposits' formation can be drawn for the plasma catalytic conversion of $CH_4$. Further investigations may be conducted to study the influence of plasma treatment, as well as deposits on the surface area and porosity of the packing. This is beyond the scope of the present article and therefore can be a topic of future research.

## 4. Conclusions

High-resolution scanning electron microscopy (HR-SEM) was utilized to characterize the deposits formed during the $CH_4$+Ar DBD plasma. A low voltage of 1 kV was applied in order to prevent the phenomenon of surface charging, which was intensively observed for high voltages ($\geq$3 kV). This could furthermore confirm the low conductivity of the deposits. For the deposits formed on the dielectric quartz tube, the SEM image exhibited an amorphous structure, where the morphology remained unchanged throughout the length of the plasma zone. EDX analysis indicated that the chemical composition of the deposits mainly consists of carbon (~91 at%), although it was noted that this chemical composition is estimated excluding the presence of hydrogen, as the EDX detector is not able to analyse hydrogen. The presence of hydrogen was identified with CHN elemental analysis, indicating that the H/C molar ratio of the deposits layer was $1.7 \pm 0.1$. This therefore indicated the polymeric nature of the formed deposits, due to a high content of H (~60 at%). The structure of $\gamma$-alumina was influenced after exposure to plasma by the formation of regionally amorphous carbon-containing agglomerates. On the contrary, the structure of Pd/$\gamma$-alumina did not exhibit the presence of carbon-rich agglomerates, indicating the resistance of the surface to deposits' formation due to the presence of Pd particles, which could promote the hydrogenation of deposit-precursors faster than their deposition to solid species. All other plasma-processed materials, including $BaTiO_3$, $MgO/Al_2O_3$, silica-SBA-15, and $\alpha$-alumina, clearly showed morphological changes after being exposed to $CH_4$+Ar plasma, not only due to the formation of deposits, but also owing to the plasma treatment of the surface, which could influence the structure of these materials.

**Supplementary Materials:** The following are available online at http://www.mdpi.com/2079-4991/9/4/589/s1, Table S1. The conversion of $CH_4$ and the selectivity of gas-phase ($C_{2+}$) products and deposits for the non-packed (Blank) and the packed DBD plasma reactor.

**Author Contributions:** Conceptualization, M.T.; Methodology, M.T.; Data Curation, M.T.; Data Analysis, M.T.; Writing Original Draft, M.T.; Writing, Review and Editing, H.G.

**Funding:** This research received no external funding.

**Conflicts of Interest:** The authors declare no conflict of interest.

## References

1. Liu, G.; Li, Y.; Chu, W.; Shi, X.; Dai, X.; Yin, Y. Plasma-assisted preparation of Ni/SiO$_2$ catalyst using atmospheric high frequency cold plasma jet. *Catal. Commun.* **2008**, *9*, 1087–1091. [CrossRef]
2. Yan, X.; Zhao, B.; Liu, Y.; Li, Y. Dielectric barrier discharge plasma for preparation of Ni-based catalysts with enhanced coke resistance: Current status and perspective. *Catal. Today* **2015**, *256*, 29–40. [CrossRef]
3. Witvrouwen, T.; Paulussen, S.; Sels, B. The use of non-equilibrium plasmas for the synthesis of heterogeneous catalysts. *Plasma Process. Polym.* **2012**, *9*, 750–760. [CrossRef]
4. Tu, X.; Gallon, H.J.; Whitehead, J.C. Plasma-assisted reduction of a NiO/Al$_2$O$_3$ catalyst in atmospheric pressure H$_2$/Ar dielectric barrier discharge. *Catal. Today* **2013**, *211*, 120–125. [CrossRef]
5. Liu, C.; Zou, J.; Yu, K.; Cheng, D.; Han, Y.; Zhan, J.; Ratanatawanate, C.; Jang, W.L. Plasma application for more environmentally friendly catalyst preparation. *Pure Appl. Chem.* **2006**, *78*, 1227–1238. [CrossRef]
6. Karuppiah, J.; Mok, Y.S. Plasma-reduced Ni/$\gamma$–Al$_2$O$_3$ and CeO$_2$–Ni/$\gamma$–Al$_2$O$_3$ catalysts for improving dry reforming of propane. *Int. J. Hydrog. Energy* **2014**, *39*, 16329–16338. [CrossRef]
7. Saoud, W.A.; Assadia, A.A.; Guizab, M.; Bouzazaa, A.; Aboussaoud, B.; Ouederni, A.; Soutrela, I.; Wolbert, D.; et al. Study of synergetic effect, catalytic poisoning and regeneration using dielectric barrier discharge and photocatalysis in a continuous reactor: Abatement of pollutants in air mixture system. *Appl. Catal. B Environ.* **2017**, *213*, 53–61. [CrossRef]
8. Assadi, A.A.; Bouzaza, A.; Vallet, C.; Wolbert, D. Use of DBD plasma, photocatalysis, and combined DBD plasma/photocatalysis in a continuous annular reactor for isovaleraldehyde elimination-synergetic effect and by products identification. *Chem. Eng. J.* **2014**, *254*, 124–132. [CrossRef]
9. Abou Ghaida, S.G.; Assadi, A.A.; Costa, G.; Bouzaza, A.; Wolbert, D. Association of surface dielectric barrier discharge and photocatalysis in continuous reactor at pilot scale: Butyraldehyde oxidation, by-products identification and ozone valorization. *Chem. Eng. J.* **2016**, *292*, 276–283. [CrossRef]
10. Assadia, A.A.; Bouzazaa, A.; Wolberta, D. Study of synergetic effect by surface discharge plasma/TiO2 combination for indoor air treatment: Sequential and continuous configurations at pilot scale. *J. Photochem. Photobiol. A Chem.* **2015**, *310*, 148–154. [CrossRef]
11. Khoja, A.H.; Tahir, M.; Amin, N.A.S. Dry reforming of methane using different dielectric materials and DBD plasma reactor configurations. *Energy Convers. Manag.* **2017**, *144*, 262–274. [CrossRef]

12. Chiremba, E.; Zhang, K.; Kazak, C.; Akay, G. Direct nonoxidative conversion of methane to hydrogen and higher hydrocarbons by dielectric barrier discharge plasma with plasma catalysis promoters. *AIChE J.* **2017**, *63*, 4418–4429. [CrossRef]
13. Nozaki, T.; Muto, N.; Kado, S.; Okazaki, K. Dissociation of vibrationally excited methane on Ni catalyst. *Catal. Today* **2004**, *89*, 57–65. [CrossRef]
14. Kado, S.; Urasaki, K.; Sekine, Y.; Fujimoto, K.; Nozaki, T.; Okazaki, K. Reaction mechanism of methane activation using non-equilibrium pulsed discharge at room temperature. *Fuel* **2003**, *82*, 2291–2297. [CrossRef]
15. Chandrashekaraiah, T.H.; Bogdanowicz, R.; Rühl, E.; Danilov, V.; Meichsner, J.; Thierbach, S.; Hippler, R. Spectroscopic study of plasma polymerized a-C:H films deposited by a dielectric barrier discharge. *Materials* **2016**, *9*, 594. [CrossRef] [PubMed]
16. Pothiraja, R.; Bibinov, N.; Awakowicz, P. Amorphous carbon film deposition on the inner surface of tubes using atmospheric pressure pulsed filamentary plasma source. *J. Phys. D Appl. Phys.* **2011**, *44*, 355206. [CrossRef]
17. Casiraghi, C.; Piazza, F.; Ferrari, A.C.; Grambole, D.; Robertson, J. Bonding in hydrogenated diamond-like carbon by raman spectroscopy. *Diam. Relat. Mater.* **2005**, *14*, 1098–1102. [CrossRef]
18. Butterworth, T.; Allen, R.W.K. Plasma-catalyst interaction studied in a single pellet DBD reactor: Dielectric constant effect on plasma dynamics. *Plasma Sources Sci. Technol.* **2017**, *26*, 065008. [CrossRef]
19. Zhang, J.; Palaniappan, A.; Su, X.; Tay, F.E.H. Mesoporous silica thin films prepared by argon plasma treatment of sol–gel derived precursor. *Appl. Surf. Sci.* **2005**, *245*, 304–309. [CrossRef]
20. Palaniappan, A.; Zhang, J.; Su, X.; Tay, F.E.H. Preparation of mesoporous silica films using sol-gel process and argon plasma treatment. *Chem. Phys. Lett.* **2004**, *395*, 70–74. [CrossRef]

*Article*

# Discharge Regimes Transition and Characteristics Evolution of Nanosecond Pulsed Dielectric Barrier Discharge

Li Zhang [1], Dezheng Yang [1,2,*], Sen Wang [1,3], Zixian Jia [4], Hao Yuan [1], Zilu Zhao [1] and Wenchun Wang [1]

[1]  Key Lab of Materials Modification, Dalian University of Technology, Ministry of Education, Dalian 116024, China; zhangli2013@mail.dlut.edu.cn (L.Z.); foreversean@126.com (S.W.); yuanhao@mail.dlut.edu.cn (H.Y.); huairuo@mail.dlut.edu.cn (Z.Z.); wangwenc@dlut.edu.cn (W.W.)
[2]  Key Laboratory of Ecophysics, College of Sciences, Shihezi University, Shihezi 832003, China
[3]  College of Electrical Engineering and Control Science, Nanjing Tech University, Nanjing 211800, China
[4]  Laboratoire des Sciences des Procédés et des Matériaux CNRS., Institut Galilée, Université Paris 13, Sorbonne Paris Cité, 93430 Villetaneuse, France; zixian.jia@lspm.cnrs.fr
*  Correspondence: yangdz@dlut.edu.cn; Tel.: +86-1332-221-4836

Received: 30 August 2019; Accepted: 24 September 2019; Published: 26 September 2019

**Abstract:** Discharge regime transition in a single pulse can present the breakdown mechanism of nanosecond pulsed dielectric barrier discharge. In this paper, regime transitions between streamer, diffuse, and surface discharges in nanosecond pulsed dielectric barrier discharge are studied experimentally using high resolution temporal–spatial spectra and instantaneous exposure images. After the triggering time of 2–10 ns, discharge was initiated with a stable initial streamer channel propagation. Then, transition of streamer-diffuse modes could be presented at the time of 10–34 ns, and a surface discharge can be formed sequentially on the dielectric plate. In order to analyze the possible reason for the varying discharge regimes in a single discharge pulse, the temporal–spatial distribution of vibrational population of molecular nitrogen $N_2$ ($C^3\Pi_u$, v = 0,1,2) and reduced electric field were calculated by the temporal–spatial emission spectra. It is found that at the initial time, a distorted high reduced electric field was formed near the needle electrode, which excited the initial streamer. With the initial streamer propagating to the dielectric plate, the electric field was rebuilt, which drives the transition from streamer to diffuse, and also the propagation of surface discharge.

**Keywords:** nanosecond pulse; temporal-spatial spectra; breakdown mechanism; reduced electric field; vibrational population

---

## 1. Introduction

As an effective method to optimize the ionization efficiency, nanosecond pulsed discharge (NPD) has become an emerging technology to generate non-thermal plasma [1]. For the sharply pulse rising time, the electrons can be accelerated effectively [2]. Therefore, high energy efficiency, excellent thermal stability, and good discharge plasma distribution can be reached in nanosecond pulsed dielectric barrier discharge (NPDBD) [3,4]. Moreover, some applications have been widely exhibited in volatile organic compounds (VOCs) removal [5], sewage treatment [6,7], polymer modification [8,9], aerospace [10], biomedical [11], etc. More recently, the nanoparticles, such as silver [12], carbon [13], and cobalt [14], have been synthetized by NPDBD. The investigation of the discharge mechanism and optimize the energy regulation could help to parameterize during the nanoparticle preparation.

For this, the dynamic investigation of NPD was studied numerically and experimentally in the last two decades [15–17], and one of the most concerned issues is the discharge mechanism in rapid electric field. Several breakdown mechanisms were investigated in NPD, for instance, streamer, Townsend, and runaway electron modes [17–26]. Townsend breakdown often involves in a diffuse manner of the electrode gap [25]. In this mode, the secondary electrons play an important role in the breakdown cycle, which were mainly produced in the process of ions transporting to the cathode. After the production of secondary electrons, the electron avalanche continues or even grows until the discharge establishes. The streamer discharge is one of the most common mechanisms of gas breakdown at high *pd* (*p* is the pressure in in standard atmosphere, and *d* is the gap distance in meters) with overvoltage, which states that the electron avalanches initiate with the seed electrons created by photoionization, and follow the positively charged trail left by the primary avalanche [25]. According to different development degrees of the streamer, there are several regimes of streamer discharge, such as, positive corona, filament in dielectric barrier discharge, spark discharge, and plasma bullet [22–24,27]. Runaway electron mechanism is appropriate for the ultrafast pulse excitation with high overvoltage at atmospheric pressure, in which X-ray can also be detected [28,29].

Generally, the NPD with the rapid gas breakdown is not in a single discharge mode. Therefore, discharge regime transitions can take place in discharge process [30–34]. Lo et al. [30] studied the streamer-to-spark transition generated by an overvoltage nanosecond pulsed discharge under atmospheric pressure air. The initial discharge in their experiment was a streamer phase with high voltage and high current, followed by a spark phase with a low voltage and a decreasing current in several hundreds of nanoseconds, and at last in the streamer-to-spark transition, the discharge contracted toward the channel axis and evolved into a highly conducting thin column. Stepanyan et al. [31,32] studied the transitions of streamer to filamentary in nanosecond surface dielectric barrier discharge in air at pressures of 1–6 bar. The discharge developed as a set of streamers at atmospheric pressure, and filament could be observed at high pressures and high voltage amplitude. Both the time- and space-resolved optical emission spectra were measured in the transition of streamer-to-filament. Glow regimes also have been observed in NPD, which developed through an initial cathode-directed streamer and followed by a return wave of potential redistribution. Pai et al. [33] found that pin electrode is beneficial to generate the wave propagation, because the electric field at the tip would be significantly higher than that for a plane electrode, and result in greater neutralization of the streamer head space charge, causing a greater potential drop to transmitting a return wave. Townsend regime was also considered as a possible regime to generated diffuse discharge. However, Townsend breakdown cannot develop sufficiently in NPD [34], because it hardly accomplished in several tens of nanoseconds, as the external applied electric field evolves much faster than the time scale for ions to move across the gap.

The temporal–spatial resolved diagnosis of plasma characteristics, including plasma spectra and ICCD images, are very important to understand the dynamics processes and rapid breakdown mechanism in NPDBD [34,35]. By these diagnoses, the reduced electric field, the vibrational population of $N_2$ ($C^3\Pi_u$), plasma dynamic evolution, can be studied to investigate plasma processes. Due to the short radiation lifetimes and high excitation rates of $N_2$ ($C^3\Pi_u$) and $N_2^+$ ($B^2\Sigma_u^+$), it is possible to calculate the reduced electric field $E/N$ ($E$ is electrical field, $N$ is gas number density, and $N = 2.68 \times 10^{19}$ cm$^{-3}$ at atmospheric air) by the ratio of intensities of $N_2$ ($C^3\Pi_u \rightarrow B^3\Pi_g$) and $N_2^+$ ($B^2\Sigma_u^+ \rightarrow X^2\Sigma_g^+$) states in different vibrational radiation bands [36]. By the refined calculation of the excitation and quenching rates of the $N_2$ ($C^3\Pi_u$, v = 0,1,2 ... ), the temporal-spatial evolution of $E/N$ can be represented by the temporal–spatial emission spectra of NPD [37]. In this paper, we concentrate our attention on the rapid breakdown mechanism and temporal–spatial evolution dynamic of NPDBD in air using needle-plate electrode. The ICCD image and the temporal–spatial resolved spectra are measured. The evolution dynamic process of the discharge and the temporal–spatial distributions of the emission intensities of $N_2$ ($C^3\Pi_u \rightarrow B^3\Pi_g$) and $N_2^+$ ($B^2\Sigma_u^+ \rightarrow X^2\Sigma_g^+$) bands are investigated. In addition, to understand the energy transition and breakdown mechanisms, the temporal–spatial distributions of vibrational population of $N_2$ ($C^3\Pi_u$, v = 0,1,2) states and the $E/N$ are calculated using the temporal–spatial emission spectra.

## 2. Experimental Setup

Figure 1 shows the schematic diagram of the experimental setup. The plasma reactor consists of a needle electrode with the curvature radius of 0.4 mm, and a grounded circular plate electrode covered by a 1 mm thick ceramic plate. The discharge is driven by a nanosecond pulsed power supply. Voltage probe (Tektronix- P6015A, Tektronix Inc, Beaverton, OR, USA) and current probe (Pearson Current Monitor-4100, Pearson Electronics Inc, Palo Alto, CA, USA) are used to measure the waveforms of pulse voltage and discharge current, which are displayed and recorded by an oscilloscope (Tektronix-TDS5054B-500 MHz, Tektronix Inc, Beaverton, OR, USA). The optical emission spectra are obtained using multichannel optical fibers and collected by a grating monochromator (Andor SR-750i, grating groove 2400 lines/mm, glancing wavelength 300 nm, Andor Technology Inc, Belfast, UK). A conjugate spectrum image system is used to acquire the spatially resolved optical spectra, where the heads of 35 parallel fibers are closely arranged in the vertical image plane of the quartz lens ($f$ = 75 mm). After the diffraction of the grating, the output spectral light can be transformed into a digital signal by an intensified charge-coupled device (ICCD) camera (Andor's iStar DH334T, Andor Technology Inc, Belfast, UK). For the time-resolved measurements, the ICCD camera is synchronized with the pulsed voltage.

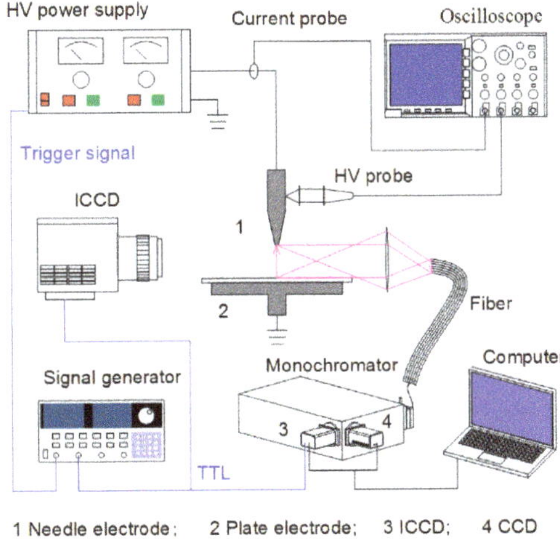


**Figure 1.** Experimental setup.

## 3. Results and Discussion

### 3.1. Dynamics Evolution of the Discharge

A series of ICCD images with the gate width of 4 ns are shown in Figure 2, presenting the dynamics process of the gas breakdown in fast pulse electric field. The pulse peak voltage, pulse repetition rate, and electrode gap distance are kept at 26 kV, 100 Hz, and 5 mm, respectively. For a suitable synchronization, the pulse voltage and ICCD are triggered by a TTL signal. The time parameters labeled in Figure 2 are shown in the waveforms of pulse voltage and discharge current, illustrated in Figure 3, where the discharge current is obtained by subtracting the displacement current from the total discharge current [38,39]. The zero time in Figures 2 and 3 is set at the initial breakdown of the gas gap. By applying a fast electric field higher than the breakdown field threshold, the local breakdown in the gap initiates from the needle electrode at 2 ns and exhibits a maximum intensity at the time of 18 ns. There are three main stages of the discharge development in NPDBD, i.e., streamer breakdown of the electrode gap, the streamer to diffuse transition, and the propagation of surface barrier discharge. Firstly, a main streamer, generated by the strong electric field, propagates from the needle to plate electrode at time t = 10 ns. It takes about 8 ns for the streamer channel to go across the gap, which means its velocity is about $6.25 \times 10^5$ m/s. When the initial streamer propagated to the dielectric plate surface, the memory charges accumulated on the dielectric plate can be erased by the conductive streamer channel. Then, the electric field in the electrode gap and on the dielectric plate surface is rebuilt, and a secondary streamer channel can be generated and propagates along the dielectric plate. A number of fine secondary streamer channels distributing around the initial streamer channel can be observed from ICCD image at t = 10 ns. Numerous secondary streamer channels propagate both in horizontal and vertical directions in a synchronous manner, the overlap of these discharge channels makes the discharge presenting a diffusive morphology at the time of 10–18 ns. A non-uniform structure with several filaments can be observed at t = 26 ns. However, the pulse voltage is already decreasing and the volume discharge in gas gap becomes weak rapidly. Then, space charges are involved in the propagation of the streamer and in the enhancement of the induced field at the streamer head. Subsequently, this feedback electric field stops the ionization in the axial direction and builds a potential electric field gradient along the dielectric surface near the breakdown region. Therefore, the surface discharge can spread to the surrounding areas over 60 ns, when the volume discharge channel has been extinguished, that is, the discharge current in Figure 3 is approximate equal to zero.

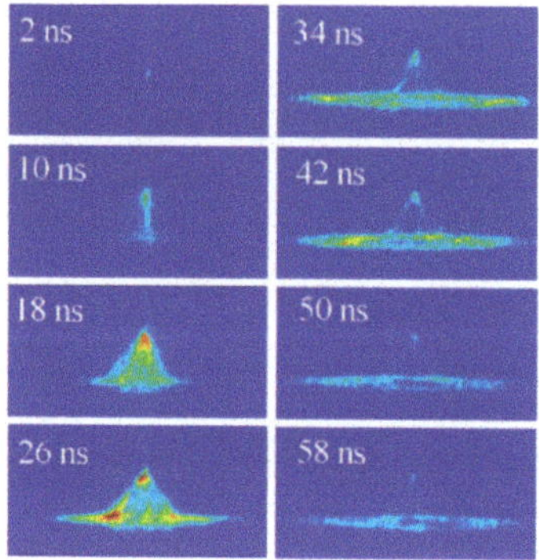


**Figure 2.** Images of discharge in air captured by an intensified charge-coupled device (ICCD) with an exposure time of 4 ns.

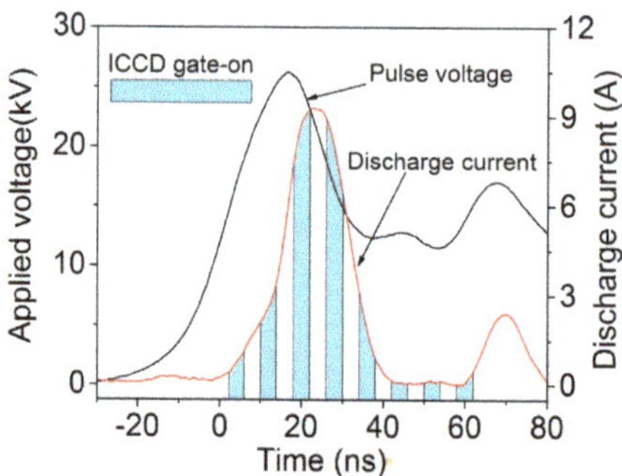


**Figure 3.** The waveforms of nanosecond pulsed voltage and discharge current.

The breakdown of the initial stage can be considered as a pulsed corona discharge in a non-uniform electric field. When the corona-like discharge is initialed at t = 2 ns, the electric field near the needle electrode can be estimated as $E_{max} = 2V/[r\ln(2d/r)] \approx 150$ kV/cm [25], where $r$ is the radius of the needle electrode, $d$ is the gap distance along the axis, and $V$ is the voltage at 2 ns. That means, $E/N$ contributed by the applied pulse voltage is about 600 Td. This high electric field is strong enough to drive the gas gap breakdown as a positive streamer mode [33]. Once the initial streamer channel propagates across the gas gap and the streamer head reaches the cathode electrode, it can erase the memory charges on the dielectric plate surface where the discharge channel touched. During the breakdown in several nanoseconds, this time is not enough for the streamer–cathode interaction to fully transform into a cathode fall [33]. In other words, the non-metal cathode and short time are not sufficient to meet the condition of secondary electron emission. Thereby, the conductive plasma channel can be considered as an anode in which the $E/N$ would become much smaller. Then, since the applied voltage is still in a high level, a newly built electric field with radial direction is formed to drive the subsequent breakdown. Meanwhile, the surrounding air can be pre-photoionized by the streamer channel. Although photoionization is orders of magnitude lower than the ionization density, it plays an important role in the propagation of the initial streamers [33,40–42]. Streamers can propagate nearly perpendicularly to the background electric field, and it can be guided by pre-ionization [40]. In addition, in the study of Nijdam [42] it is found that when the streamers do not follow the background field lines, they are usually repelled by neighboring streamers and follow the new local electric field direction. As a consequence, abundant of fine secondary streamer channels distributing around the initial streamer can be formed by pre-photoionization (as shown in Figure 2 at t = 10 ns), which contribute the transition to a diffuse regime.

### 3.2. Optical Emission Spectra of Nanosecond Pulsed Dielectric Barrier Discharge (NPDBD)

Figure 4a,b shows the optical emission spectra (OES) of NPDBD in the ranges of 333–339 nm and 390–401 nm at atmospheric air as a function of the distance on needle-plate axial direction. For the measurements in the experiment, the pulse peak voltage, pulse repetition rate, and electrode gap distance were kept at 26 kV, 100 Hz, and 5 mm, respectively. It can be seen in Figure 4 that both the emission intensities of $N_2^+(B^2\Sigma_u^+ \rightarrow X^2\Sigma_g^+, 0\text{–}0)$ and $N_2$ $(C^3\Pi_u \rightarrow B^3\Pi_g, 0\text{–}0)$ exhibit the maximum value at the position about 1 mm from the needle tip and decrease as the distance from the needle tip increases. A significant difference between their spatial distribution is the spectra of $N_2^+(B^2\Sigma_u^+ \rightarrow X^2\Sigma_g^+)$ mainly emitted from the region near the needle tip, which is much smaller than that of $N_2$ $(C^3\Pi_u \rightarrow B^3\Pi_g)$.

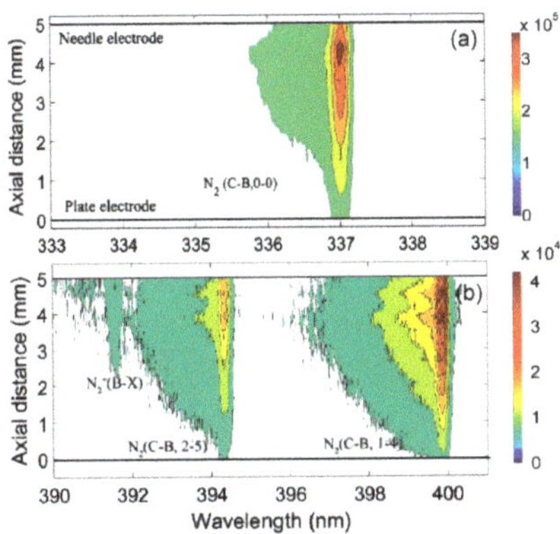


**Figure 4.** Spatial resolved spectra of nanosecond pulsed dielectric barrier discharge (NPDBD) (**a**) in range of 333–339 nm; (**b**) in range of 389–401 nm.

According to the spatially resolved OES shown in Figure 4, spatial–temporal resolved spectra can be obtained by an ICCD detector. Figure 5 shows the transient spatial resolved OES in the range of 389–401 nm at four different times. During the measurement, the gate width of ICCD was kept at

5 ns, and the pulse peak voltage, pulse repetition rate, and electrode gap distance are kept at 26 kV, 100 Hz, and 5 mm, respectively. It shows that there is an obvious difference between the distribution of emission intensities of $N_2^+(B^2\Sigma_u^+ \to X^2\Sigma_g^+, 0–0)$ and $N_2 (C^3\Pi_u \to B^3\Pi_g)$. When the discharge initiates from the needle electrode, the emission intensity of $N_2^+(B^2\Sigma_u^+ \to X^2\Sigma_g^+, 0–0)$, which stands for the high reduced electric field, is approximated equal to the intensity of $N_2 (C^3\Pi_u \to B^3\Pi_g, 2–5)$. At the time of 30–40 ns, the bands of $N_2^+(B^2\Sigma_u^+ \to X^2\Sigma_g^+, 0–0)$ become very weak.

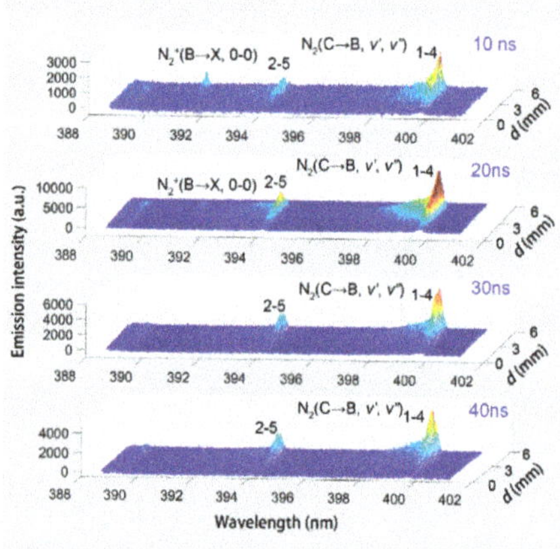

**Figure 5.** Spatial resolved spectra in the range of 389–401 nm at four different times.

Figure 6a,b shows the emission intensities of $N_2^+(B^2\Sigma_u^+ \to X^2\Sigma_g^+, 0–0)$ and $N_2 (C^3\Pi_u \to B^3\Pi_g, 2–5)$ at 0 mm, 2.5 mm, and 5 mm from the plate electrode as functions of time in one single pulse, by integrating the spatial–temporal resolved spectra. The NPDBD was operated at atmospheric air at 26 kV pulse peak voltage, 100 Hz pulse repetition rate, and 5 mm electrode gap distance. The gate width of ICCD was set as 5 ns and the zero time is set at the initial breakdown of the gas gap.

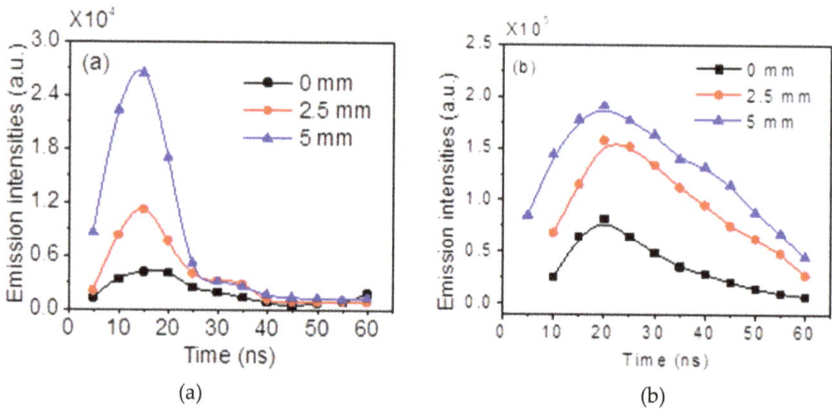

**Figure 6.** Spatiotemporal distribution of the emission intensities: (**a**) $N_2^+(B^2\Sigma_u^+ \to X^2\Sigma_g^+, 0–0, 391.4\,nm)$; (**b**) $N_2 (C^3\Pi_u \to B^3\Pi_g, 2–5, 394.3\,nm)$.

It can be seen from Figure 6a that once the breakdown initiated, the spectra of $N_2^+(B^2\Sigma_u^+ \to X^2\Sigma_g^+, 0–0)$ is firstly emitted from the region near the needle electrode. Its emission intensity increases sharply at first 15 ns and then decrease, exhibiting a maximum at the time of 15 ns. The existence time of $N_2^+$ ($B^2\Sigma_u^+ \to X^2\Sigma_g^+$, 0–0) is about 20 ns, which is only about 1/3 of the discharge duration time in a single pulse (in a characteristic time of 50–60 ns). Distance from the plate electrode has an obvious influence on the emission intensity of $N_2^+$($B^2\Sigma_u^+ \to X^2\Sigma_g^+$, 0–0). When the distance from the plate electrode decreases from 5 mm to 2.5 mm, which means the detection region moves to the middle of the electrode gap from the region near the needle electrode, the emission intensity decreases sharply to about $1.2 \times 10^4$ (a.u.). There is a little distinction in the evolution of $N_2^+$($B^2\Sigma_u^+ \to X^2\Sigma_g^+$, 0–0) near the plate electrode. When the volume discharge is extinguished at 55–60 ns, a slight increase of the emission intensity of $N_2^+$($B^2\Sigma_u^+ \to X^2\Sigma_g^+$, 0–0) can be observed (from 2156 to 9182).

As shown in Figure 6b, the temporal evolutions of $N_2$ ($C^3\Pi_u \to B^3\Pi_g$, 2–5) spectral bands present an obvious different tendencies both on spatial and temporal dimensionality. Firstly, the time period for the emission of $N_2$ ($C^3\Pi_u \to B^3\Pi_g$, 2–5) is much longer, it can almost be detected during the whole time from 10–60 ns. The maximum values of emission intensity of $N_2$ ($C^3\Pi_u \to B^3\Pi_g$, 2–5) appears at the time of 20 ns, which slightly lag behind the $N_2^+$($B^2\Sigma_u^+ \to X^2\Sigma_g^+$, 0–0) band. Secondly, the attenuation gradients of $N_2$ ($C^3\Pi_u \to B^3\Pi_g$, 2–5) is much smaller. The emission intensity of $N_2$ ($C^3\Pi_u \to B^3\Pi_g$, 2–5) at 5 mm from the plate electrode is about 1.2 times of the emission intensity at 2.5 mm, while the emission intensity of the $N_2^+$($B^2\Sigma_u^+ \to X^2\Sigma_g^+$, 0–0) band at 5 mm is about 2.2 times of the emission intensity at 2.5 mm.

As observed in Figure 2, the discharge initialed from the needle electrode as a streamer mode. When the positive streamer propagates to the plate electrode, the streamer head with a distorted high electrical field can produce high energetic electrons with the main energy of about 2–20 eV [34], which can ionize and excite the $N_2$ molecules to excited ions $N_2^+$($B^2\Sigma_u^+$). The relative strong bands of $N_2^+$($B^2\Sigma_u^+ \to X^2\Sigma_g^+$) can be emitted from the region near the needle electrode at the time of 10–20 ns. When the initial positive streamer develops sufficiently, the main streamer changes to diffuse mode. At this time, the discharge intensity arrives to the maximum (25 ns). However, since the local strong electrical field caused by the streamer breakdown would not exist, the $E/N$ decrease to a low level, the electron energy is insufficient to ionize the $N_2$ molecules, and the emission intensity of $N_2^+$($B^2\Sigma_u^+ \to X^2\Sigma_g^+$, 0–0) drops very low after t = 25 ns.

### 3.3. Time and Space Distribution of Vibrational Population of $N_2$ ($C^3\Pi_u$, $v = 0,1,2$)

In atmospheric pressure air plasma, some energy can be transfused and stored in the molecule vibrational energy levels by the electron impact vibrational excitation process. Since the energy level gap of the molecules vibrations is in the scale of one-tenth eV (for $N_2$ ($C^3\Pi_u$, v = 0) to $N_2$ ($C^3\Pi_u$, v = 1) is about 0.22 eV), which is in the intermediate state between electronic energy (in the scale of several eV) and rotation energy, the vibrational energies of $N_2$, $O_2$, etc., play an important role in plasma energy transfer and plasma chemical processes [43].

For the specified vibrational transition from upper level v′ to the lower level v″, the emission intensity $I_{v'v''}$ is proportional to the density of photon $n_f$, which can be obtained by following Equations [27]:

$$I_{v'v''} \propto dn_f/dt = N_{v'} (FC)_{v'v''} (R_e)_{v'v''} v_{v'v''} \tag{1}$$

$$I_{0\text{-}3}/I_{1\text{-}4} = N_0 (FC)_{0\text{-}3} (Re)_{0\text{-}3} v_{0\text{-}3}/N_1 (FC)_{1\text{-}4} (Re)_{1\text{-}4} v_{1\text{-}4} \tag{2}$$

Since $(Re)_{v'v''}$ is almost constant, the relative populations of $N_2$ ($C^3\Pi_u$, v = 1) can be expressed as:

$$N_1/N_0 = I_{1\text{-}4}(FC)_{1\text{-}4}\, v_{1\text{-}4}/I_{0\text{-}3}(FC)_{0\text{-}3} v_{0\text{-}3} \tag{3}$$

The relative populations of $N_2$ ($C^3\Pi_u$, v = 2) can also be calculated similarly.

Figure 7 shows the temporal distribution of the relative vibrational populations of $N_2$ ($C^3\Pi_u$, v = 1), $N_2$ ($C^3\Pi_u$, v = 2) with the population of $N_2$ ($C^3\Pi_u$, v = 0) normalized to 1. The discharge was operated at the pulse peak voltage, pulse repetition rate, and electrode gap distance of 26 kV, 100 Hz, and 5 mm, respectively. The time in the Figure is counted from the initial moment of the discharge.

During the discharge in each pulse, the populations of higher vibrational level $N_2$ ($C^3\Pi_u$) exhibits an obvious increase with the discharge time. At the time of 55 ns, the population of $N_2$ ($C^3\Pi_u$, v = 1) and $N_2$ ($C^3\Pi_u$, v = 2) increase to 0.39 and 0.14 (at 0 ns they are 0.34 and 0.096), respectively. Obviously, the increase of the population of $N_2$ ($C^3\Pi_u$, v = 2) between the times t = 0 and t = 55 ns is larger than that of $N_2$ ($C^3\Pi_u$, v = 1).

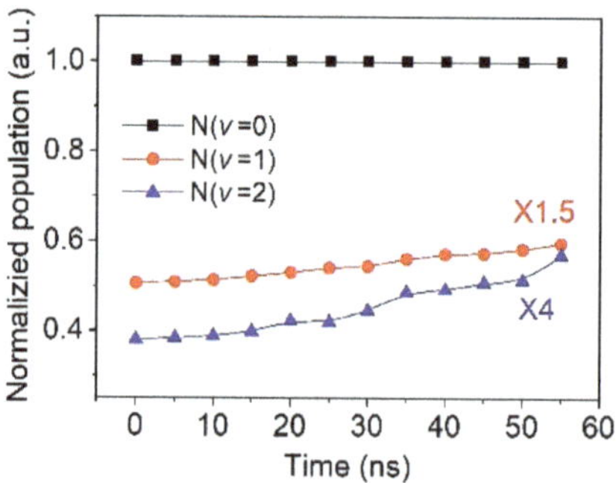


**Figure 7.** Temporal dependence of the relative vibrational populations of $N_2$ ($C^3\Pi_u$, v = 0), $N_2$ ($C^3\Pi_u$, v = 1), and $N_2$ ($C^3\Pi_u$, v = 2).

During the discharge, the excited molecules $N_2$ ($C^3\Pi_u$, v = 0, v = 1, v = 2) are generated by the electron impact excitation processes with ground state nitrogen molecules $N_2$ ($X^1\Sigma_g^+$).

$$N_2(X^1\Sigma_g^+) + e \rightarrow N_2(C^3\Pi_u, v = 0\text{--}3)\, k_{v'} \tag{4}$$

The reaction rate constants $k_{v'}$ for different vibrational levels v = 0, v = 1, v = 2 can be expressed as a function of the cross sections, which is influenced by the *E/N*. It should be noted that the distribution of the vibrational levels does not reach the equilibrium after the fast electron impact reaction. The overpopulations of high vibrational levels exist for the high electron temperature in NPDBD [44].

During the energy transfer processes, there are two approaches leading the obvious increase of vibrational populations of $N_2$ ($C^3\Pi_u$, v = 1) and $N_2$ ($C^3\Pi_u$, v = 2), i.e., the electron impact vibrational excitation (Equation (5)) and vibrational–vibrational (V–V) energy exchange (Equations (6) and (7)). Since the vibrational–rotational (V–R) energy exchange has a time scale of the order of microsecond to millisecond, it can be neglected during the discharge [45].

$$e + N_2\ (C^3\Pi_u, v = 0) \rightarrow e + N_2\ (C^3\Pi_u, v > 0) \tag{5}$$

$$N_2\ (C^3\Pi_u, v = 0) + N_2\ (v > 1) \rightarrow N_2\ (C^3\Pi_u, v = 1) + N_2\ (v\text{-}1) \tag{6}$$

$$N_2\ (C^3\Pi_u, v = 1) + N_2\ (v > 2) \rightarrow N_2\ (C^3\Pi_u, v = 2) + N_2\ (v\text{-}1) \tag{7}$$

Fundamentally, the energy of vibration excitation state of $N_2$ ($C^3\Pi_u$, v) is granted from the electron energy, so the population of vibrational excited state of nitrogen molecules can be availabel for some insight into electron temperature for an integrated time [34,46]. Figure 8 shows the spatial distribution of the relative vibrational populations of $N_2$ ($C^3\Pi_u$, v = 0), $N_2$ ($C^3\Pi_u$, v = 1), $N_2$ ($C^3\Pi_u$, v = 2). For a better comparison, the relative vibrational population of $N_2$ ($C^3\Pi_u$, v = 0) is normalized to 1. The positions of needle electrode and dielectric plate are 5 mm and 0 mm, marked with different color lump in the figure. The discharge is operated at 26 kV pulse peak voltage, 100 Hz pulse repetition rate, and 5 mm electrode gap distance.

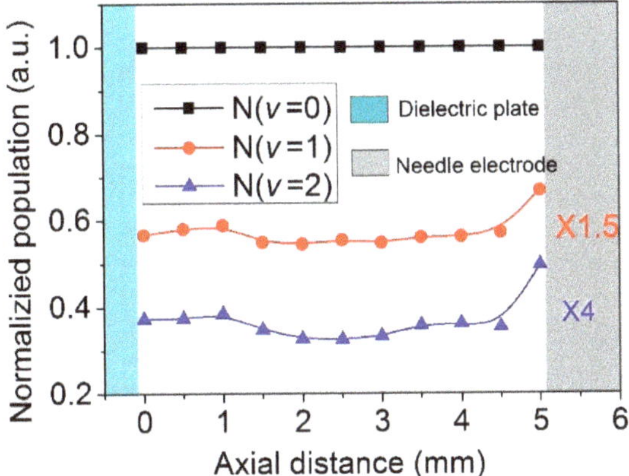

**Figure 8.** Spatial distribution of the relative vibrational populations of $N_2$ ($C^3\Pi_u$, v = 0), $N_2$ ($C^3\Pi_u$, v = 1), and $N_2$ ($C^3\Pi_u$, v = 2).

From the spatial distribution of the relative vibrational populations, it is shown that both the relative populations of $N_2$ ($C^3\Pi_u$, v' = 1) and $N_2$ ($C^3\Pi_u$, v' = 2) exhibit a maximum near the needle tip and decrease gradually with the distance from the needle electrode. However, near the plate electrode, the populations of $N_2$ ($C^3\Pi_u$, v' = 1) and $N_2$ ($C^3\Pi_u$, v' = 2) present a slight increase. Since the vibration distribution of $N_2$ is mainly caused by the electron impact vibration excitation in our experiment, the populations of $N_2$ ($C^3\Pi_u$, v' = 1) and $N_2$ ($C^3\Pi_u$, v' = 2) can show the maps of *E/N*. In addition, near the needle electrode, the high *E/N* is beneficial to the breakdown of the gas gap and the slight increase of the electric field near the dielectric plate, which can excite the surface discharge on the dielectric plate.

### 3.4. Calculation of Reduced Electric Field in NPDBD

In NPDBD at atmospheric air, it is proposed that the *E/N* can be calculated using the intensity ratio of second positive system (SPS) of $N_2$ to first negative system (FNS) of $N_2^+$ [36]. Therefore, the evolution of *E/N* can be represented by the temporal–spatial emission spectra ratio of $N_2$ ($C^3\Pi_u \to B^3\Pi_g$) and $N_2^+$($B^2\Sigma_u^+ \to X^2\Sigma_g^+$) [37]. For the detected spectra of $N_2$ ($C^3\Pi_u \to B^3\Pi_g$, v', v'') and $N_2^+$($B^2\Sigma_u^+ \to X^2\Sigma_g^+$,0–0, 391.4 nm), the corresponding population of upper vibrational excited state $N_2$ ($C^3\Pi_u$, v = 0–3) and $N_2^+$($B^2\Sigma_u^+$, v = 0) are determined by the electron impact vibrational excitation from ground state nitrogen molecule $N_2$ ($X^1\Sigma_g^+$) by the Equations (4) and (8):

$$e + N_2(X^1\Sigma_g^+) \to N_2^+(B^2\Pi_u^+, v = 0)\, k_i \tag{8}$$

The reaction rates $k_{v'}$ (v' = 0–3) and $k_i$ can be calculated by the cross sections from BOLSIG$^+$ and database LXcat as Equation (9):

$$k = \gamma \int_0^\infty \varepsilon\sigma(\varepsilon)F_0 d\varepsilon, \tag{9}$$

where $\gamma = (2e/m)^{1/2}$ is a constant, $\varepsilon = (v/\gamma)^2$ is the electron energy in eV, in which v is electron velocity, and the function $F_0$ is the isotropic part of electron distribution $F$. For the cross sections $\sigma(\varepsilon)$, it can be obtained by the relationship with the reduce electric field (*E/N*) in function (8):

$$-\frac{\gamma e^2}{3}\left(\frac{E}{N}\right)^2 \frac{d}{d\varepsilon}\left(\frac{\varepsilon}{Q}\frac{dF_0}{d\varepsilon}\right) = C_0[F_0, \{\sigma(\varepsilon)\}] - \frac{\Omega}{NC}\varepsilon^{1/2}F_0, \tag{10}$$

where the momentum cross section $Q$ can be defined as $Q = \sigma(\varepsilon) + \Omega/N\gamma\varepsilon^{1/2}$, the energy distribution $F_0$ is constant in time and space, and $C_0$ is the change in $F_0$ due to collisions [47]. For an accurate calculation, the cross sections of the vibrational structure of $N_2$ ($C^3\Pi_u$, $v = 0$–3) was calculated using the Frank Condon factors of the $N_2$ ($C^3\Pi_u \rightarrow B^3\Pi_g$) listed in Table 1.

**Table 1.** Franck-Condon factors for the $C^3\Pi_u$-$B^3\Pi_g$ (Second Positive System) [45].

| v″ | v′ = 0 | v′ = 1 | v′ = 2 | v′ = 4 | v′ = 5 |
|---|---|---|---|---|---|
| 0 | $4.55 \times 10^{-1}$ | $3.88 \times 10^{-1}$ | $1.34 \times 10^{-1}$ | $2.16 \times 10^{-2}$ | $1.16 \times 10^{-3}$ |
| 1 | $3.31 \times 10^{-1}$ | $2.92 \times 10^{-2}$ | $3.35 \times 10^{-1}$ | $2.52 \times 10^{-1}$ | $5.66 \times 10^{-2}$ |
| 2 | $1.45 \times 10^{-1}$ | $2.12 \times 10^{-1}$ | $2.30 \times 10^{-2}$ | $2.04 \times 10^{-1}$ | $3.26 \times 10^{-1}$ |
| 3 | $4.94 \times 10^{-2}$ | $2.02 \times 10^{-1}$ | $6.91 \times 10^{-2}$ | $8.81 \times 10^{-2}$ | $1.13 \times 10^{-1}$ |
| 4 | $1.45 \times 10^{-2}$ | $1.09 \times 10^{-1}$ | $1.69 \times 10^{-1}$ | $6.56 \times 10^{-3}$ | $1.16 \times 10^{-1}$ |
| 5 | $3.87 \times 10^{-3}$ | $4.43 \times 10^{-2}$ | $1.41 \times 10^{-1}$ | $1.02 \times 10^{-1}$ | $2.45 \times 10^{-3}$ |
| 6 | $9.68 \times 10^{-4}$ | $1.52 \times 10^{-2}$ | $7.72 \times 10^{-2}$ | $1.37 \times 10^{-1}$ | $4.70 \times 10^{-2}$ |

For the depopulation, the $N_2$ ($C^3\Pi_u$, $v = 0$–3) and $N_2^+$ ($B^2\Sigma_u^+$, $v = 0$) can be quenched by the spontaneous radiative depopulation and collisions with heavy particles by Equations (11)–(16):

$$N_2(C^3\Pi_u, v\prime) \rightarrow N_2(B^3\Pi_u, v\prime\prime) + h\nu_{v\prime v\prime\prime} \quad k = 1/\tau_{v\prime v\prime\prime} \tag{11}$$

$$N_2(C^3\Pi_u, v\prime = 0 - 3) + N_2 \rightarrow \text{products } k_{q,N_2} \tag{12}$$

$$N_2(C^3\Pi_u, v\prime = 0 - 3) + O_2 \rightarrow \text{products } k_{q,O_2} \tag{13}$$

$$N_2^+(B^2\Sigma_u^+, v\prime = 0) + N_2 \rightarrow \text{products } k_{q,N_2} \tag{14}$$

$$N_2^+(B^2\Sigma_u^+, v\prime = 0) + O_2 \rightarrow \text{products } k_{q,O_2} \tag{15}$$

$$N_2^+(B^2\Sigma_u^+, v\prime = 0) + N_2 + M \rightarrow N_4^+ + M \ k_{conv} \tag{16}$$

The radiative lifetimes and deactivation rate constants of different separated vibrational state of $N_2$ ($C^3\Pi_u$) and $N_2^+$ ($B^2\Sigma_u^+$) are shown in Table 2:

**Table 2.** Radiative and deactivation parameters of $N_2$ ($C^3\Pi_u$) and $N_2^+$ ($B^2\Sigma_u^+$).

| | $N_2$ ($C^3\Pi_u$, v = 0) | $N_2$ ($C^3\Pi_u$, v = 1) | $N_2$ ($C^3\Pi_u$, v = 2) | $N_2$ ($C^3\Pi_u$, v = 3) | $N_2^+$ ($B^2\Sigma_u^+$, v = 0) |
|---|---|---|---|---|---|
| $\tau$ (ns) [37] | 42 | 41 | 39 | 41 | 62 |
| $A_{v\prime v\prime\prime}$ | 0.05 (0–3) | 0.11 (1–4) | 0.14 (2–5) | 0.14 (3–6) | 0.72 (0–0) |
| $k_{q,N2}$ ($10^{-10}$ cm$^3$·s$^{-1}$) | 0.13 | 0.29 | 0.46 | 0.43 | 2.1 |
| $k_{q,N2}$ ($10^{-10}$ cm$^3$·s$^{-1}$) | 3.0 | 3.1 | 3.7 | 4.3 | 5.1 |
| $k_{conv,}$ ($10^{-29}$ cm$^3$·s$^{-1}$) | - | - | - | - | 5.0 |
| $g$ | 0.012 | $1.0 \times 10^{-2}$ | $8.3 \times 10^{-3}$ | $7.3 \times 10^{-3}$ | $2.2 \times 10^{-3}$ |

Therefore, the change in the excited state particle [$N_{exc}$] concentrations of $N_2$ ($C^3\Pi_u$, $v\prime = 0,1,2$) and $N_2^+$ ($B^2\Sigma_u^+$) can be expressed as the production of the electron impact processes and the depopulation as (17a) and (17b), where the associative conversion by three-body collisions (Equation (16)) is considered in the quenching of the nitrogen ions:

$$\frac{d[N_{exc}]}{dt} = k_{exc}n_e[N_2] - \frac{1}{\tau}[N_{exc}] - k_{q,N_2}[N_2][N_{exc}] - k_{q,O_2}[O_2][N_{exc}] \tag{17a}$$

$$\frac{d[N_{exc}]}{dt} = k_{exc}n_e[N_2] - \frac{1}{\tau}[N_{exc}] - k_{q,N_2}[N_2][N_{exc}] - k_{q,O_2}[O_2][N_{exc}] \\ - k_{conv}[N_2][M][N_{exc}] \tag{17b}$$

For the selected vibrational transition of SPS of $N_2$ and FNS of $N_2{}^+$, the emission intensity expresses as:

$$I_{v'v''} = h\nu_{v'v''}N_{v'v''}\tau_{v'v''}A_{v'v''} = h\nu_{v'v''}n_e[N_2]gA_{v'v''}, \tag{18}$$

where the probability of $A_{v'v''}$ is calculated by $q_{v'v''}$ using function $A_{v'v''} = \frac{q_{v'v''}v_{v'v''}^3}{\Sigma_{v''}q_{v'v''}v_{v'v''}^3}$, the corrected g-function can be expressed by $g = (1 - \tau(k_{q,N_2}[N_2] + k_{q,O_2}[O_2])[M]\{+k_{conv}[N_2][M]^2\})^{-1}$, and the number density of $[N_2]$ and $[O_2]$ are estimated to be equal to $2.12 \times 10^{19}$ cm$^{-3}$ and $5.6 \times 10^{18}$ cm$^{-3}$ at atmospheric air.

Thus, the intensity ratio $R_{391}/R_{v'v''}$ is expressed as:

$$R_{391}/R_{v'v''} = I_{391}/I_{v'v''} = \left(\frac{\lambda_{391}}{\lambda_{v'v''}}\right)^{-1} \frac{A_{391}}{A_{v'v''}} \frac{g_{391}}{g_{v'v''}} \frac{k_{391}}{k_{v'v''}} \tag{19}$$

As described in Equations (8) and (9), the excitation rate constants of electron impact processes depend on the $E/N$ only, that means, the intensity ratios of $I_{391}/I_{405}$, $I_{391}/I_{400}$, and $I_{391}/I_{394}$, etc., can be used to calculate the reduce electric field. The measured intensity ratios for these vibrational bands at a position of 5 mm from the needle electron tips and at the time of 15 ns are marked in Figure 9, which are about 420 Td, 400 Td, and 395 Td, respectively.

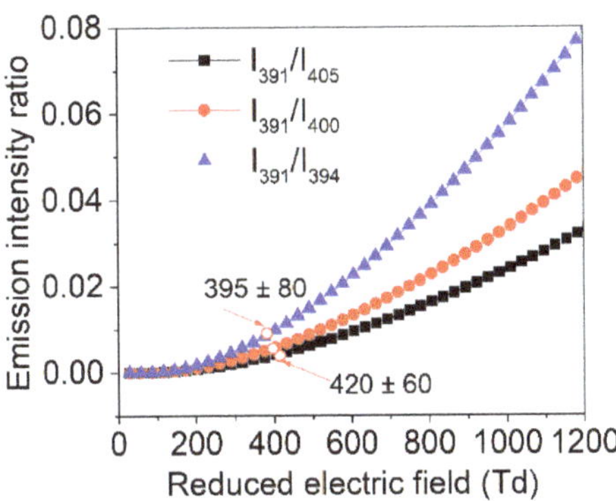


**Figure 9.** The intensity ratios of $I_{391}/I_{405}$, $I_{391}/I_{400}$, and $I_{391}/I_{394}$ as function of $E/N$.

*3.5. Temporal Evolution of Reduced Electric Field E/N*

Figure 10 shows the temporal evolution of $E/N$ at various positions (0 mm, 2.5 mm, 5 mm from the plate electrode) in NPDBD. In the measurement, the pulse peak voltage, pulse repetition rate, and electrode gap are also kept at 26 kV, 100 Hz, and 5 mm, respectively. It shows that the curves of $E/N$ present different tendencies compared with the waveform of pulse voltage. In the region near the needle tip (5 mm from plate electrode), the $E/N$ presents a maximum at 5 ns, which is about 590 ± 80 Td. When the initial streamer channel was formed, the $E/N$ decreases sharply in the discharge duration accompanied with the sharply increase of plasma optical emission intensity. In the period of 15–35 ns, the $E/N$ is about 270–420 Td, which is only about 1/2–2/3 compared with the $E/N$ at 5 ns. In the study of Fridman [48], the $E/N$ of diffuse regime in NPDBD is only about 1/2 of the streamer regime. Therefore, the low $E/N$ at t = 15 ns indicates that mode transition from streamer regime to diffuse regime is accomplished. At the central position of the electrode gap (2.5 mm from plate electrode), the $E/N$ is much weaker compared with that near the needle tip. The distorted high $E/N$ cannot be observed, instead, the maximum $E/N$ is about 370 Td at the time of 10 ns, then the $E/N$ decreases gradually with the transition from streamer to diffuse discharge. At the time of 25 ns, the $E/N$ at the central position is almost kept consistent with the $E/N$ at the position near the needle tip. Near the surface of dielectric plate (0 mm), the $E/N$ presents a completely different evolution tendency. For the discharge duration of 5–25 ns, it decreases gradually and keeps a low value (220–350 Td). However, once the diffuse discharge extinguished and the surface discharge begins to propagate to the outside direction (t = 30–50 ns), the $E/N$ increases with the duration of discharge time obviously. At the time of 50 ns, the $E/N$ on the surface of dielectric plate is about 365 Td, which is about 100 Td higher than the $E/N$ at the positions of needle tip and 50 Td higher than that at central place of electrode gap.

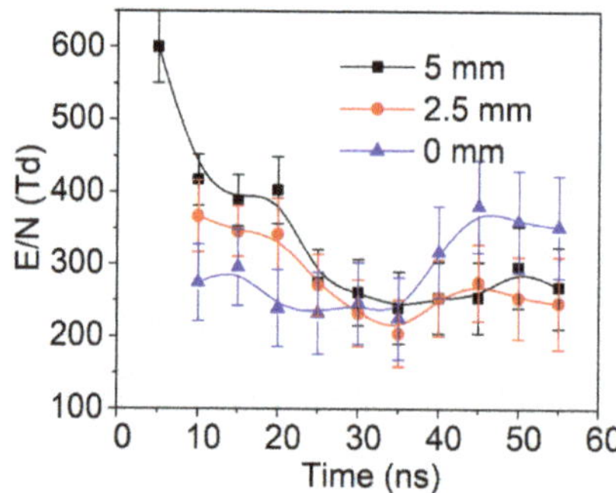

**Figure 10.** The temporal evolution of $E/N$ at 0 mm, 2.5 mm, 5 mm from the plate electrode in NPDBD.

In the NPDBD, the $E/N$ is determined by the overlap of applied pulsed electric field and the built-in electric field formed by the space charge in the plasma region and memory charge on the surface of dielectric plate. At the initial time, a distorted high $E/N$ is formed near the needle electrode due to the extremely asymmetrical electrode configuration, which excites the initial streamer from needle electrode to plate electrode. Caused by the high conductivity in the streamer channel, the $E/N$ decreases sharply when the gas gap is broken down. Once the diffuse discharge forms, the $E/N$ in the whole electrode gap is rebuilt, it almost equally distributes in the axial distance. The memory charges on the surface of dielectric plate can be erased by the plasma, which can build an electric field along dielectric plate. This horizontal electric field drives the surface barrier discharge propagating in the radial direction after the volume discharge extinguished.

## 4. Conclusions

Evolution dynamic process in a discharge pulse is observed by one-shot ICCD images. Three main stages in NPDBD are distinguished, which are the streamer breakdown from needle tip to plate electrode, the regime transition from streamer to diffuse, and the propagation of surface discharge on the plate electrode surface. At the beginning of the discharge, the $E/N$ near the needle tip can be estimated to about 590 Td. This high $E/N$ excites the initial breakdown as a positive streamer regime. The streamer builds up a new electric field with radial direction and pre-photoionizes the surrounding air to drive the subsequent breakdown. Hence, an abundance of fine secondary streamer channels around the initial streamer form at t = 10 ns, contributing to the transition to diffuse regime. By measuring the temporal–spatial resolved spectra, it is found that the spectra of $N_2^+(B^2\Sigma_u^+ \rightarrow X^2\Sigma_g^+)$ and $N_2$ ($C^3\Pi_u \rightarrow B^3\Pi_g$) present obvious different evolution tendencies. The band of $N_2^+(B^2\Sigma_u^+ \rightarrow X^2\Sigma_g^+)$, indicator of high $E/N$, is mainly emitted from the region near the needle tip in the initial period of the breakdown process. The energy distribution, the relative vibration population of $N_2$ ($C^3\Pi_u$, $v = 0,1,2$), and the $E/N$ are calculated. The populations of $N_2$ ($C^3\Pi_u$, v = 1,2) increase with the discharge duration time. The increase of $N_2$ ($C^3\Pi_u$, v = 2) is larger than that of $N_2$ ($C^3\Pi_u$, v = 1). The evolutions of $E/N$ are calculated using the temporal–spatial resolved spectra of $N_2^+(B^2\Sigma_u^+ \rightarrow X^2\Sigma_g^+)$ and $N_2$ ($C^3\Pi_u \rightarrow B^3\Pi_g$). It is found that $E/N$ near the plate electrode (0 mm), at the middle of the electrode gap (2.5 mm), near the needle electrode (5 mm), present different tendencies with the waveform of pulse voltage. A distorted high $E/N$ can be observed near the needle electrode at 5 ns. At the time of 10–25 ns, the $E/N$ decreases to about 270–320 Td, which indicates that mode transition from streamer regime to diffuse regime is accomplished. Near the surface of dielectric plate, the $E/N$ decreases gradually and keeps a low value at the time of 10–25 ns, but increases obviously after the diffuse discharge extinguished (at the time of about 30–50 ns). At the time of 50 ns, the $E/N$ on the surface of dielectric plate is about 100 Td higher than the $E/N$ at the positions of needle tip and 50 Td higher than that at central place of electrode gap. It drives the surface barrier discharge propagating in the radial direction along the dielectric plate.

**Author Contributions:** Conceptualization, D.Y. and L.Z.; methodology, L.Z.; formal analysis, S.W. and H.Y.; investigation, L.Z., D.Y. and Z.J; data curation, Z.Z. and S.W.; writing—original draft preparation, L.Z. and D.Y; writing—review and editing, Z.J., D.Y. and W.W.

**Funding:** This research was funded by National Key R&D Program of China (2016YFC0207201), the Science and Technology on High Power Microwave Laboratory Fund (JCKYS2018212036), the National Natural Science Foundation of China (Grant No. 11965018), and Fundamental Research Funds for the Central Universities (Grant No. DUT18LK42).

**Conflicts of Interest:** The authors declare no conflict of interest.

## References


1. Pai, D.Z.; Lacoste, D.A.; Laux, C.O. Transitions between corona, glow, and spark regimes of nanosecond repetitively pulsed discharges in air at atmospheric pressure. *J. Appl. Phys.* **2010**, *107*, 093303. [CrossRef]
2. Ito, T.; Kobayashi, K.; Czarnetzki, U.; Hamaguchi, S. Rapid formation of electric field profiles in repetitively pulsed high-voltage high-pressure nanosecond discharges. *J. Phys. D Appl. Phys.* **2010**, *43*, 062001. [CrossRef]
3. Zhang, L.; Yang, D.; Wang, W.; Wang, S.; Yuan, H.; Zhao, Z.; Sang, C.; Jia, L. Needle-array to Plate DBD Plasma Using Sine AC and Nanosecond Pulse Excitations for Purpose of Improving Indoor Air Quality. *Sci. Rep.* **2016**, *6*, 25242. [CrossRef] [PubMed]
4. Kosarev, I.N.; Khorunzhenko, V.I.; Mintoussov, E.I.; Sagulenko, P.N.; Popov, N.A.; Starikovskaia, S.M. A nanosecond surface dielectric barrier discharge at elevated pressures: Time-resolved electric field and efficiency of initiation of combustion. *Plasma Sources Sci. Technol.* **2012**, *21*, 45012. [CrossRef]
5. Blin-Simiand, N.; Pasquiers, S.; Jorand, F.; Postel, C.; Vacher, J.R. Removal of formaldehyde in nitrogen and in dry air by a DBD: Importance of temperature and role of nitrogen metastable states. *J. Phys. D Appl. Phys.* **2009**, *42*, 122003. [CrossRef]
6. Wang, S.; Yang, D.Z.; Wang, W.C.; Zhang, S.; Liu, Z.J.; Tang, K.; Song, Y. An atmospheric air gas-liquid diffuse discharge excited by bipolar nanosecond pulse in quartz container used for water sterilization. *Appl. Phys. Lett.* **2013**, *103*, 2011–2015. [CrossRef]
7. Bubnov, A.G.; Grinevich, V.I.; Kuvykin, N.A.; Maslova, O.N. The Kinetics of Plasma-Induced Degradation of Organic Pollutants in Sewage Water. *High Energy Chem.* **2004**, *38*, 41–45. [CrossRef]
8. Miron, C.; Hulubei, C.; Sava, I.; Quade, A.; Steuer, A.; Weltmann, K.; Kolb, J.F. Polyimide Film Surface Modification by Nanosecond High Voltage Pulse Driven Electrical Discharges in Water. *Plasma Process. Polym.* **2015**, *12*, 734–745. [CrossRef]
9. Yuan, H.; Wang, W.; Yang, D.; Zhou, X.; Zhao, Z.; Zhang, L.; Wang, S.; Feng, J. Hydrophilicity modification of aramid fiber using a linear shape plasma excited by nanosecond pulse. *Surf. Coat. Technol.* **2018**, *344*, 614–620. [CrossRef]
10. Pendleton, S.J.; Kastner, J.; Gutmark, E.; Gundersen, M.A. Surface Streamer Discharge for Plasma Flow Control Using Nanosecond Pulsed Power. *IEEE Trans. Plasma Sci.* **2011**, *39*, 2072–2073. [CrossRef]
11. Gherardi, M.; Turrini, E.; Laurita, R.; Gianni, E.D.; Ferruzzi, L.; Liguori, A.; Stancampiano, A.; Colombo, V.; Fimognari, A. Atmospheric Non-Equilibrium Plasma Promotes Cell Death and Cell-Cycle Arrest in a Lymphoma Cell Line. *Plasma Process. Polym.* **2016**, *12*, 1354–1363. [CrossRef]
12. Lu, P.; Kim, D.W.; Park, D.W. Silver nanoparticle-loaded filter paper: Innovative assembly method by nonthermal plasma and facile application for the reduction of methylene blue. *Surf. Coat. Technol.* **2019**, *366*, 7–14. [CrossRef]

13. Sun, D.L.; Hong, R.Y.; Wang, F.; Liu, J.Y.; Rajesh Kumar, M. Synthesis and modification of carbon nanomaterials via AC arc and dielectric barrier discharge plasma. *Chem. Eng. J.* **2016**, *283*, 9–20. [CrossRef]

14. Wang, L.; Yi, Y.H.; Guo, H.C.; Du, X.M.; Zhu, B.; Zhu, Y.M. Highly dispersed co nanoparticles prepared by an improved method for plasma-driven NH 3 decomposition to produce H 2. *Catalysts* **2019**, *9*, 1–13.

15. Yang, D.Z.; Yang, Y.; Li, S.Z.; Nie, D.X.; Zhang, S.; Wang, W.C. A homogeneous dielectric barrier discharge plasma excited by a bipolar nanosecond pulse in nitrogen and air. *Plasma Sources Sci. Technol.* **2012**, *21*, 035004. [CrossRef]

16. Tarasenko, V.F. Nanosecond discharge in air at atmospheric pressure as an X-ray source with high pulse repetition rates. *Appl. Phys. Lett.* **2006**, *88*, 601. [CrossRef]

17. Aleksandrov, N.L.; Kindysheva, S.V.; Nudnova, M.M.; Starikovskiy, A.Y. Mechanism of ultra-fast heating in a non-equilibrium weakly ionized air discharge plasma in high electric fields. *J. Phys. D Appl. Phys.* **2010**, *43*, 255201. [CrossRef]

18. Tarasenko, V.F.; Baksht, E.K.; Burahenko, A.G.; Shut'ko, Y.V. Diffuse discharge, runaway electron, and X-ray in atmospheric pressure air in an inhomogeneous electrical field in repetitive pulsed modes. *Appl. Phys. Lett.* **2011**, *98*, 021503. [CrossRef]

19. Anikin, N.B.; Zavialova, N.A.; Starikovskaia, S.M.; Starikovskii, A.Y. Nanosecond-Discharge Development in Long Tubes. *IEEE Trans. Plasma Sci.* **2008**, *36*, 902–903. [CrossRef]

20. Yatom, S.; Shlapakovski, A.; Beilin, L.; Stambulchik, E.; Tskhai, S.; Krasik, Y.E. Recent studies on nanosecond-timescale pressurized gas discharges. *Plasma Sources Sci. Technol.* **2016**, *25*, 064001. [CrossRef]

21. Yan, K.; Li, R.; Zhu, T.; Zhang, H.; Hu, X.; Jiang, X.; Liang, H.; Qiu, R.; Wang, Y. A semi-wet technological process for flue gas desulfurization by corona discharges at an industrial scale. *Chem. Eng. J.* **2006**, *116*, 139–147. [CrossRef]

22. Zhang, Y.; Wang, H.; Jiang, W.; Bogaerts, A. Two-dimensional particle-in cell/Monte Carlo simulations of a packed-bed dielectric barrier discharge in air at atmospheric pressure. *New. J. Phys.* **2015**, *17*, 083056. [CrossRef]

23. Shao, T.; Tarasenko, V.F.; Zhang, C.; Lomaev, M.I.; Sorokin, D.A.; Yan, P.; Kozyrev, A.V.; Baksht, E.K. Spark discharge formation in an inhomogeneous electric field under conditions of runaway electron generation. *J. Appl. Phys.* **2012**, *111*, 023304.

24. Lu, X.; Naidis, G.V.; Laroussi, M.; Ostrikov, K. Guided ionization waves: Theory and experiments. *Phys. Rep.* **2014**, *540*, 123–166. [CrossRef]

25. Raizer, Y.P. *Gas. Discharge Physics*; Springer-Verlag: Berlin, Germany, 1991.

26. Pai, D.Z.; Lacoste, D.A.; Laux, C.O. Nanosecond repetitively pulsed discharges in air at atmospheric pressure-the spark regime. *Plasma Sources Sci. Technol.* **2010**, *19*, 065015. [CrossRef]

27. Wang, W.; Wang, S.; Liu, F.; Zheng, W.; Wang, D. Optical study of OH radical in a wire-plate pulsed corona discharge. *Spectrochim. Acta Part A Mol. Biomol. Spectrosc.* **2006**, *63*, 477–482. [CrossRef]

28. Babich, L.P.; Loĭko, T.V.; Tsukerman, V.A. High-voltage nanosecond discharge in a dense gas at a high overvoltage with runaway electrons. *Sov. Phys. Uspekhi* **1990**, *33*, 521–540. [CrossRef]

29. Levko, D.; Yatom, S.; Vekselman, V.; Gleizer, J.Z.; Gurovich, V.T.; Krasik, Y.E. Numerical simulations of runaway electron generation in pressurized gases. *J. Appl. Phys.* **2012**, *111*, R265. [CrossRef]

30. Lo, A.; Cessou, A.; Lacour, C.; Lecordier, B.; Boubert, P.; Xu, D.A.; Laux, C.O.; Vervisch, P. Streamer-to-spark transition initiated by a nanosecond overvoltage pulsed discharge in air. *Plasma Sources Sci. Technol.* **2017**, *26*, 045012. [CrossRef]

31. Stepanyan, S.A.; Starikovskiy, A.Y.; Popov, N.A.; Starikovskaia, S.M. A nanosecond surface dielectric barrier discharge in air at high pressures and different polarities of applied pulses: Transition to filamentary mode. *Plasma Sources Sci. Technol.* **2017**, *23*, 045003. [CrossRef]

32. Shcherbanev, S.A.; Khomenko, A.Y.; Stepanyan, S.A.; Popov, N.A.; Starikovskaia, S.M. Optical emission spectrum of filamentary nanosecond surface dielectric barrier discharge. *Plasma Sources Sci. Technol.* **2017**, *26*, 02LT01. [CrossRef]

33. Pai, D.Z.; Stancu, G.D.; Lacoste, D.A.; Laux, C.O. Nanosecond repetitively pulsed discharges in air at atmospheric pressure-the glow regime. *Plasma Sources Sci. Technol.* **2009**, *18*, 045030. [CrossRef]

34. Ito, T.; Kanazawa, T.; Hamaguchi, S. Rapid Breakdown Mechanisms of Open Air Nanosecond Dielectric Barrier Discharges. *Phys. Rev. Lett.* **2011**, *107*, 065002. [CrossRef] [PubMed]

*Nanomaterials* **2019**, *9*, 1381

35. Zhang, Y.; Li, J.; Jiang, N.; Shang, K.F.; Lu, N.; Wu, Y. Optical characteristics of the filamentary and diffuse modes in surface dielectric barrier discharge. *Spectrochim. Acta Part A Mol. Biomol. Spectrosc.* **2016**, *168*, 230–234. [CrossRef] [PubMed]

36. Paris, P.; Aints, M.; Valk, F.; Plank, T.; Haljaste, A.; Kozlov, K.V.; Wagner, H.E. REPLY: Reply to comments on "Intensity ratio of spectral bands of nitrogen as a measure of electric field strength in plasmas". *J. Phys. D Appl. Phys.* **2006**, *38*, 3894–3899. [CrossRef]

37. Pancheshnyi, S.V.; Starikovskaia, S.M.; Starikovskii, A.Y. Collisional deactivation of $N_2(C_3\Pi$ u, v = 0, 1, 2, 3) states by $N_2$, $O_2$, $H_2$ and $H_2O$ molecules. *Chem. Phys.* **2000**, *262*, 349–357. [CrossRef]

38. Shao, T.; Long, K.; Zhang, C.; Yan, P.; Zhang, S.; Pan, R. Experimental study on repetitive unipolar nanosecond-pulse dielectric barrier discharge in air at atmospheric pressure. *J. Phys. D Appl. Phys.* **2008**, *41*, 215203.

39. Liu, S.H.; Neiger, M. Excitation of dielectric barrier discharges by unipolar submicrosecond square pulses. *J. Phys. D Appl. Phys.* **2001**, *34*, 1632–1638. [CrossRef]

40. Wu, S.; Lu, X.; Liu, D.; Yang, Y.; Pan, Y.; Ostrikov, K. Photo-ionization and residual electron effects in guided streamers. *Phys. Plasmas* **2014**, *21*, 103508. [CrossRef]

41. Nijdam, S.; Geurts, C.G.C.; Van Veldhuizen, E.M.; Ebert, U. Reconnection and merging of positive streamers in air. *J. Phys. D Appl. Phys.* **2009**, *42*, 045201. [CrossRef]

42. Nijdam, S.; Takahashi, E.; Teunissen, J.; Ebert, U. Streamer discharges can move perpendicularly to the electric field. *New. J. Phys.* **2014**, *16*, 103038. [CrossRef]

43. Shkurenkov, I.; Adamovich, I.V. Energy balance in nanosecond pulse discharges in nitrogen and air. *Plasma Sources Sci. Technol.* **2016**, *25*, 015021. [CrossRef]

44. Shkurenkov, I.; Burnette, D.; Lempert, W.R.; Adamovich, I.V. Kinetics of excited states and radicals in a nanosecond pulse discharge and afterglow in nitrogen and air. *Plasma Sources Sci. Technol.* **2014**, *23*, 065003. [CrossRef]

45. Suchard, S.N.; Melzer, J.E. (Eds.) *Spectroscopic Data*; Springer: New York, NY, USA, 1976.

46. Zhang, S.; Wang, W.; Jia, L.; Liu, Z.; Yang, Y.; Dai, L. Rotational, Vibrational, and Excitation Temperatures in Bipolar Nanosecond-Pulsed Diffuse Dielectric-Barrier-Discharge Plasma at Atmospheric Pressure. *IEEE Trans. Plasma Sci.* **2013**, *41*, 350–354. [CrossRef]

47. Hagelaar, G.J.M.; Pitchford, L.C. Solving the Boltzmann equation to obtain electron transport coefficients and rate coefficients for fluid models. *Plasma Sources Sci. Technol.* **2005**, *14*, 722–733. [CrossRef]

48. Liu, C.; Dobrynin, D.; Fridman, A. Uniform and non-uniform modes of nanosecond-pulsed dielectric barrier discharge in atmospheric air: Fast imaging and spectroscopic measurements of electric field. *J. Phys. D Appl. Phys.* **2014**, *47*, 252003. [CrossRef] [PubMed]

*Review*

# A Review on the Promising Plasma-Assisted Preparation of Electrocatalysts

**Feng Yu [1], Mincong Liu [1], Cunhua Ma [1], Lanbo Di [2], Bin Dai [1],* and Lili Zhang [3],***

[1]  School of Chemistry and Chemical Engineering, Shihezi University, Shihezi 832003, China;
    yufeng05@mail.ipc.ac.cn (F.Y.); qdulmc@163.com (M.L.); mchua@shzu.edu.cn (C.M.)
[2]  College of Physical Science and Technology, Dalian University, Dalian 116622, China; dilanbo@163.com
[3]  Institute of Chemical and Engineering Sciences, Agency for Science, Technology and Research,
    Jurong Island 627833, Singapore
*   Correspondence: db_tea@shzu.edu.cn (B.D.); zhang_lili@ices.a-star.edu.sg (L.Z.); Tel.: +86-993-205-7272 (B.D.)

Received: 29 August 2019; Accepted: 3 October 2019; Published: 10 October 2019

**Abstract:** Electrocatalysts are becoming increasingly important for both energy conversion and environmental catalysis. Plasma technology can realize surface etching and heteroatom doping, and generate highly dispersed components and redox species to increase the exposure of the active edge sites so as to improve the surface utilization and catalytic activity. This review summarizes the recent plasma-assisted preparation methods of noble metal catalysts, non-noble metal catalysts, non-metal catalysts, and other electrochemical catalysts, with emphasis on the characteristics of plasma-assisted methods. The influence of the morphology, structure, defect, dopant, and other factors on the catalytic performance of electrocatalysts is discussed.

**Keywords:** electrocatalyst; plasma; defect rich; surface etching; heteroatom doping

## 1. Introduction

Tremendous research efforts have been put to address global energy and environmental concerns. One of the promising energy solutions is electrocatalytic technology, such as fuel cells, the hydrogen production reaction, $CO_2$ recycling, and ammonia synthesis (Figure 1) [1,2].

Electrocatalytic technology mainly involves two categories: (a) the electrocatalytic oxidation reaction; i.e., the oxygen evolution reaction (OER), hydrogenation reaction (HOR), and methanol oxidation reaction (MOR); and (b) the electrocatalytic reduction reaction; i.e., the oxygen reduction reaction (ORR), $CO_2$ reduction reaction ($CO_2$RR), $N_2$ reduction reaction (NRR), hydrogen evolution reaction (HER), etc. Electrocatalytic reactions can be carried out at atmospheric temperature and pressure. For example, sustainable ammonia synthesis is achieved under carbon-free conditions using $N_2$ and $H_2O$ as the raw materials; the electrochemical reduction of $CO_2$ is used to prepare chemicals with high energy densities, such as C1 and Cn, electrochemically converting electrical energy into an efficient and clean energy carrier-hydrogen energy [3,4]. In addition, chemical energy can be converted into electrical energy using HOR, MOR, or ORR in a fuel cell. ORR and OER can also be used as reversible half-reactions for rechargeable metal-oxygen batteries, using metals, such as Li, Mg, and Zn as energy carriers rather than hydrogen [5].

Currently, noble metal catalysts (such as Au, Pt, Ru, and Ag) are the most important materials for electrocatalysts. They are characterized by easy adsorption of reactants on the surface, the ability to form active intermediates, excellent acid-base corrosion resistance, and outstanding catalytic activity, selectivity, and stability. Compared with noble metal electrocatalysts, inexpensive non-noble metal electrocatalysts also have excellent electrocatalytic performance, especially for transition metal alloys and compounds (oxides, sulfides, nitrides, phosphides, and carbides) [6]. Transition metals can bond with the reaction molecules to form a transition state with lower energy barrier, thus reducing the activation energy of the whole reaction path and accelerating the chemical reaction [7]. Considering that there are many kinds of non-noble metal electrocatalysts, electrocatalysts with special chemical compositions and physical structures can also be designed and prepared to meet the needs of different electrocatalytic reactions [8]. The excellent acid-base corrosion resistance of non-metallic electrocatalysts have attracted extensive attention of researchers, especially for carbon-based electrocatalysts. Conventional carbon-based molecules have no hollow orbitals, which makes it difficult to participate in the electrocatalytic reaction. However, the electrocatalytic performance of carbon-based materials can be effectively improved by introducing other heteroatoms or functional groups on the surface of carbon materials [9].

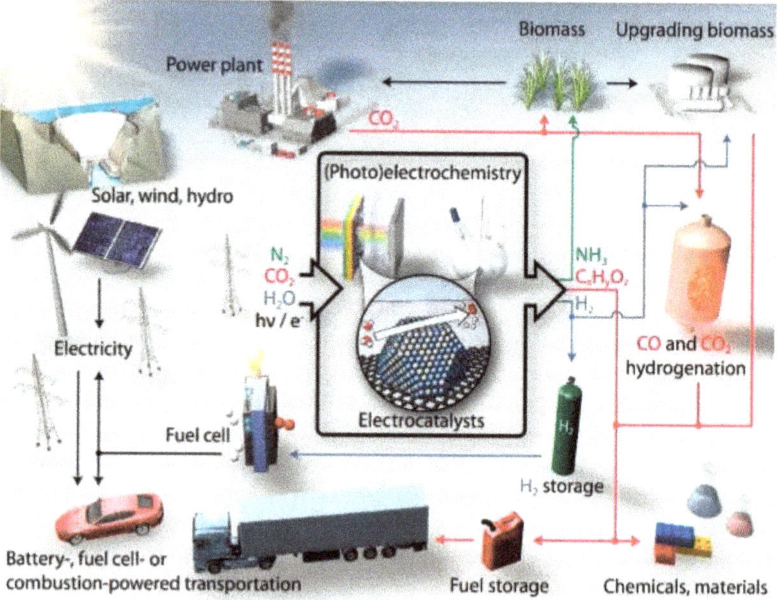


**Figure 1.** The roadmap of electrocatalytic reaction technology for sustainable energy use in the future [1]. (Reproduced with permission from [1]. American Association for the Advancement of Science, 2017).

In recent years, the plasma-assisted preparation method has attracted more and more research interest and has been widely used in the synthesis and modification of electrocatalyst materials. Plasma is partially ionized gas consisting of electrons, ions, molecules, free radicals, photons, and excited species, all of which are active species for the preparation and treatment of catalysts [10]. Differently from the traditional preparation methods, plasma can generate redox species, surface etching, and element doping during the nucleation of catalysts and crystal growth, so as to prepare highly dispersed, small particle size, defect-rich electrocatalysts [11]. This paper reviews the recent progress in plasma-assisted preparation methods, and discusses the effect of plasma on the enhanced catalytic activity through improved dispersion of active species and active sites. Finally, challenges and a future perspective on the development of plasma-assisted preparation of electrocatalysts are provided.

## 2. Noble Metal Electrocatalysts

### 2.1. Plasma Enhanced Deposition

In 1991, Keijser et al. [12], Dutch scientists, first proposed plasma enhanced atomic layer deposition (PEALD) technology. As shown in Figure 2a–c, the common PEALD technologies include direct plasma-enhanced atomic layer deposition, remote plasma-enhanced atomic layer deposition and radical-enhanced atomic layer deposition [13,14]. In 2015, Ting et al. [15] used PEALD technology to deposit Pt nanoparticles with different particle sizes on the $TiO_2$ surface layer of Ti thin films, and prepared $Pt/TiO_2$ catalysts. It was found that the average particle size of Pt nanoparticles increased from 3 nm to 7 μm with the increase of deposition times (Figure 2d–i). Pt nanoparticles with a size of ~5 nm exhibited excellent catalytic activity for MOR with the best CO tolerance and electrochemical stability (Table 1), which had potential applications in methanol fuel cells.

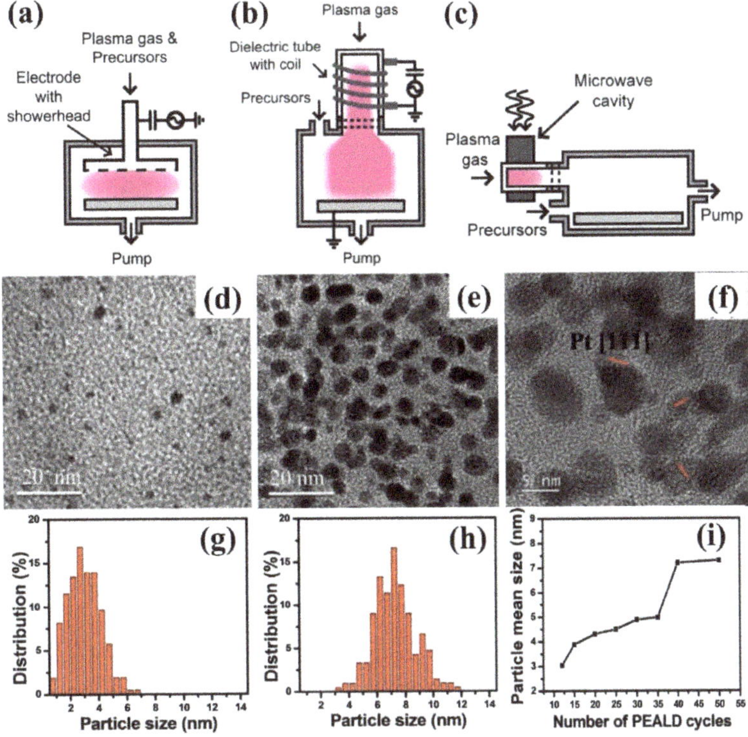


**Figure 2.** Plasma enhanced atomic layer deposition (PEALD) technology (**a**) direct plasma enhanced atomic layer deposition; (**b**) remote plasma enhanced atomic layer deposition; (**c**) radical-enhanced atomic layer deposition [13] (Reproduced with permission from [13], American Vacuum Society, 2007.) HRTEM images of samples with different PEALD deposition times: (**d**) 12 times, (**e**) 50 times, and (**f**) 40 times. Corresponding particle size distribution maps: (**g**) 12 times and (**h**) 50 times. (**i**) The graph of relationship between the average particle size distribution and the number of depositions [15]. (Reproduced with permission from [15]. Elsevier B.V., 2015).

In the same year, Yoshiaki et al. [16] produced highly ionized metal plasma by coaxial pulse arc plasma deposition (CAPD), which deposited Pt nanoparticles with average particle diameter of 2.5 nm on the carbon carrier (Ketjenblack carbon). Compared with the commercial 20% Pt/C, the catalytic activity and stability of the catalysts were significantly improved for MOR. Moreover, the catalysts exhibited excellent initial potential and half-wave potential for ORR performance. The half-wave potential was 0.87 V, which was higher than that of 5% Pt/C (0.78 V) and 20% Pt/C catalysts (0.84 V).

Plasma sputtering (IPS) is also used to prepare noble metal electrocatalysts. Grigoriev et al. [17] used Cabot carbon black (Vulcan XC-72), carbon nanotubes and nanofibers as substrates to deposit Pt and PtPd nanoparticles, which had potential electrochemical activity in fuel cells, water electrolysis cells and dual-function fuel cells. Falch et al. [18] deposited then $Pt_xPd_y$ film on $SiO_2$ substrates by magnetron enhanced plasma sputtering. It was found that Pt3Pd2 and PtPd4 exhibited good catalytic activity in the electrocatalytic oxidation of $SO_2$. Compared with pure platinum (0.598 + 0.011 V, standard hydrogen electrode (SHE)), those two thin film materials had lower initial potential (0.587 + 0.004 V, SHE), which showed their potential application in $SO_2$ oxidation.

## 2.2. Gas Plasma-Assisted Preparation

Plasma is a high energy state gas with electrons, ions, and free radicals. Common gases can be used to form plasma, such as $H_2$, $O_2$, $CO_2$, Ar, $NH_3$, and $N_2$. When high energy plasma gas gets into contact with the surface of the material, it will lead to the physical and chemical changes of the material surface, such as chemical reduction, surface etching, the generation of surface active groups, etc. [19] In 2018, Ma et al. [20] irradiated Pt-based catalysts with high-energy electrons and activated ions in $H_2$ plasma, which could simultaneously reduce Pt ions and graphene oxide (GO), as shown in Figure 3a. By adjusting the mass ratio of GO to multi-walled carbon nanotubes (MWCNT), Pt nanoparticles with different particle size distribution were obtained (Figure 3b–e). Pt/GNT had excellent MOR performance and anti-poisoning performance. The current density of Pt/GNT was 97.9 mA/mg, which was 2.2 times higher than that of commercial Pt/C (44.1 mA/mg). The current density of Pt/GNT was 691.1 mA/mg, which was much higher than that of commercial Pt/C (368.2 mA/mg).

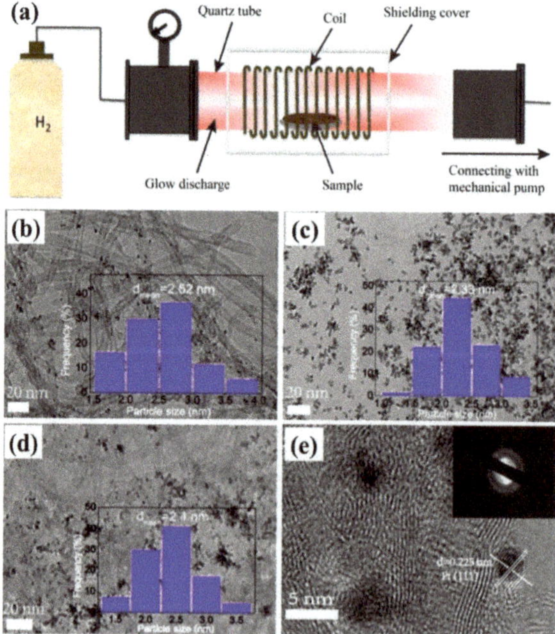


**Figure 3.** (**a**) The sketch of device for sample treatment by $H_2$ plasma. TEM images and corresponding Pt particle size distribution images: (**b**) Pt/carbon nanotubes (CNTs), (**c**) Pt/RGO, and (**d**) Pt/GNT (graphene oxide (GO):CNTs = 1:2, mass ratio). (**e**) HRTEM and SAED images of Pt/GNT [20]. (Reproduced with permission from [20]. Elsevier B.V., 2018).

By contrast, Xu et al. [21] found that Pt nanoparticles in Pt/CNTs-HP prepared by $H_2$ plasma reduction of Pt-based precursor showed more uniform distribution on CNT with the particle size of about 2 nm. They had better MOR electrocatalytic activity than a traditional hydrogen reduced catalyst (Pt/CNTs-H) and $NaBH_4$ reduced catalyst (Pt/CNTs-N). Precursors of noble metal catalysts can also be reduced by Ar plasma. Sui et al. [22] obtained Pt-Ir/TiC bimetallic electrocatalysts by Ar plasma reduction method, which had finer metal crystals and higher metal dispersion. The bimetallic electrocatalyst showed higher ORR/OER catalytic activity than that of the Pt-Ir/TiC electrocatalysts prepared by chemical reduction method.

Gas plasma can also be used to treat noble metal catalysts directly. Ipshita et al. [23] treated the Au@Pt catalyst with Pt loading of 1.75 ug/cm$^2$ (corresponding to two layers of Pt atoms covering a gold core with a diameter of 5 nm) by Ar plasma. It was found that Ar plasma can enhance the formation of highly active Pt (110) crystal face. The total surface area was as high as $48 \pm 3$ m$^2$/g, which was equivalent to 44% of the utilization rate of Pt atoms. The catalysts exhibited excellent CO poisoning resistance and outstanding MOR and ORR properties. Koh et al. [24] prepared "Au Islands" catalysts by treating gold foil with O$_2$ plasma, which showed excellent CO$_2$RR performance. Compared with polycrystalline gold electrode, the "Au Islands" catalysts had excellent CO selectivity, and their Faraday efficiency was more than 95%, which attracted people's attention.

### 2.3. Solution Plasma Sputtering

Solution plasma sputtering (SPS) can occur between electrodes and solutions, providing a new plasma-liquid interface that initiates a variety of physical and chemical processes, as shown in Figure 4a,b. The unique interaction can be used to prepare noble metal nanoparticles and to promote carbon aggregation, so as to produce excellent electrocatalysts [25,26]. Kim et al. [27] obtained metal filtrate through corrosion of electrodes by discharging in water with pulsed plasma. Pt and Pt-M (M = Cu, Ag and Pd) bimetallic nanoparticles were prepared, which had no element segregation or phase segregation during the alloying of Pt and M. Among them, Pt-Ag bimetallic nanoparticles exhibited excellent MOR electrocatalytic activity, stability, and durability. At the same time, Kim et al. [28] pointed out that the corrosion of anode electrode was more aggressive than that of cathode electrode during plasma discharge, and the composition of Pt-Pd bimetallic nanoparticles can be varied with power and electrode structures. Cho et al. [29] found that Pt$_{69}$Pd$_{31}$ had excellent catalytic activity for MOR, with a current density of 6.81 mA/cm$^2$, making it an ideal candidate catalyst for methanol fuel cell.

Metal-carbon composites can be obtained by adding carbon materials to the metal filtrate prepared by SPS. In 2017, Zhang et al. [30] used SPS technology to synthesize PtPd filtrate directly from Pt and Pd wire, and then Koqin black carbon material was added to obtain PtPd/C. As shown in Figure 4c–f, PtPd/KB-2 prepared in a mixture of methanol and water had better dispersion of metal particles than that prepared in water. The particle diameter was about 2–5 nm. The MOR activity of PtPd/KB-2 (12 wt% Pt) was four times better than that of commercial Pt/C catalyst, and the electrochemical surface area was 2.5 times of that of commercial Pt/C catalyst. Meanwhile, the mass activity of PtPd/KB-2 after 300 cycles can reach 43%. In 2018, Horiguchi et al. [31] designed a SPS mobile cell, which can continuously add Vulcan XC72R to the solution containing Pt to prepare Pt/XC72 electrocatalysts, providing a new route for the continuous production of electrocatalysts.

Metal-carbon composites can also be obtained by placing metal nanoparticles prepared by SPS in uniformly dispersed carbon-containing suspension. In 2017, Huang et al. [32] acquired Pt/CoPt/MWCNTs composite catalysts by dispersing Pt/CoPt composite nanoparticles by SPS in a uniform aqueous solution of multi-walled carbon nanotubes (MWCNT). The electrocatalysts exhibited excellent MOR catalytic activity and good stability, which exhibited an activity of 1719 mA/mg$_{Pt}$ 3.16 times higher than that of commercial Pt/C. It can be used as a catalyst for direct methanol fuel cell. Su et al. [33] prepared Pt nanoparticles with particle size of 2 nm by SPS in a XC72 suspension. Pt/C/TiO$_2$ electrocatalysts were then synthesized by mixing the obtained Pt/C composites with TiO$_2$ nanotubes under ultrasound. The catalysts showed good electrocatalytic activity, and the CO toxicity resistance of the current density was 315.2 mA/mg, which is 1.73 times higher than that of commercial Pt/C.

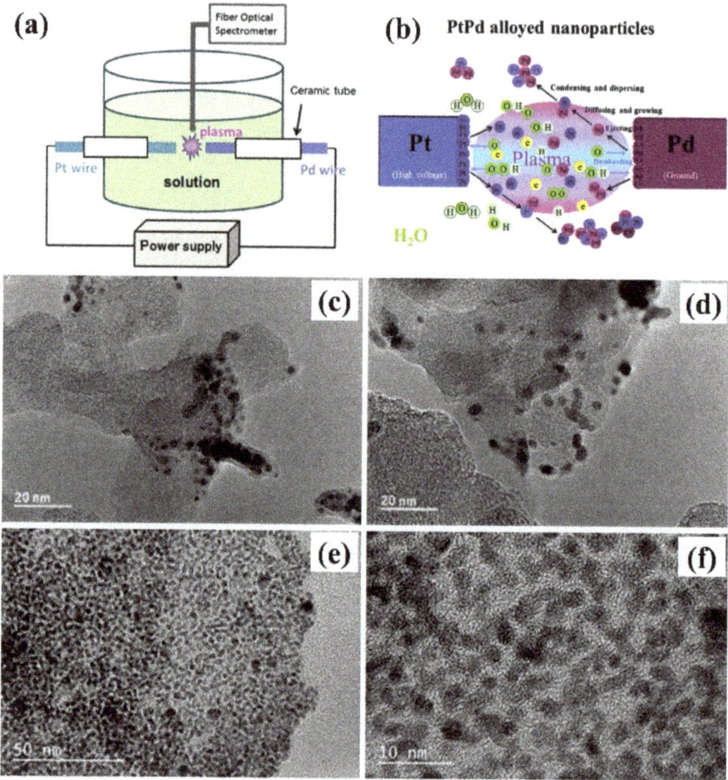



**Figure 4.** (**a**) The experimental device and (**b**) the schematic diagram of PtPb nanoparticles synthesized by solution plasma sputtering. Catalysts formed in different solutions: (**c,d**) PtPd/KB in aqueous solution; (**e,f**) PtPd/KB-2 in water-methanol mixture [30]. (Reproduced with permission from [30]. Elsevier B.V., 2017).

Besides, the carbon-containing suspension can be used instead of water or alcohol solvents to prepare carbon-supported precious metal electrocatalysts. Hu et al. [34] successfully prepared PdAu/KB electrocatalysts by plasma discharge in the suspension of Keqin black carbon material using Pd and Au wires as electrodes. PdAu alloy nanoparticles with an average particle size of 2–5 nm were uniformly distributed on KB, and exhibited outstanding ORR activity in an acidic solution (0.5 M $H_2SO_4$, 240 cycles) and an alkaline solution (0.5 M NaOH, 700 cycles). As electrocatalysts, PdAu alloy nanoparticles have potential applications in future fuel cells or metal air batteries.

In addition to water or alcohol solvents, carbon-supported noble metal electrocatalysts can be directly obtained by using organic solutions as liquid plasma discharge media. As shown in Figure 5a–b, carbon nanospheres (CNS) can be formed when benzene is used as an organic solvent in plasma discharge. At the same time, metal nanoparticles (such as Au and Pt) produced by metal electrode sputtering are deposited on carbon nanospheres to obtain highly active electrocatalysts, loaded with precious metal nanoparticles. At the same time, the metal nanoparticles (Au, Pt, etc.) produced by metal electrode sputtering can be loaded on carbon nanospheres to obtain highly active carbon-supported noble metal nanoparticles electrocatalysts [35,36]. The carbon nanospheres obtained by this method exhibited a diameter range of 20 to 30 nm and a pore size range of 13 to 16 nm. The Au nanoparticles had a particle diameter of less than 10 nm and were uniformly dispersed on the carbon nanospheres (Figure 5c–e), making them an ideal carbon-supported, noble metal electrocatalyst.

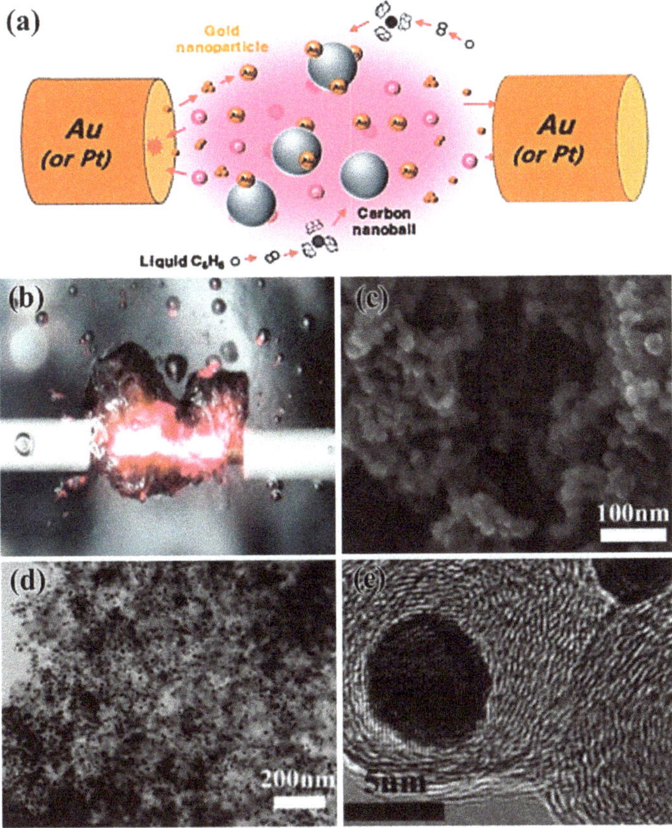



**Figure 5.** (**a**) The schematic diagram and (**b**) the physical diagram of carbon-loaded noble metal materials prepared by organic solution plasma. (**c**) Field emission scanning electron microscope (FESEM), (**d**) scanning transmission electron microscope (STEM), and (**e**) HRTEM diagrams of Au/CNBs catalysts [35]. (Reproduced with permission from [35]. Royal Society of Chemistry, 2013).

Solution plasma process can also directly reduce noble metals to prepare electrocatalysts [37]. Lee et al. [38] synthesized Pt/C electrocatalysts with Pt nanoparticles being reduced by $H_2PtCl_6·6H_2O$, while carbon supports were formed by the corrosion of carbon electrodes during the solution plasma process. The Pt nanoparticles with a diameter of about 38.14 nm were composed of many primary particles with a diameter of about 1.85 nm, which exposed more (111) crystal faces and exhibited good HOR catalytic activity. In 2018, Hussain et al. [39] used He/H$_2$ plasma to reduce $H_2PtCl_6$ to Pt nanoparticles which were then loaded on nitrogen-doped reduced graphene oxide (rGO-N). The obtained Pt/rGO-N electrocatalysts showed a comparable high ORR electrocatalytic activity and superior stability to the commercial Pt/C. It was found that the ORR activity in the electrolytes of 0.1 M KOH and 0.05 M $H_2SO_4$ were three and two times higher than that in commercial Pt/C, respectively. Cui et al. [40] prepared $PtO_aPdO_b@Ti_3C_2T_x$ catalysts by loading PtPd bimetallic oxide nanoparticles onto two-dimensional MXene carrier by solution plasma reduction, as shown in Figure 6. The particle size of PtPb nanoparticles increased with the extension of plasma reaction time. The catalyst obtained by plasma reaction for 3 min showed good catalytic activity for HER and OER, exhibiting the activation potential of HER in 0.5 M $H_2SO_4$ of 57 mV, which was the same for that of OER in a 0.1 M KOH solution of 1.54 V at the current density of 10 mA/cm$^2$. In particular, it had superior water electrolysis performance in an alkaline solution of 1.0 M KOH, which showed that the electrolysis water voltage at the current density of 10 mA/cm$^2$ was 1.53 V.

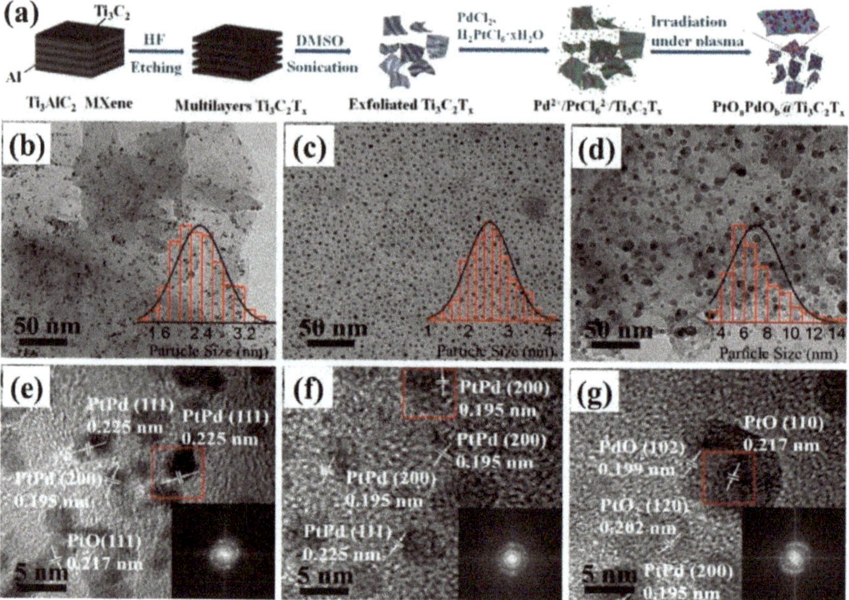

**Figure 6.** (**a**) The flow chart of PtO$_a$PdO$_b$@Ti$_3$C$_2$T$_x$ catalysts prepared by liquid plasma reduction. TEM and HRTEM images with 200 W plasma for (**b,e**) 1 min, (**c,f**) 3 min, and (**d,g**) 5 min. (**e–g**) corresponding fast fourier transform (FFT) diffraction patterns [40]. (Reproduced with permission from [40]. American Chemical Society, 2018).

### 2.4. Plasma Prepared and Modified Electrocatalysts Support

Plasma technology is also used to prepare and modify the catalyst carrier that achieves hydrophilicity and element doping, and promote the dispersion of the catalyst's active components and the improvement of its catalytic activity [41–43]. Chetty et al. [44] formed abundant CO and COO-functional groups on the surface of multi-walled carbon nanotubes (CNT) by O$_2$ plasma, which improved the dispersion of Pt-Ru nanoparticles and the catalytic activity toward MOR. Ding et al. [45] prepared N-doped graphene-coated Pt nanocrystals (N-GPN) by N$_2$ plasma treatment of GPN (Figure 7a–f). From Figure 7g–h, it can be seen that nitrogen doping can produce more active sites and improve the electrocatalytic performance for ORR, which can effectively regulate the activity of the electrocatalysts. Loganathan et al. [46] prepared different Pt/C(N) and Pt/C(AA) catalysts with excellent ORR catalytic activities by modifying carbon black used for the fuel cell catalyst support system with N$_2$ and allylamine plasma, respectively. Du et al. [47] realized in-situ N-doping and surface modification on carbon paper carrier surface by 25% N$_2$ + 75% H$_2$ plasma treatment, promoting the growth of Pt nanowire arrays (NW). The self-supporting Pt NW/GDL (large area gas diffusion layer) electrocatalysts with an area of 5 cm$^2$, could be directly used as catalyst electrodes for fuel cells, which showed excellent charge transfer and mass transfer properties, and their power density was twice as high as that of traditional Pt nanoparticles.

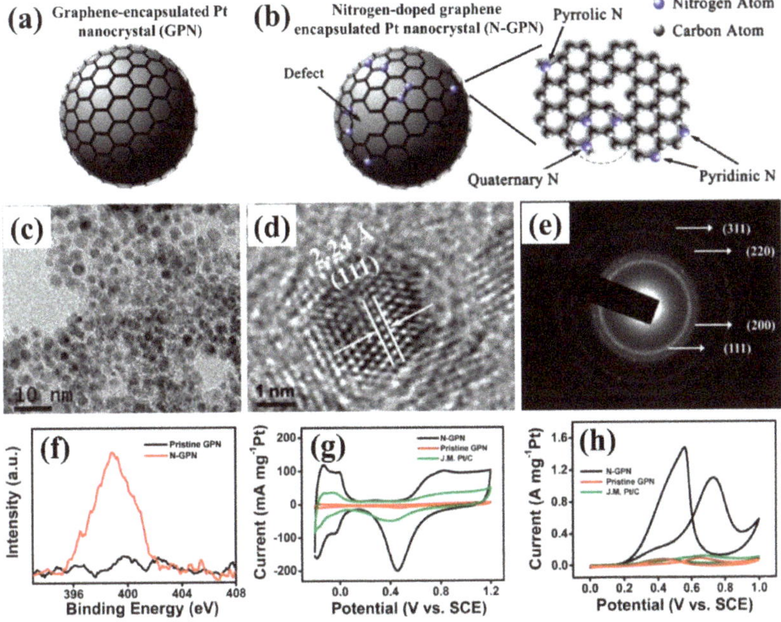



**Figure 7.** The schematic diagram of (**a**) graphene-supported Pt nanoparticles (GPN) and (**b**) N-doped GPN (N-GPN); The images of (**c**) TEM, (**d**) HRTEM, and (**e**) selected area electron diffraction (SAED) for GPN; (**f**) XPS image; (**g**) chemical vapor (CV)—0.5 M $H_2SO_4$, and (**h**) CV—0.5 M $H_2SO_4$ + 1.0 M $CH_3OH$ for GPN and N-GPN [45]. (Reproduced with permission from [45]. American Chemical Society, 2014).

In 2017, Hu et al. [48] obtained Pt/ZnO/KB photocatalysts with solution plasma technology. They first synthesized ZnO nanowires using zinc electrode wires, and then replaced them with Pt electrode wires, loaded Pt nanoparticles onto zinc oxide, and finally, placed Pt/ZnO in Keqin black suspension to obtain aa Pt/ZnO/KB photocatalyst. Pt/ZnO/KB exhibited excellent MOR photocatalytic activity due to the better electron transport performance and smaller resistance of ZnO nanowires. The catalytic activity of Pt/ZnO/KB was 964 mA/mg, which was more than three times that of Pt/KB (306 mA/mg). Meanwhile Pt/ZnO/KB showed excellent CO toxicity resistance and stability, which made it possible for methanol fuel cells to work all-weather in light or dark.

## 3. Non-Noble Metal Electrocatalysts

### 3.1. Metal and Alloy Electrocatalysts

In recent years, non-noble metal catalysts have attracted more and more attention. In particular, transition metals with empty orbits combined with molecules containing lone pairs of electrons through coordination bonds to form transition states with lower barriers reduce the activation energy of the whole reaction path and accelerate the chemical reaction. In 2016, Flis-Kabulska et al. [49] prepared Ni–Fe–C electrocatalysts by introducing carbon into NiFe alloy using $CH_4$+$H_2$ plasma carburization. Although the solubility of carbon in NiFe alloy was very low, a carbide layer with thickness of about 2 μm was formed on the surface, which greatly improved the surface hardness and corrosion resistance. Ni–Fe–C alloy exhibited good HER catalytic activity in 25 wt.% KOH solution at 80 °C. Jin et al. [50] used the microwave plasma chemical vapor deposition (MPECVD) method to coat carbon nanoparticles on carbon cloth, which prepared a three-dimensional CoNPs@C with excellent conductivity and catalytic activity. It was found that a low overpotential of 153 mV can be generated at the current density of

10 mA/cm$^2$ for HER, and the overpotential was 270 mV for OER, and the decomposition voltage was 1.65 V when composing a full hydrolyzed cell.

Cu-based catalysts are commonly used in the preparation of hydrocarbons by CO$_2$RR, such as carbon monoxide, methane, ethylene, ethanol, and n-propanol [51]. Mistry et al. [52] used plasma to treat oxidized Cu-based catalysts for the catalytic reduction of CO$_2$ at low overpotential. It can be seen from Figure 8a–h that Cu$_2$O was not completely reduced during the reaction, and Cu$^+$ ions were still on the surface. Cu-based catalysts derived from oxides only had a minimal effect on catalytic performance, while Cu$^+$ played an important role in reducing the initial potential and improving the selectivity of ethylene. Among the catalysts, Cu foil treated by O$_2$ plasma had 60% ethylene selectivity at −0.9 V (versus reversible hydrogen electrode (RHE)) potential (Figure 8i,j). In 2017, Gao et al. [53] adjusted Cu (100) crystal face and the content of oxygen/chloride ion by treatment of Cu nanoparticles with H$_2$, O$_2$, and Ar plasma, respectively. It was found that the oxygen content on the surface and subsurface of Cu nanoparticles was the key to achieve high activity and selectivity of hydrocarbons/alcohols, even more important than the existence of the Cu (100) crystal plane. Cu nanoparticles treated by O$_2$ plasma had lower overpotential and higher selectivity for ethylene, ethanol, and n-propanol, of which the selectivity for ethylene and ethanol was 45% and 22%, and the maximum Faradaic efficiency (FE) of C2 and C3 products can reach 73%.

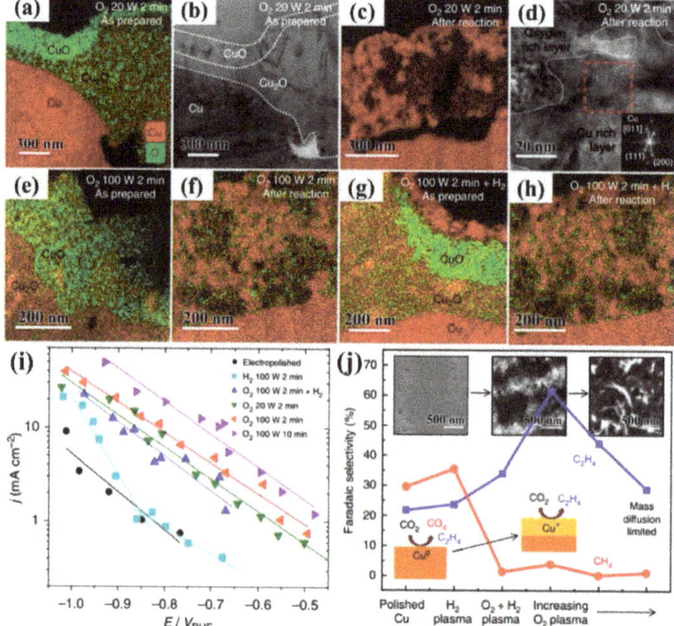

**Figure 8.** (a,c,e–h) EDS images and (b,d) HRTEM images of different plasma activated copper foil before and after reaction; (i) electrochemical activity during CO$_2$RR process; (j) hydrocarbon selectivity of plasma treated Cu foil. The corresponding inset SEM images were after the reaction: H$_2$ plasma treated metal Cu foil, O$_2$ plasma treated with Cu foil at 20 W for 2 min, and O$_2$ plasma treated at 100 W for 10 min, respectively [52]. (Reproduced with permission from [52]. Nature Publish Group, 2016).

### 3.2. Transition Metal Oxides

#### 3.2.1. Cobalt Oxides

Cobalt oxide ($CoO_x$) has attracted much attention for its superior catalytic activity in electrocatalysts [54,55]. In 2017, Kim et al. [56] prepared $Ag@Co_3O_4$ core-shell hybrid nanocrystals in aqueous solution by treating Ag and Co electrode wires with solution plasma technology. Compared with Ag and Ag–Co electrocatalysts prepared in alcohol solution, it showed a better electronic effect and geometric effect, which was beneficial to the fracture of the O–O bond for ORR, resulting in the catalytic activities being 5.2 and 2.6 times higher, respectively. The modification of the carrier can also effectively improve the catalytic activity. Taiwo et al. [57] synthesized three-dimensional nanoporous graphene sheets with highly specific surface areas as a catalyst carrier by microwave-induced plasma, and then doped them with nitrogen and supported $Co_3O_4$ nanoparticles to prepare $15Co_3O_4$/N-AP/800 electrocatalysts. The catalysts exhibited excellent electrical conductivity and ORR performance, showing that the Tafel slope was 42 mV/dec, which was better than 82 mV/dec of Pt/C. In 2018, Lang et al. [58] used DC arc discharge plasma to synthesize CoO nanospheres mixed with $La_2O_3$ and Pt/C to prepare highly active CO–La–Pt ternary ORR catalysts for Li–O batteries. The specific capacity and energy density of the electrode reached 3250.2 mAh/g and 8574.2 Wh/kg at 0.025 $mA/cm_2$, and the capacity retention rate reached 38.3% after 62 cycles.

Transition metal oxides with abundant defects and vacancies were prepared by plasma etching, which exhibited excellent electrocatalytic performance [59]. In 2016, Xu et al. [60] used Ar plasma to treat $Co_3O_4$ nanosheets, which produced oxygen vacancies on the surface of $Co_3O_4$ nanosheets, which had higher current density and lower initial potential than untreated $Co_3O_4$ nanosheets. The catalysts showed good OER performance and the current density was increased by 10 times at a voltage of 1.6 V. Wang et al. [61] introduced oxygen vacancies in cobalt oxyhydroxide (CoOOH) by Ar plasma technique. It was proven by density functional theory that Vo–COOH catalysts with oxygen vacancies could initiate additional reaction steps to accelerate the oxidation of $H_2O$—$H_2O^* \leftrightarrow (HO + H)^*$ and $(HO + H)^* \leftrightarrow HO^* + H^+ + e^-$, and had a lower kinetic barrier. At the same time, the experimental observation confirmed that the two sites could promote the OER catalytic activity of $V_0$–CoOOH, which accelerated the deprotonation process and enhanced water oxidation due to their synergistic catalysis. Meanwhile, this method can also be extended to other OER catalysts, which has broad application prospects. In 2018, Ma et al. [62] applied $Co_3O_{4-x}$ with oxygen-rich vacancies, treated by Ar plasma, to the catalytic reactions of ORR (half-wave potential 0.84 V) and OER (overpotential of 330 mV and Tafel slope of 58 mV/dec), resulting in excellent catalytic activity, as shown in Figure 9b. The Zn-based hybrid battery (Figure 9a) prepared on this basis can simultaneously carry out two kinds of electrochemical reactions (OER and ORR electrocatalysis, and reversible Co–O $\leftrightarrow$ Co–O–OH redox reaction), with a high power density of 3200 W/kg and a high energy density of 1060 Wh/kg. The battery showed good water resistance and washing resistance. Its capacity retention rate was 99.2% after 20 h of immersion test and 93.2% after 1 h of washing test. More interestingly, when exposed to air from underwater, the battery can automatically restore the power output, with excellent electrochemical performance, good environmental adaptability, and "air recyclability," so was considered to have potential to become the next generation of energy storage device and be widely used in flexible wearable electronic devices and other technical fields (Figure 9c).

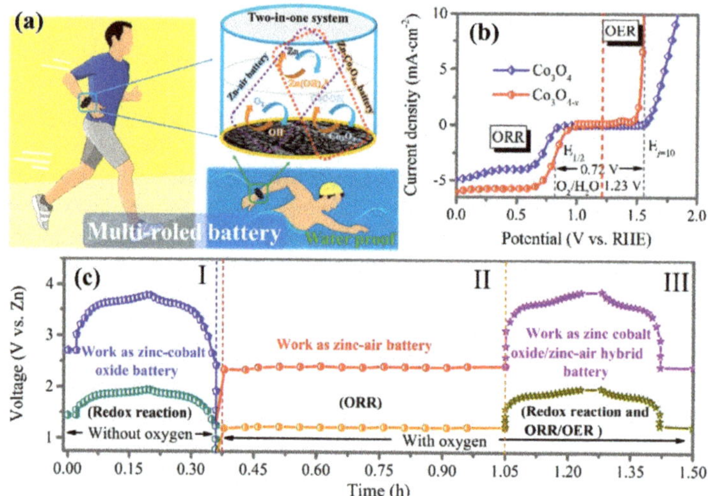



**Figure 9.** (**a**) The schematic diagram of Zn–Co₃O₄₋ₓ/Zn-air hybrid battery; (**b**) the overall polarization curve of Co₃O₄ nanorods and Co₃O₄₋ₓ; (**c**) the constant current charge/discharge curve of Zn–Co₃O₄₋ₓ /Zn-air hybrid battery. (I) Work of the Zn–Co₃O₄₋ₓ battery in an anaerobic environment; (II) Work of the zinc-air battery in an aerobic environment; (III) Work of the hybrid battery in an aerobic environment [62]. (Reproduced with permission from [62]. American Chemical Society, 2018).

The doping of heteroatoms can also improve the electrocatalytic activity of $CoO_x$, and the plasma technology can be used to dope nitrogen [63]. In 2017, Xu et al. [64] found that surface etching and nitrogen doping were realized on $Co_3O_4$ nanosheets by $N_2$ plasma. The sample $N–Co_3O_4$ had more active sites due to N-doping and oxygen vacancies, leading to a significant improvement in electronic conductivity. Compared to $Co_3O_4$ nanosheets (1.79 V and 234 mV/dec), it showed a lower potential (1.54 V) at a current density of 10 mA/cm² and a Tafel slope of 59 mV/dec. Uhlig et al. [65] directly treated the CoAc/C precursor with $N_2$ plasma, that simultaneously achieved nitrogen doping of the carrier and active components. The product showed excellent ORR catalytic activity in 0.1 M KOH and 0.1 M $K_2CO_3$ solutions. However, cobalt oxide produced more by-products, which may have resulted in membrane damage to the fuel cell, which remains unimproved upon.

Elemental P can be doped into an oxide by plasma [66]. In 2017, Xiao et al. [67] found that P atoms were immediately transported and filled into the position of an oxygen vacancy defect, which was a structural defect caused by surface oxygen etching of materials by Ar plasma airflow treatment (Figure 10a–f). The 3p orbital and 3d orbital of P atoms have lone pair of electrons, so the $V_0$ of $Co_3O_4$ filled with P atoms can be used to effectively stabilize vacancies, which can induce local charge density and adjust surface charge state, and also can be used to adjust the relative proportion of $Co^{2+}/Co^{3+}$, improving the adsorption and electrocatalytic properties of materials [68]. As can be seen in Figure 10g–i, $V_0$-$Co_3O_4$ had a relatively lower coordination number than $Co_3O_4$ due to the formation of oxygen vacancies, and $P–Co_3O_4$ had a much higher coordination number than $V_0–Co_3O_4$ and $Co_3O_4$, due to the filling of P atoms. When P occupied the vacancy sites, electrons migrated out of the 3d orbital of Co, and more $Co^{2+}$ (Td) remained in $P–Co_3O_4$, which exhibited excellent catalytic activity for HER and OER. $P–Co_3O_4$ required only an overpotential of 120 mV in the HER process and the overpotential of 280 mV in the OER process at the current density of 10 mA/cm². The Tafel slopes for HER and OER were 52 mV/dec and 51.6 mV/dec, respectively. $P–Co_3O_4$ can effectively catalyze full water-splitting in a 5 M KOH solution (80 °C), which showed an overpotential of 420 mV at a current density of 100 mA/cm².

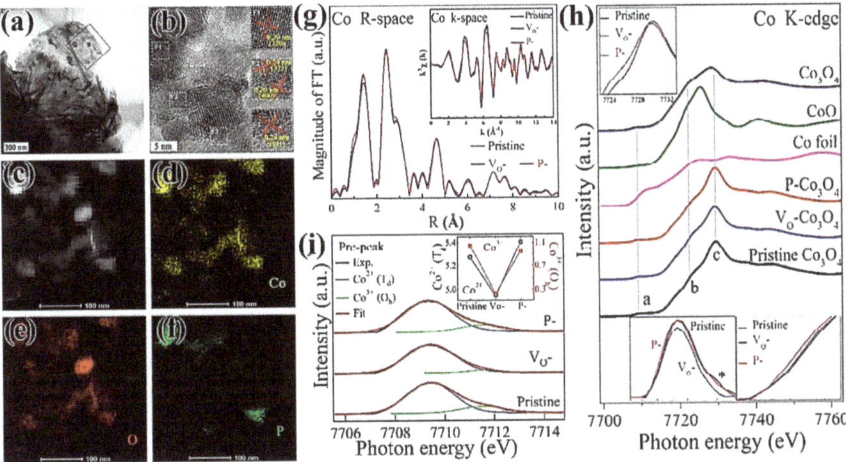

**Figure 10.** (**a**) TEM images, (**b**) HRTEM images, and (**c–f**) STEM-EDX elemental mapping images of P–Co$_3$O$_4$; (**g**) Co K-side extended x-ray absorption fine structure (EXAFS). The inset is the Fourier-transformed EXAFS oscillations; (**h**) Co K-edge XANES spectra of pristine V$_0$–Co$_3$O$_4$ and P–Co$_3$O$_4$. Top inset magnifies the main peak region. Bottom insets magnify the pre-peak region; (**i**) deconvoluted pre-peak of Co K-edge XANES. Inset compares the amount of Co$^{2+}$(Td) and Co$^{3+}$(Oh) states of pristine, V$_0$–Co$_3$O$_4$, and P–Co$_3$O$_4$ [67]. (Reproduced with permission from [67]. Royal Society of Chemistry, 2017).

In addition, excellent electrocatalytic performance can be obtained by reducing the size of the catalysts to expose more active sites. Dou et al. [69] prepared atomic-scale CoO$_x$-ZIF (Zeolitic Imidazolate Framework, ZIF) catalysts by directly treating ZIF-67 precursors with O$_2$ plasma. ZIF-67 has an abundant pore structure, which provided abundant channels for O$_2$ to enter ZIF to activate Co$^{2+}$ to obtain atomic-scale CoO$_x$. After treatment with O$_2$ plasma, the porous structure and large surface area of ZIF-67 were still retained, which were very conducive to the transport of materials toward OER. The prepared atomic-scale CoO$_x$-ZIF catalysts can provide abundant active sites for oxygen evolution reaction, and its catalytic performance for OER was better than that of noble metal catalyst RuO$_2$. The atomic scale electrocatalysts were obtained directly in-situ in MOFs (metal organic frameworks, MOFs) by that method for the first time, which provided a new idea for the convenient preparation of atomic scale electrocatalysts.

### 3.2.2. Perovskite-Type Oxides

Perovskite-type oxide ABO$_3$ is a new type of inorganic nonmetallic material with unique physical and chemical properties. A-site is usually the ion of a rare earth or alkaline earth element, and B-site is the ion of transition element. A-site and B-site can be partially replaced by other metal ions with similar radii and keep their crystal structures unchanged. Therefore, ABO$_3$ is an ideal sample for studying the surface and catalytic performance of catalysts in theory, which has great development potential in the fields of environmental protection and industrial catalysis [70]. In particular, unlike the anion vacancies in other metal oxides, ABO$_3$ can also achieve cation vacancies. Chen et al. [71] used Ar plasma technology to surface-treat the SnCo$_{0.9}$Fe$_{0.1}$(OH)$_6$ (SnCoFe) precursor and adjusted the electrocatalytic activity of OER by defect engineering. Because Sn(OH)$_4$ had weak chemical bonds and its lattice energy was much lower than that of cobalt or iron hydroxides, Sn vacancies preferentially formed in the process of forming various cationic vacancies, thus a perovskite-type hydroxide (SnCoFe–Ar) electrocatalyst with rich Sn vacancies has been developed. The abundant Sn vacancies made the catalyst materials have a larger specific surface area and superhydrophilicity, exposed more CoFe active sites, and created more 1–2 nm channels on the surfaces of the catalyst materials, which improved the material transport capacity and electronic transport capacity, and optimized the adsorption capacity of the reactants. Compared with the overpotential of 420 mV and Tafel slope of 77.0 mV/dec for SnCoFe, the overpotential and the Tafel slope of the SnCoFe–Ar were 300 mV and 42.3 mV/dec at the current density of 10 mA/cm$^2$, which exhibited superior catalytic performance for OER.

Transition metal ions also affect the catalytic performance of the electrocatalysts. Hayden et al. [72] successfully prepared $SrTi_{1-x}Fe_xO_{3-y}$ (STFO) perovskite-type composite gradient electrocatalysts by high-throughput physical vapor deposition (HT-PVD) with $O_2$ plasma. With the increase of x value, the lattice parameters of STFO increased from 0.392 + 0.001 nm of $SrTiO_3$ to 0.386 + 0.001 nm of $SrFeO_3$, and the conductivity increased as well. When x > 0.75, the conductivity reached $\rho = 0.041$ S/cm. The corresponding OER reduced the initial potential at current 100 μA from 1.52 VRHE (x = 0.2) to 1.40 VRHE (x = 0.85), but the high OER activity was accompanied with low stability. Therefore, $SrTi_{0.5}Fe_{0.5}O_{3-y}$ showed the best ORR activity and electrode stability.

In order to develop efficient electrocatalysts with extraordinary mass activity and stability, Chen et al. [73] deposited amorphous $SrCo_{0.85}Fe_{0.1}P_{0.05}O_{3-\delta}$ (SCFP) nano-thin films with weak chemical bonds on the conductive nickel foam (NF) substrate by high-energy Ar plasma. The rapidly reconstituted SCFP-NF bifunctional catalyst exhibited good electron conductivity, ultra-high activity, and stability, due to the destruction of strong chemical bonds by plasma in the crystalline SCFP target. The ultra-high quality activity (overpotential 550 mV) of 1000 mA/mg in 1.0 M KOH solution was about 2.1 times higher than that of $RuO_x$-NF coupled Pt-NF electrode, as shown in Figure 11a,b. At the same time, the catalysts exhibited outstanding catalytic activity with a water decomposition stability of 650 h (current density of 10 mA/cm$^2$), which was significantly superior to the current $RuO_x$-NF coupled Pt-NF electrode, which represented a major breakthrough in water cracking, as shown in Figure 11c–e. The simple reconstruction strategy was expected to be used in other advanced energy conversion and storage devices to develop new and efficient catalysts.

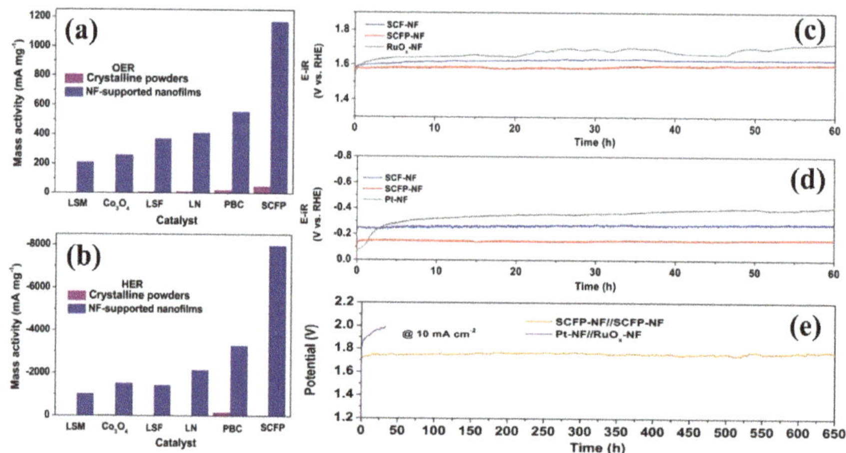

**Figure 11.** Comparison of the mass activity at an overpotential of 0.35 V between crystalline powders and NF-supported nanofilms for the oxygen evolution reaction (OER) (**a**) and hydrogen evolution reaction (HER) (**b**); the stability tests of SCFP-nickel foam (NF) and control samples for OER (**c**) and HER (**d**); (**e**) The stability tests for the water splitting of bifunctional SCFP-NF catalysts and Pt-NF coupled RuO-NF [73]. (Reproduced with permission from [73]. Wiley-VCH, 2018).

### 3.2.3. Two-Dimensional (2D) Layered Double Hydroxides

Transition-metal layered double hydroxides (LDH) have unique 2D structures, large specific surface areas and special electronic structures, showing good electrocatalytic performance [74,75]. The LDHs prepared by the conventional methods have a high number of layers, which limits the exposure of the electrocatalytic active sites, and thus inhibits the electrocatalytic activity. At present, the method of liquid phase exfoliation to prepare LDHs generally uses organic solvents, which has the disadvantages of substantial time-consumption, toxicity, and easy adsorption of organic solvents. Compared with traditional liquid phase exfoliation, plasma exfoliation has the advantages of cleanliness, rapidity, and efficiency, and avoids the adsorption of organic solvent molecules. In 2017, Liu et al. [76] prepared 2D ultra-thin nanosheets (CoFe-LDHs–$H_2O$) by treating bulk CoFe bimetallic hydroxide nanosheets (CoFe-LDHs) with water plasma technology, which can be stabilized in the form of powder, as shown in Figure 12a,b. Compared with bulk CoFe-LDHs, the thickness of CoFe-LDHs–H2O nanosheets decreased significantly (Figure 12c,g), and the 2D base surface became rough (Figure 12d,h). From Figure 12e,f,I,j, it can be seen that water plasma had a very good peeling effect on CoFe-LDHs nanosheets, as the thickness of nanosheets decreased from 25.2 nm to 1.54 nm. The obtained CoFe-LDHs–$H_2O$ had larger specific surface area, resulting in many cobalt vacancies, iron vacancies, and oxygen vacancies on its surface, which enhanced the adsorption of OER intermediates and further boosted the catalytic activity. Tafel slope and charge transfer impedance decreased significantly, such that the overpotential potential was only 36 mV/dec.

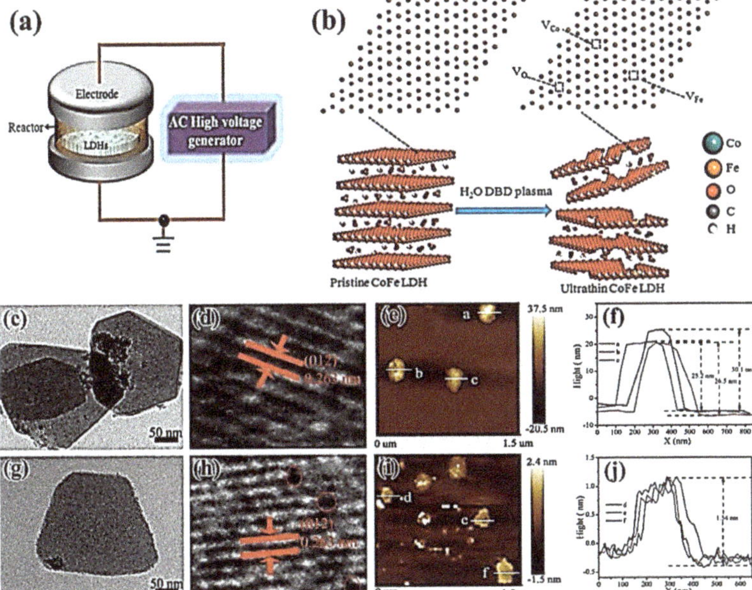


**Figure 12.** The schematic diagram of (**a**) a dielectric barrier discharge (DBD) reactor and (**b**) water DBD plasma-activated stripping of CoFe-LDHs nanosheets; TEM, HRTEM, and atomic force microscope (AFM) images and corresponding nanosheet thickness images of bulk CoFe-LDHs (**c**–**f**) and plasma-treated CoFe-LDHs–H2O samples (**g**–**j**) [74]. (Reproduced with permission from [74]. Royal Society of Chemistry, 2017).

In the meantime, Wang et al. [77,78] obtained 2D ultra-thin CoFe-LDHs–Ar and CoFe-LDHs–$N_2$ nanosheets by treating CoFe-LDHs with Ar plasma and $N_2$ plasma, respectively. Gas plasma technology was not only non-toxic, clean and time-saving, but CoFe-LDHs–Ar and CoFe-LDHs–$N_2$

exfoliated by the dry method also had abundant oxygen vacancies, cobalt vacancies, and iron vacancies. The formation of multiple vacancies was conducive to regulating the electronic structure of the material surface, reducing the coordination number around cobalt and iron atoms, increasing the degree of chaos around atoms, and facilitating the adsorption of the oxygen evolution reaction intermediate, which led to improving the electrocatalytic performance of the material. CoFe-LDHs–Ar nanosheets exfoliated by Ar plasma exhibited good oxygen precipitation performance. The overpotential of CoFe-LDHs–Ar nanosheets was only 266 mV at the current density of 10 mA/cm$^2$, showing good activity. CoFe-LDHs–N$_2$ nanosheets stripped by N$_2$ plasma not only had abundant metal vacancies and oxygen vacancies, but also realized nitrogen doping. The doping of elemental nitrogen was beneficial to changing the electronic structure around the reaction site and improving the catalytic activity by adsorbing oxygen precipitation intermediates, and the resulting defect-rich 2D ultra-thin CoFe-LDHs–N$_2$ nanosheets made it easy to expose more electrocatalytic active sites to increase their oxygen evolution performance. The overpotential was only 233 mV at the current density of 10 mA/cm$^2$, and the Tafel slope and charge transfer impedance were also reduced, which had very good stability. The new method of stripping 2D layered materials to achieve nitrogen doping and defect-richness simultaneously can be used for reference to other similar materials.

### 3.2.4. Other Metal Oxides

In addition, other transition metal oxides are also used as electrocatalysts. For example, in 2017, Oturan et al. [79] prepared sub-stoichiometric titanium oxide (Ti$_4$O$_7$) by plasma deposition and electrocatalytic oxidation to treat the pollutant antibiotic amoxicillin (AMX), which can rapidly oxidize and degrade 0.1 mM (36.5 mg/L) AMX in a short time. Luo et al. [80] prepared FeO$_x$/C electrocatalysts by uniformly depositing iron oxide clusters on porous carbon substrates with high ionized iron plasma generated by arc discharge. It was found that the catalysts showed good ORR and OER properties, and an excellent rate performance and cycle life in a Li-O$_2$ battery. A discharge capacity of 500 mAh/g was retained after 37 cycles at the current density of 100 mA/cm$^2$. In 2018, Peng et al. [81] prepared MnO$_x$@C-D electrocatalysts by dielectric barrier discharge technology for ORR. Because of the effect of plasma, the catalysts allowed manganese ions with different valences to coexist, and had high oxygen adsorption capacities. The aggregation of nanometer catalysts was inhibited at a medium temperature, which showed a higher reaction activity than that of the catalysts prepared by traditional calcination method. Jiang et al. [82] induced abundant defects on the surface of MnO$_2$ nanowires by Ar plasma etching, and observed that Ar–MnO$_2$ had abundant edges and oxygen vacancies to produce more active sites, showing excellent ORR performance. At the current density of 157 mA/cm$^2$, the power density of Al-air battery based on A–MnO$_2$ catalysts can reach 159 mW/cm$^2$, which is much higher than that of untreated MnO$_2$ catalysts (115 mW/cm$^2$).

In the process of HER catalysis, pre-reduction of transition metal oxides is an effective way to improve the catalytic activity of HER. Zhang et al. [83] treated NiMoO$_4$ nanowire arrays by carbon plasma for 30 s, which formed Ni$_4$Mo nanoclusters on the surface and deposited a layer of graphitized carbon, as shown in Figure 13a–d. Compared with NiMoO$_4$ nanowire arrays treated by H$_2$ reduction, the sample (C-60s) exhibited excellent catalytic activity for hydrogen evolution (Figure 13e–g), and could maintain its array morphology, chemical composition, and catalytic activity during long-term intermittent hydrogen evolution. It opened up a new way for simultaneous activation and stabilization of transition metal oxide electrocatalysts. In 2018, Geng et al. [84] used H$_2$ plasma to surface-treatment of 2D ZnO nanosheets, which produced abundant oxygen vacancy defects on the surface. Although these oxygen vacancy defects did not increase the number of active centers, they can lead to the efficient activation of CO$_2$ molecules in the electron-rich state and enhance the electrochemical catalytic reduction activity of CO$_2$. At the voltage of −1.1 V compared to RHE, the oxygen-rich ZnO nanosheets with vacancies can effectively reduce the CO$_2$ to a CO product with a current density of −16.1 mA/cm$^2$ and a Faraday efficiency of about 83%, exhibiting excellent catalytic activity for CO$_2$RR.

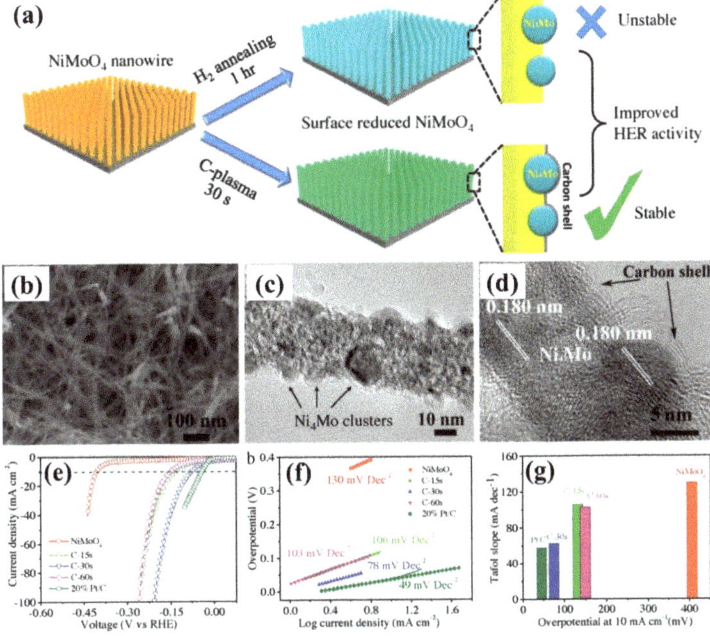

**Figure 13.** (**a**) The preparation of C-plasma treated and $H_2$ annealed $NiMoO_4$; (**b**) SEM and (**c,d**) TEM images of the sample C-30s; the electrochemical performance tests of the sample: (**e**) linear sweep voltammetry (LSV) curve, (**f**) Tafel slope, and (**g**) the relationship between overpotential and Tafel slope at current density of 10 mA/s [83]. (Reproduced with permission from [83]. Wiley-VCH, 2018).

### 3.3. Transition Metal Sulfides

Transition metal sulfides which have received extensive attention in recent years, have become new and highly efficient electrocatalysts, particularly HER catalysts. The hydrogen evolution reaction mainly occurs at the boundary of 2D materials in which the atom is incompletely coordinated. Therefore, the preparation of 2D transition metal sulfides with incompletely coordinated atomic boundaries is the key to whether the material can replace the noble metal platinum. Li et al. [85] etched 2D $TaS_2$ nanosheets by using $O_2$ plasma technology. It was found that the plasma could control the processing of high-density atomic-scale pores on three-dimensional (3D) two-layer crystals, which can increase the catalytic sites and effectively improve the hydrogen evolution catalytic activity of 2D materials, as shown in Figure 14a. As can be seen from Figure 14b–g, the atomic defect density increased with the increase of processing time, which exhibited excellent HER catalytic activity of treatment at 15 min.

2D $MoS_2$ is also used for electrocatalytic reactions, which can also be used to enhance the activity by Ar, $O_2$, and $H_2$ plasma. Tao et al. [86] prepared $Ar–MoS_2$ and $O_2–MoS_2$ thin film materials for OER catalytic reaction using Ar and $O_2$ plasma treatment, respectively. The hydrophilic contact angles of Ar and $O_2–MoS_2$ were found to be 61.8° and 48.4°, which were significantly better than untreated $MoS_2$ (96.5°). The Tafel slope of $O_2–MoS_2$ was up to 105 mV/dec, which was significantly better than $Ar–MoS_2$ (117 mV/dec) and $MoS_2$ (160 mV/dec). In 2018, Zhang et al. [87] and Huang et al. [88] found that $MoO_3$ species were formed at the edge and plane positions of $MoS_2$ by using $O_2$ plasma to treat $MoS_2$ and $H_3Mo_{12}O_{40}P/MoS_2$ composite intercalation compounds respectively. The $MoO_3$ can also be reduced and decomposed from the lattice of $MoS_2$ in the HER catalytic process, effectively improving the HER catalytic activity of $MoS_2$. In addition, the presence of S defects also enhanced the catalytic activity of $MoS_2$ for HER. Cheng et al. [89] treated $MoS_2$ with $H_2$ plasma, and the sample $MoS_2$ treated for 15 min produced high density S vacancies on the base plane of single-layer crystalline

MoS$_2$, whose overpotential could reach 240 mV (versus RHE) at the current density of 10 mA/cm$^2$. The amorphous a–MoS$_2$ treated by H$_2$ plasma had abundant sulfur vacancies, resulting in a decrease of overpotential from 206 mV of a–MoS$_2$ to 143 mV of MoS$_{1.7}$ [90].

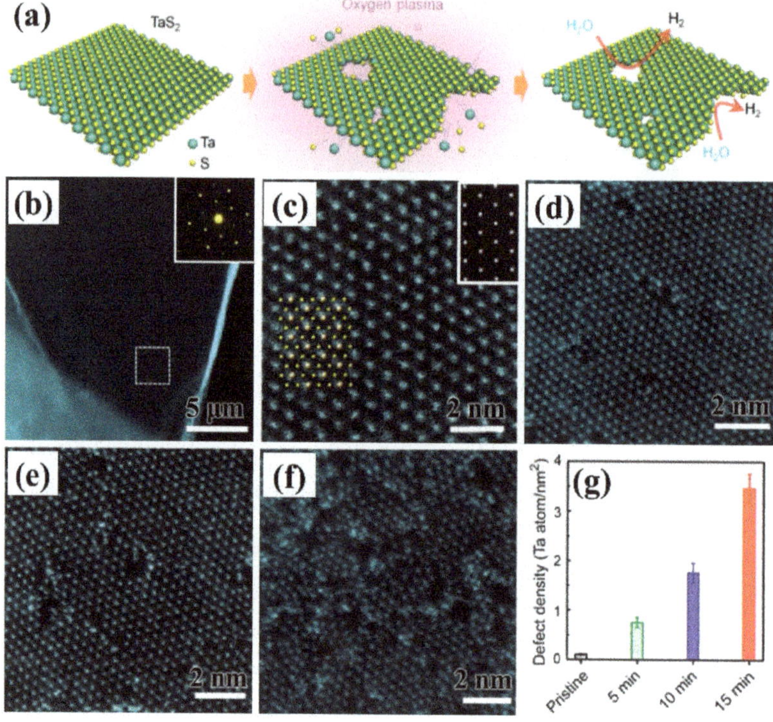


**Figure 14.** (**a**) Mechanism diagram of O$_2$ plasma treated TaS$_2$ nanosheets for HER; (**b**) TEM images and (c) HAADF-STEM images of TaS$_2$; (**d–f**) HAADF-STEM images by plasma treatment for 5 min, 10 min, and 15 min; (**g**) the density of edge Ta atoms [85]. (Reproduced with permission from [85]. Wiley-VCH, 2016).

Cobalt sulfide also has good electrocatalytic activity. Dou et al. [91] prepared bifunctional catalysts with catalytic ORR and OER by simultaneously etching and doping the catalyst's surface and using NH$_3$ plasma. Co$_9$S$_8$ nanoparticles (Co$_9$S$_8$/G) loaded on graphene were treated by ammonia plasma. Nitrogen was successfully doped into the lattices of Co$_9$S$_8$ and graphene by the treatment of Co$_9$S$_8$ nanoparticles loaded on graphene (Co$_9$S$_8$/G) with NH$_3$ plasma. Moreover, partial etching also occurred on the surface of Co$_9$S$_8$ and graphene. The doping of heteroatoms can effectively adjust the electronic structure of Co$_9$S$_8$ and graphene, while the etching of the surface can expose the catalysts to more catalytically active sites, which can obtain a high activity bifunctional electrocatalysts with similar ORR catalytic performance to commercial Pt/C, and OER catalytic performance superior to RuO$_2$. In 2018, Zhang et al. [92] prepared a sulfur-rich, Co$_3$S$_4$, ultrathin porous nanosheet with abundant sulfur vacancies (Co$_3$S$_4$ PNS$_{vac}$) using Ar plasma to treat the Co$_3$S$_4$/TETA precursor. When the hydrogen evolution potential was 200 mV, the mass activity was as high as 1056.6 A/g, which is 107 and 14 times that of Co$_3$S$_4$ nanoparticles and nanosheets respectively, which was also better than the HER electrocatalytic activity of Pt/C. In addition, the simple and rapid chemical conversion strategy can be extended to the synthesis of ultra-thin porous CoSe$_2$ and NiSe$_2$ sheets with anion-rich defects, due to their universality, which opened up new ways to control the electrocatalytic performance of 2D nanomaterials through defect engineering and ultra-thin porous structure engineering.

In recent years, self-supporting electrode materials have attracted widespread attention. In 2018, He et al. [93] synthesized CuI nanosheet arrays by using dielectric barrier discharge (DBD) plasma with iodine vapor on copper foam, and then prepared $Cu_2S$ nanosheet arrays via sulfur ion exchange. The $Cu_2S$/CF showed excellent OER catalytic activity in 1 M KOH solution with an overpotential of only 290 mV at the current density of 10 mA/cm$^2$. Yeo et al. [94] used a PEALD technique to grow a tungsten disulfide ($WS_2$) film on a Ni mesh ($WS_2$/Ni-foam), which exhibited superior HER activity. In acidic electrolyte, the overpotential and the Tafel slope of the catalysts were 280 mV and 63 mV/dec at a high working current density of 100 mA/cm$^2$. Daniel et al. [95] modified $WS_2$ and $MoS_2$ by $SF_6$/$C_4F_8$ plasma. It was found that their overpotentials shifted to 100 mV and 200 mV, and the Tafel slopes decreased by 50 mV/dec and 120 mV/dec, respectively. Especially for $WS_2$ samples treated by plasma for 31s, the Tafel slope was only 81 mV/dec. Plasma can also be used to treat the carrier. Qu et al. [96] used a plasma-treated Ni–Fe foam (PNFF) carrier to form a PNFF with many micro-grooves on the surface. Then, $CoS$/$Ni_3S_2$–FeS nanoflowers were formed by vulcanization, after Co nanoparticles were deposited on PNFF carrier by electrodeposition, which showed the overpotentials toward HER and OER of 75 mV and 136 mV at the current density of 10 mA/cm$^2$, respectively. The amount of $H_2$ and $O_2$ produced were 680 μmol/h and 1230 μmol/h, respectively, showing high electrocatalytic activity and overall water splitting performance.

*3.4. Transition Metal Selenides*

Transition metal selenides also exhibit excellent electrocatalytic performance, especially for the HER catalytic reaction. [97] HER has become the bottleneck of hydrogen production from electrolytic water due to the slow reaction kinetics. The lower overpotential and the Tafel slope identify the benefits of HER catalytic activity; plus the HER can react at a lower applied voltage. For example, the Tafel slope of a Pt-based catalysts can be reached 30 mV/dec, which is close to the theoretical limit of 29 mV/dec, but its reserves and price limit wider applications. In 2017, Deng et al. [98] used a 3D porous vertical graphene array (VG) prepared by MPECVD as a conductive substrate, and prepared a $MoSe_2$/VG/array by hydrothermal method treated by ammonia gas heat treatment modification. Next, the 2H to 1T (2H-1T) phase transition of $MoSe_2$ was initiated by N doping to form N–$MoSe_2$/VG having a 2H-1T composite phase. The introduction of 1T phase reduced the band width of $MoSe_2$ and improved the electron transport performance. At the same time, N doping increased the hydrogen evolution active site at the edge of $MoSe_2$ sheet, which showed that the Tafel slope can reach 49 mV/dec, demonstrating a good hydrogen evolution performance. Qu et al. [99] used the glangcing angle deposition (GLAD) method combined with the $N_2$/$H_2$ plasma technology to prepare the $MoSe_2$/Mo composite, as shown in Figure 15. The plasma-assisted selenization process gave $MoSe_2$/Mo a large surface area, while accelerating the charge transfer of the metal phase of $MoS_2$ exposed to the edge. The Tafel slope of catalysts can reach 34.7 mV/dec, which exhibits excellent HER catalytic activity.

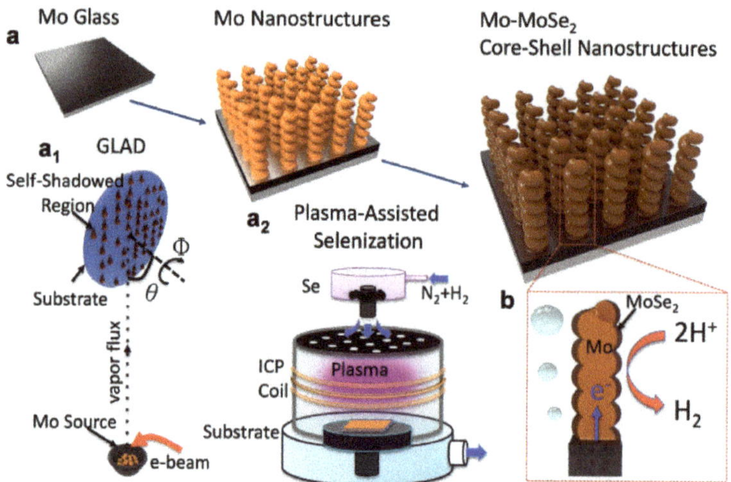



**Figure 15.** (**a**) Plasma-assisted selenization process: inset of (**a1**) MoSe$_2$/Mo composite prepared by GLAD method; inset of (**a2**) plasma-assisted selenization treatment. (**b**) Schematic diagram of HER and charge transfer [99]. (Reproduced with permission from [99]. Wiley-VCH, 2016).

### 3.5. Transition Metal Nitrides

Transition metal nitrides are often used as electrocatalysts, known as "quasi-platinum catalysts," providing a new way to prepare economical and efficient electrocatalysts. Zhang et al. [100] prepared a three-dimensional porous NiMoN material on carbon cloth with N$_2$ plasma treatment technology, which displayed high roughness and electron transport capability. The overpotential can reach 109 mV at a current density of 10 mA/cm$^2$ due to its excellent synergistic effects from Ni, Mo, and N for HER performance. Ouyang et al. [101] used N$_2$ plasma treatment to grow a 3D structure of nitriding (hNi$_3$N) on Ni foam (NF), which showed excellent OER catalytic activity and laid the foundation for the development of a high-performance, metal nitride, energy storage switching electrode. In 2018, Liu et al. [102] produced nickel nitride (Ni$_3$N$_{1-x}$) rich in nitrogen vacancies by nitriding Ni foam (NF) by microwave plasma-generation (Figure 16a,b). The presence of nitrogen vacancies effectively promoted the adsorption of water molecules, improved the adsorption-desorption behavior of the intermediate adsorbed hydrogen, and enhanced the HER activity of Ni$_3$N$_{1-x}$, as shown in Figure 16c–f. The self-supporting Ni$_3$N$_{1-x}$/NF electrode showed an overpotential of 55 mV and a Tafel slope of 54 mV/dec in an alkaline solution at the current density of 10 mA/cm$^2$.

Some nitride also exhibits excellent electrocatalytic properties in terms of ORR catalytic performance. Wang et al. [103] deposited high-density discrete Cu$_3$N nanocrystals on XC-72 carbon black by plasma enhanced atomic layer deposition (PEALD). It was found that the work function of Cu$_3$N nanocrystals was 5.04 eV, which was lower than Pt (5.60 eV). Cu$_3$N had stronger electron transfer performance than typical Pt catalysts. At the same time, the synergistic coupling effect between Cu$_3$N nanocrystals and carbon support made the Cu$_3$N$_{200}$/C sample exhibit a smaller $\pi$ (= 4.34 eV) than the pure Cu$_3$N nanocrystals. It displayed excellent ORR catalytic activity, significantly improved quality activity, and greater durability. Panomsuwan et al. [104] prepared iron-nitrogen-doped carbon nanoparticle–carbon fiber (Fe–N–CNP–CNF) by solution plasma. Because of the synergistic effect of the high graphitization of CNF, the mesoporous/macroporous CNP, the catalytic active center of ORR (graphite N and Fe-N bond), and the carbon-encapsulated Fe/Fe$_3$C particles, the Fe–N–CNP–CNF obtained showed excellent catalytic activity, durability and methanol resistance toward oxygen reduction (ORR) in an alkaline solution. Zhong et al. [105] found that the use of air plasma etching of iron-nitrogen co-doped porous carbon (Fe–N/C) electrocatalysts can remove sp$^3$ Cs with poor stability and amorphous $sp^2$ Cs, exposing more active catalytic $FeN_4$ centers, and transforming a small number of Fe-based nanoparticles into $FeN_4$ species, which significantly improved the catalytic activity of Fe–N/C for the oxygen reduction reaction (ORR) in acidic and alkaline electrolytes.

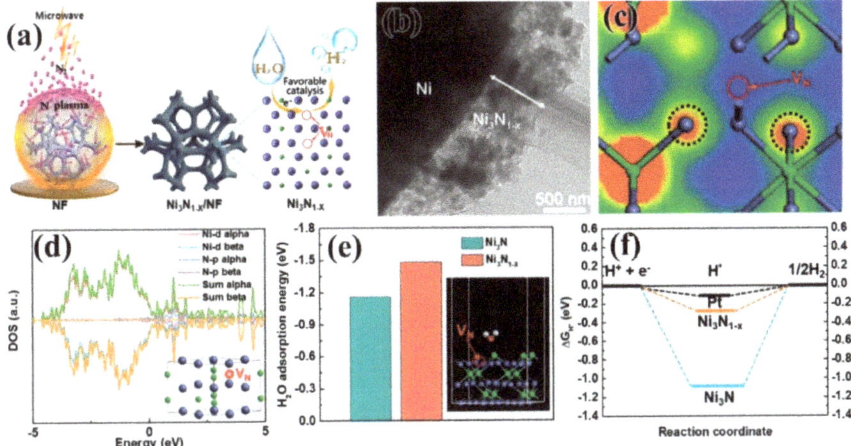

**Figure 16.** (**a**) Schematic diagram of preparation of self-supporting $Ni_3N_{1-x}$/NF electrode by microwave plasma-generation; (**b**) TEM images, (**c**) charge density distribution, and (**d**) the total electron density and partial electron density (TDOS and PDOS) of $Ni_3N_{1-x}$/NF; (**e**) adsorption energy of $H_2O$ molecules on the surface of $Ni_3N$ and $Ni_3N_{1-x}$; (**f**) free energy of H adsorption by various substances at equilibrium potential [102] (open access).

### 3.6. Transition Metal Phosphides

Transition metal phosphides have attracted great interest as electrocatalytic catalysts, particularly for HER and OER catalysts for electrocatalytic water splitting [106]. Currently, bimetallic phosphides exhibit higher catalytic activity than monometallic compounds [107] Liang et al. [108] realized the transition of NiCo–OH to NiCoP on nickel foam by $PH_3$ plasma, and prepared NiCoP electrocatalysts, as shown in Figure 17a, which exhibited excellent HER and OER electrocatalytic activity under alkaline conditions (Figure 17b–d). Zhang et al. [109] used plasma-enhanced chemical vapor deposition (PECVD) to form phosphate and phosphide groups on foamed nickel in the presence of $PH_3$, $CO_2$, and $H_2$ to form NiFePi/P. The change in the surrounding electronic environment of metal ions, due to the strong synergistic effect between phosphate and phosphide, not only increased the active center, but also improved the wettability and metal properties of the catalyst, the high conductivity, the wettability, and the active sites, which lead to excellent OER performance in an alkaline solution.

The doping of metal elements can also effectively balance the adsorption/desorption of oxygen and hydrogen-containing intermediates of metal phosphides, while enhancing density of electronic states' (DOS) strength and enhancing charge transfer kinetics. Dinh et al. [110] etched metal V-doped $Ni_2P$ nanosheets (V-$Ni_2P$) by $O_2$ plasma, and found that V substitution in the prepared OV–$Ni_2P$ may introduce defects and crystal dislocations, thus increasing the number of active sites and the carrier concentration. The strength of the oxygen bonding and the charge transfer kinetics were significantly enhanced after vanadium doping. Due to the $O_2$ plasma treatment, the phosphide surface was partially oxidized to produce a phosphate–phosphide region which balanced the adsorption/desorption of hydrogen and oxygen-containing intermediates. After the plasma, the Brunauer-Emmett-Teller (BET) surface area and the hydrophilicity of the material increased significantly. O–V–$Ni_2P$ had a large surface area (168.2 $m^2$/g) and excellent hydrophilicity (contact angle of 16.8°), leading to a higher electroactive surface area and a lower charge transfer resistance. The study found that the 10% V-doped $Ni_2P$

nanosheets ($O_3$–$V_{10}$–$Ni_2P$) treated by $O_2$ plasma for 3 min gave the best performance. In the 1.0 M KOH solution, η10-HER and η10-OER were 108 mV and 257 mV, respectively. In addition, a smaller Tafel slope was obtained for HER (72.3 mV/dec) and OER (43.5 mV/dec). As comprehensive, dual-function catalysts for water electrolysis, $O_3$–$V_{10}$–$Ni_2P$||$O_3$–$V_{10}$–$Ni_2P$ batteries required η10 to be only 1.563 V, which was superior to the most advanced $IrO_2$||Pt/C (1.687 V). The catalysts also had significant durability and 20 h of operational stability. These excellent properties were attributed to high-valence vanadium doping, which acted as a good electron acceptor and contributed to OER performance. Peng et al. [111] prepared nickel-doped amorphous FeP nanoparticle-supported titanium nitride nanowire array (Ni–FeP/TiN/CC) composites by high-energy metal ion implantation, which showed excellent HER performance in alkaline systems. The overpotential was 75 mV at a current density of 10 mA/cm$^2$, and the value remained at 93% of the initial value after continuous hydrogen evolution for 10 h at a high overpotential of 300 mV.

Cobalt phosphide is a new type of catalyst that has been developed in recent years to replace platinum metal-based electrocatalysts. However, the stability of cobalt phosphide is not as good as platinum, which limits the large-scale application of $CoP_x$ in water electrolyzers. Goryachev et al. [112] prepared $Co_3O_4$ films by plasma enhanced atomic layer deposition method, and obtained $CoP_x$ electrodes by thermal phosphating ($PH_3$), which has CoP and surface rich P (P/CO>1). It was found that the surface activation energy and the exchange current density was 81 ± 15 kJ/mol and $j_0$ = −8.9 × 105 A/cm$^2$. Liu et al. [113] obtained plasma-activated (PA)–CoPO by the treatment of $Co_3(PO_4)_2$ nanosheet arrays on a nickel network with $H_2$ plasma, which had larger surface area, enhanced conductivity, rich coordination unsaturated $Co^{3+}$, and numerous oxygen vacancies. The catalysts showed excellent OER and HER performance. The voltages of oxygen evolution and hydrogen evolution were 240 mV and 50 mV at a current density of 10 mA/cm$^2$, and the Tafel slopes were 53 mV/dec and 35 mV/dec, respectively. Using a PA–CoPO nanosheet array as the anode and cathode, a full water split of 10 mA/cm$^2$ was achieved at a low voltage of 1.48 V, which was superior to $IrO_2$/C–Pt/C at a sufficiently high overpotential. That provides a path for the design of high performance electrocatalysts for total hydrolysis, as shown in Figure 17e,f.

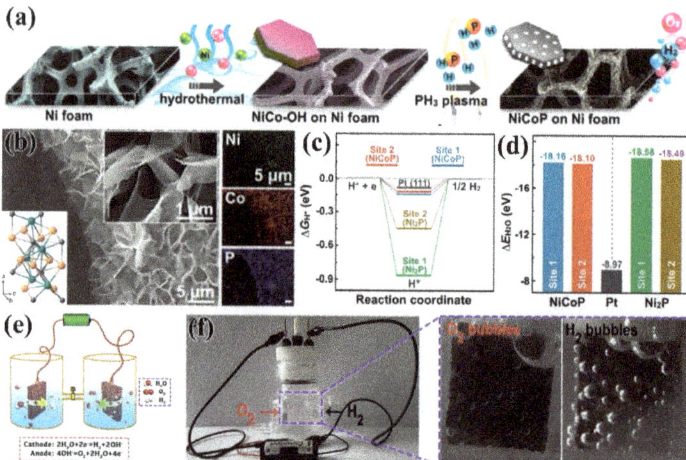

**Figure 17.** (**a**) Schematic diagram of the conversion of NiCo–OH nanosheets to NiCoP by $PH_3$ plasma treatment; (**b**) SEM image, energy spectrum, and crystal structure of NiCoP; (**c**) free-energy diagram for $H_2$ adsorption on the $Ni_2P$, NiCoP (0001), and Pt (111) surfaces; (**d**) adsorption energy of water [108] (Reproduced with permission from [108]. American Chemical Society, 2016.) (**e**) Schematic diagram of the bifunctional electrocatalysts OER and HER; (**f**) photograph of the water splitting in the alkaline solution at 1.5 V [113] (Reproduced with permission from [113]. Elsevier B.V., 2018).

### 3.7. Transition Metal Carbides

Nickel carbide ($Ni_3C_x$) films are prepared by the $H_2$ plasma atomic layer deposition technique. They are polycrystalline and highly homogeneous, with a rhombic $Ni_3C$ crystal structure, and without any nanographite or amorphous carbon [114]. Xiong et al. [115] used the process to conformally coat a uniform thin layer of $Ni_3C$ on carbon nanotubes (CNTs), in order to obtain core-shell nanostructured $Ni_3C$/CNT composites. The ALD-prepared $Ni_3C$/CNT composite exhibited excellent performance in supercapacitors and electrocatalytic hydrogen evolution. Ko et al. [116] used plasma-assisted deposition to grow WC nanowalls from bottom to top on a silicon wafer, which were highly crystalline and showed superior HER performance in a 0.5 M $H_2SO_4$ solution. Not only was the Tafel slope 67 mV/dec, but also oxidation would not occur even after 10,000 cycles of recycling, which displayed good durability.

### 3.8. Other Compounds

In 2018, Guo et al. [117] performed a non-destructive modification of the Prussian blue analog (PBA) structure by air plasma. The reactive oxygen species produced by plasma selectively bound to the metal in the framework, while retaining the porous structure of the framework with the high dispersion and orderliness of metal sites in the framework. Porous catalyst Co–PBA-plasma 2 h was obtained with extremely high oxygen evolution activity. Prussian blue was a Fe/Co double metal cyanide skeleton composed of cyanide bridged Fe and Co cations (Figure 18a–c). The reactive oxygen species generated in the plasma were bonded to the open sites of Co, promoting the oxidation state transition of Co to Co(III) (Figure 18d). Due to the highly reactive metal catalytic sites in the nanoporous framework, the OER catalysts showed a low overpotential characteristic of only 330 mV at a high current density of 100 mA/cm$^2$. This value is close to the actual operating conditions of industrial alkaline materials, implying a very promising technology for developing high performance catalysts. Yan et al. [118] used Ar plasma to prepare phytic acid–$Co^{2+}$ (P-Phy–$Co^{2+}$) with coordination unsaturation, which had an oxygen evolution potential of 306 mV at a current density of 10 mA/cm$^2$. The method can be extended to CoFe bimetallic system. The oxygen evolution potential of P-Phy–$Co^{2+}$/$Fe^{3+}$ prepared at the current density of 10 mA/cm$^2$ was 265 mV, and the Tafel slope was less than 36.51 mV/dec.

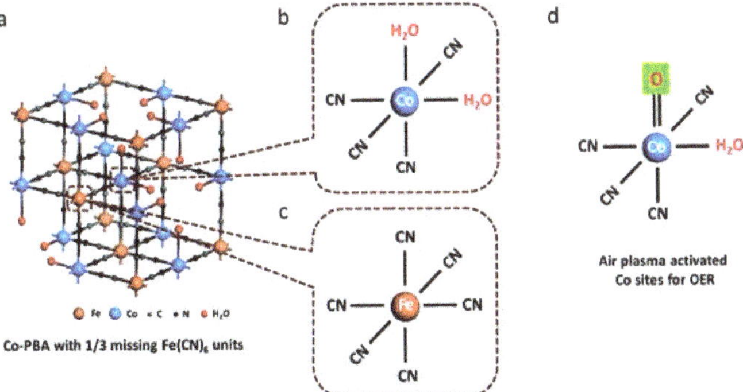


**Figure 18.** (**a**) The structure of Co–PBA with the composition of $Co_3(Fe(CN)_6)_2$; the coordination structure of (**b**) Co and (**c**) Fe sites. Each Co center has two open sites, which are occupied by coordinated water molecules, while the Fe center is completely coordinated by six CN groups; (**d**) schematic diagram of metal sites in Prussian blue's structure modified by air plasma [117]. (Reproduced with permission from [117]. Wiley-VCH, 2018).

## 4. Carbon Based Electrocatalysts

Carbon is one of the most abundant and important elements in nature. Carbon-based catalysts have become one of the most popular electrocatalysts in recent years [119]. Carbon-based materials have more advantages than traditional materials, such as low costs, various structures, and good electrical and thermal conductivity. The excellent catalytic performance can be obtained by tailoring carbon-based materials with specific sizes, doping types, contents, morphologies and structures, promoting active site exposure, increasing the transport of reaction-related substances, and enhancing the transfer of electrons throughout the electrode [120,121]. At the same time, more defects become the active sites of the electrocatalyst after heteroatom doping. In addition, intrinsically defective carbon electrode catalysts also exhibit catalytic activity comparable to carbon materials doped with heteroatoms (e.g., F, S, P, and B) [122].

### 4.1. Defective Carbon Materials

Defective carbon materials have also attracted attention for enhancing electrocatalytic activity. Generally, defect-rich materials are prepared at elevated temperatures or using template precursors to form more edge and hole defects [123]. In 2016, Tao et al. [124] obtained defect-rich graphene and carbon nanotubes by using Ar plasma to etch the surface of graphene and carbon nanotubes, as shown in Figure 19a–e. The P-CNTs acquired with plasma treatment had also many defects.

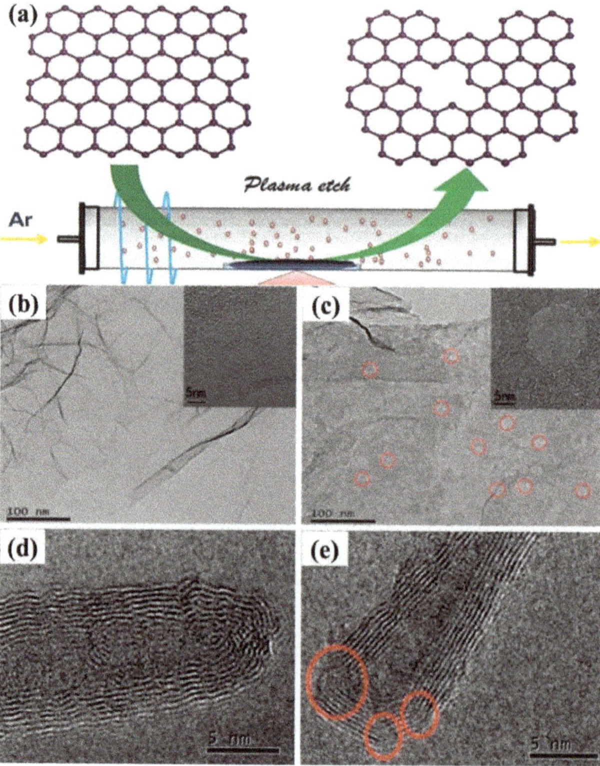


**Figure 19.** (a) Schematic diagram of the preparation of Ar plasma etching graphene and carbon nanotubes surface by Ar plasma; TEM images of (b) graphene and (c) Ar plasma treated graphene; TEM images of (d) carbon nanotubes and (e) Ar plasma treated carbon nanotubes [124]. (Reproduced with permission from [124]. Royal Society of Chemistry, 2016).

Subsequently, Liu et al. [125] directly treated commercial carbon cloth with Ar plasma technology and exposed it to air, which not only made the surface of carbon fiber more rough and porous, but also exposed more graphene-like nanosheets on the surface. And the deficient, oxygen-doped graphene was also produced in situ on its surface. Compared with pure carbon cloth, the distance between carbon nanosheets was about 0.37 nm after simple plasma etching of carbon cloth P-CC, which was larger than that of graphite $d_{002}$ = 0.34 nm. The value of $SP^2/SP^3$ of carbon was significantly reduced, which means that plasma treatment can produce more defects. The carbon fibers after plasma etching exhibited a larger specific surface area, exposing more active sites, and the treated carbon fibers had better conductivity, making them more conducive to material transfer, resulting in better catalytic performance in OER and ORR. An amorphous, edge-rich/defective graphene was generated in situ on the surface of the carbon fibers by argon plasma etching, while the dangling bonds of these defect sites were exposed to air and reacted with oxygen or water to achieve oxygen functionalization. In 2018, Lehmann et al. [126] used plasma-enhanced chemical vapor deposition to prepare defect-rich layered carbon nanowalls (hCNW) with readily accessible graphite edge locations at the top of the wall and many defect locations within the porous sidewalls, which was considered to be the ideal sites for adsorption and electron transfer. And the ORR initiation potential of hCNW-60 in 0.1 M KOH solution was 830 mV with a two electron transfer.

### 4.2. Nitrogen-Doped Carbon Materials

The doping of nitrogen atoms in graphene is considered to be a good way to improve the electrocatalytic performance. Wang et al. [127] mainly prepared N-doped graphene oxide (N–PEGO) by low temperature plasma technology. Ammonium carbonate was used as the activator and nitrogen source in the preparation of N–PEGO by low temperature plasma technology, which can effectively achieve the exfoliation of graphene oxide and nitrogen doping in one step (Figure 20). The N–PEGO prepared by this method has a nitrogen doping amount of 5.3 at%, a specific surface area of 380.0 $m^2/g$, an initial potential of 0.89 V (versus RHE), and an excellent oxygen reduction reaction (ORR) catalysis, which also showed better stability and methanol resistance than the commercial noble metal Pt/C catalysts. Zhu et al. [128] achieved nitrogen doping of graphene foam by $N_2$ plasma technology (NGF–CFP), which exhibited excellent OER performance.

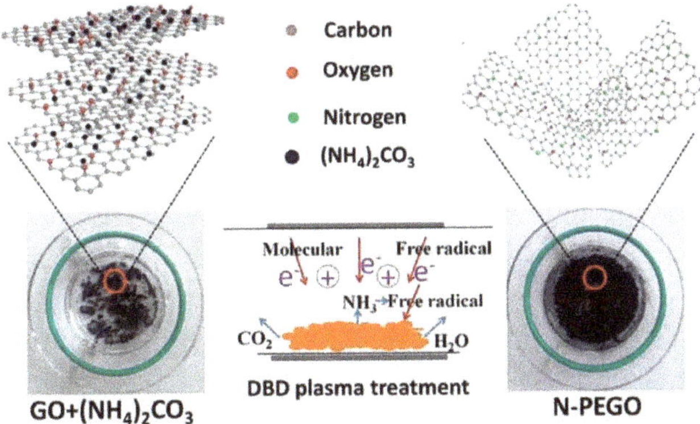


**Figure 20.** Schematic diagram of N-doped graphene oxide prepared by low temperature plasma technology [127]. (Reproduced with permission from [127]. Royal Society of Chemistry, 2018).

N-doped carbon nanotubes are also used as electrocatalysts [129]. In 2015, Du et al. [130] successfully synthesized nitrogen-doped carbon nanotubes (NCNTs) with a nitrogen content of 5.38 at.% via microwave plasma chemical vapor deposition (MPCVD). Because of the doping of nitrogen, NCNT exhibited the same ORR electrocatalytic activity as Pt–CNT, which was 0.87 V of the initial point and 4.1 of the electron transfer number, and had the highest ORR performance at a load of 729 $\mu g/cm^2$. Subramanian et al. [131] prepared carbon nanotube carpets (VA-NCNTs) with nitrogen-doped vertical alignment by treating carbon nanotubes with $N_2$ plasma. Compared with undoped nitrogen VA-CNTs, VA-NCNTs exhibited well electrocatalytic performance under alkaline conditions. Zhang et al. [132] obtained NCNT/glass carbon (GC) electrode by treating carbon nanotubes with $NH_3$ plasma, which can reduce $CO_2$ to formate in water without using metal catalysts. If polyethyleneimine (PEI) is coated on NCNT/GC electrode, the catalytic overpotential can be significantly reduced, and the current density and efficiency can be increased, which is helpful for stabilizing the intermediate of the $CO_2$ reduction of PEI.

The plasma deposition method and the solution plasma method are considered to be effective methods for one-step synthesis of nitrogen-doped carbon nanoparticles (NCNP) [133,134]. NCNPs can be synthesized in situ by solution plasma, and the type of C–N bond can be controlled by the structure of precursors and additives (Figure 21a–c). Li et al. [135] prepared different NCNPs using pyridine and acrylonitrile as heterocyclic and linear structure precursors, and using ruthenium as additives, which can realize the control of nitrogen elements (Figure 21d–g). It was found that the current density was proportional to the content of graphite-N. The existence of graphite-N promoted the direct four-electron transfer pathway of ORR, while the higher percentage of amino-N made the ORR initiation potential move to a positive value. Amino-N and graphite-N played a synergistic role in improving ORR activity. Li et al. [136] produced N-doped carbon nanoparticles by solution plasma containing pyridine-N, amino-N, and graphite-N bonds, which provided a simple and effective method to study the relationship between C–N bonding structure and the electrochemical performance of N-doped carbon catalysts. At the same time, Panomsuwan et al. [137,138] obtained NCNPs samples with different nitrogen doping content (0.63–1.94 at.%) by changing the difference of C/N molar ratio in organic precursors, using organic liquid mixtures such as benzene and pyrazine as precursors. It was found that the initial potential and current density of the ORR electrochemical properties of NCNP were improved with the increase of nitrogen doping content, which was mainly due to the graphite-N and pyridine-N on NCNPs. Compared to commercial Pt/C catalysts, NCNP showed superior long-term durability and strong methanol tolerance.

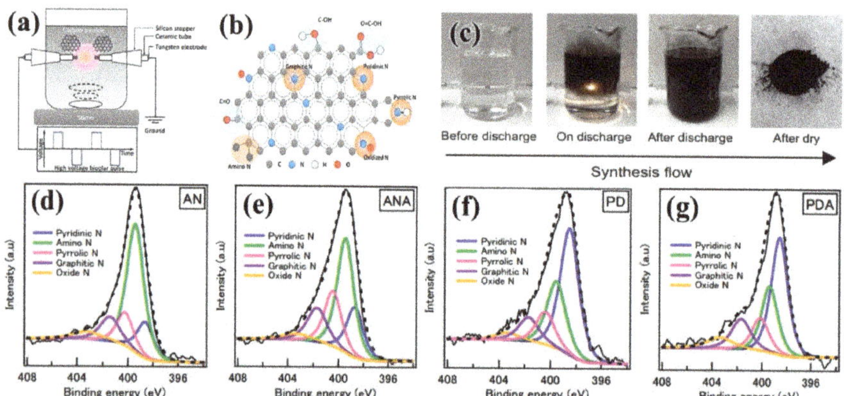

Figure 21. (a) Schematic diagram of preparation of N-doped carbon nanoparticles (NCNP) by solution plasma method; (b) species of doped single atoms [135] (copyright @ 2017 Royal Society of Chemistry.) (c) Process for preparing nano-carbon materials by solution plasma method [139] (open access.) (d–g) Distribution of nitrogen elements of NCNP prepared from acrylonitrile (AN), acrylonitrile + ANA, pyridine (PD), and pyridine + hydrazine (PDA) as precursors [135]. (Reproduced with permission from [135]. Royal Society of Chemistry, 2017).

### 4.3. Oxygen-Doped Carbon Materials

The solution plasma method can also be used to synthesize carbon nanoparticles containing oxygen in one step. Ishizaki et al. [140] successfully synthesized oxygen-containing carbon nanomaterials using a mixture of benzene (BZ) and 1,4-dioxane (DO). Although DO content did not affect the initial potential, it did influence the current density of ORR. With the increase of DO content, the O contents in the samples increased, such that the order of current density of ORR carbon nanosamples was as follows: BZ90 + DO10 > BZ100 > BZ70 + DO30 > BZ50 + DO50. Kondratowicz et al. [141] used oxygen plasma to adjust the properties of reduced graphene oxide (rGO), which was an easy-to-control and eco-friendly method. Oxygen plasma treatment can improve the adsorption of enzymes on rGO electrodes by introducing oxygen groups and increasing porosity. With different plasma treatment times, different oxygen groups (such as carboxyl and hydroxyl groups) can be introduced on the surface of rGO to change the wettability of rGO, and other functional groups (such as quinones and lactones) can also be produced in a longer treatment time. In addition, the external surface of rGO was partially etched, resulting in an increase in surface area and porosity of the material. The current density of rGO treated for 10 min was twice as high as that of untreated rGO.

### 4.4. Sulfur-Doped Carbon Materials

Sulphur atom doping can effectively change the electronic and chemical properties of graphene, making it possible for graphene to be used as the electrocatalyst. In 2016, Wang et al. [142] realized the reduction of graphene oxide and doping of S with the treatment of microwave-assisted stripping and hydrogen sulfide plasma. Sulfur-doped graphene was used for an ORR electrocatalytic reaction under alkaline conditions, showing excellent electrochemical performance. In 2017, Ting et al. [143] etched sulfur-doped graphene by Ar plasma, resulting in more topological defects, while maintaining the original doping structure of SG (Figure 22a). Benefiting from the synergistic coupling of S-doping and plasma-induced topographic defects, SG–P had greatly enhanced HER activity and good stability in acidic media, demonstrated by its low overpotential of 178 mVat the current density of 10 mA/cm$^2$, and Tafel slope of 86 mV/dec. The optimum HER activity of SG–P can be obtained by combining thiophene-rich substances with appropriate plasma-induced topological effects (Figure 22b–g). In 2018, Zhou et al. [144] obtained the self-supporting electrode 3DSG-Ar by Ar plasma treatment of three-dimensional sulfur-doped graphene (3DSG), which showed excellent HER electrocatalytic activity. After 2000 cycles, it still had good electrocatalytic stability, and the Tafel slope was 64 mV/dec.

### 4.5. Boron-Doped Carbon Materials

Panomsuwan et al. [145] prepared boron-doped carbon nanoparticles (BCNP) by solution plasma process using benzene and triphenyl borate as precursors (Figure 23). Compared with undoped carbon nanoparticles, the electrocatalytic activity of BCNP for oxygen reduction reaction (ORR) in alkaline solution was improved in terms of initial potential and current density. In addition, BCNP showed excellent long-term durability and methanol oxidation resistance in ORR. Li et al. [146] obtained porous BDD/Ta multilayers by etching polycrystalline boron-doped diamond (BDD) on tantalum substrates with H$_2$/Ar plasma. It was found that the effective electroactive surface area and charge transfer ability of etched microcrystals at liquid crystal interface were improved. The porous BDD/TA electrodes were applied to electro-Fenton method for rapid degradation of methylene blue.

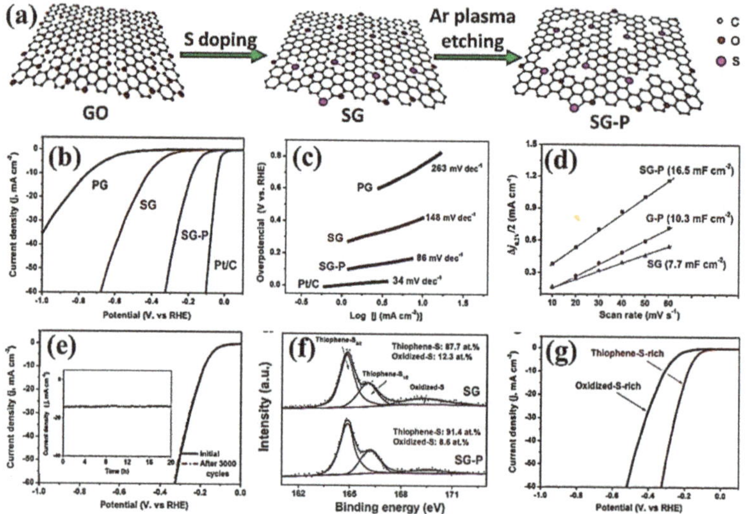

**Figure 22.** (**a**) Schematic diagram of the treatment of S-doped graphene BY Ar plasma. (**b**) Polarization potential and (**c**) Tafel slope at a scan speed of 5 mV/s in a 0.5 M H2SO4 solution of PG, SG, SG–P, and commercial 20% Pt/C. (**d**) Current densities of SG, G–P, and SG-P. (**e**) Cycle performance of SG–P. (**f**) PG, SG, SG–P, and commercial 20% Pt/C of S's 2p for SG and SG–P. (**g**) Different polarization curves for SG–P enriched in thiophene sulfur and oxidized sulfur [143]. (Reproduced with permission from [143]. Elsevier B.V., 2017).

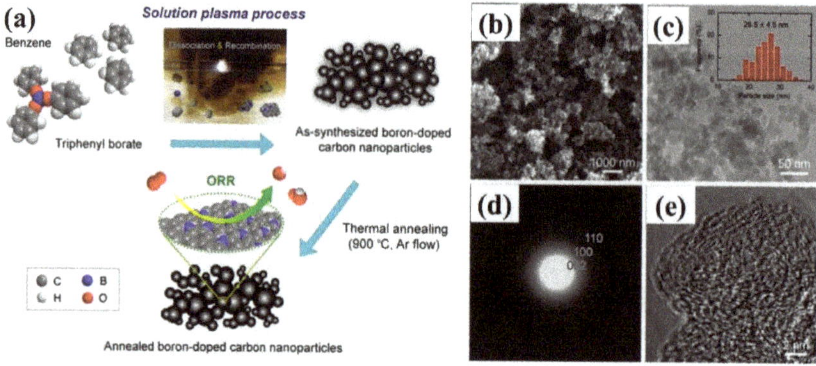

**Figure 23.** (**a**) Schematic diagram of the preparation of B-doped nanoparticles (BCNP) by solution plasma. (**b**) SEM image, (**c**) TEM images, and particle size distribution image; (**d**) SAED diffraction pattern; and (**e**) HRTEM images of BCNP [145]. (Reproduced with permission from [145]. Elsevier B.V., 2015).

**Table 1.** Preparation of electrocatalysts by plasma technology.

| Reaction Type | Samples | Methods | Electrochemical Performance | Ref. |
| --- | --- | --- | --- | --- |
| MOR | $Pt/TiO_2$ | PEALD | the MOR current density drops to a small value after 1500 s with NCALD < 30. | [15] |
| | Pt/C | CAPD | The calculated ECSAs of 75.4 $m^2$/g | [16] |
| | Pt/GNT | $H_2$ plasma | The current density of 97.9 mA/mg and mass activity of 691.1 mA/mg Pt | [20] |
| | Pt/CNTs-HP | $H_2$ plasma | The current density of 15.8 mA/mg | [21] |
| | Au@Pt | Ar plasma | Mass activity up to $48 \pm 3$ $m^2$/g | [23] |
| | Pt-Ag | SPS | The calculated ECSAs of 28.15 $m^2$/g | [27] |
| | $Pt_{40}Pd_{60}$ | SPS | The calculated ECSAs of 28.15 $m^2$/g | [28] |
| | $Pt_{69}Pd_{31}$ | SPS | The catalytic activity of 6.81 mA/$cm^2$ | [29] |
| | PtPd/KB-2 | SPS | The electrocatalytic activity was more than 4 times of that of commercial Pt/C | [30] |
| | Pt/CoPt/MWCNS | SPS | Mass activities of 1719 mA/mg Pt | [32] |
| | $Pt/C/TiO_2$-2 | SPS | Mass activities of 315.2 mA/mg Pt | [33] |
| | Pt-Ru/OCNT | $O_2$ plasma | The onset potential of 0.3 V | [44] |
| | Pt/ZnO/KB | SPS | Catalytic activity of 964 mA/mgPt | [48] |
| OER | Pt-Ir/TiC | Ar plasma | The current density of 2.5 mA/$cm^2$ at 1 V | [22] |
| | $PtO_aPdO_b@Ti_3C_2T_x$ | SPS | Activation potential of 1.5 V in 0.1 M KOH | [40] |
| | Ag/1400-15 | Spark plasma sintering | The value of 61 mV | [41] |
| | CoNPs@ C | MPECVD | Overpotential of 270 mV | [50] |
| | CoO–N/C | Cold plasma deposition | The oxygen evolution potential of 378 mV | [55] |
| | $Co_3O_4$ nanosheets | Ar plasma | The current density of 44.44 mA/$cm^2$ at 1.6 V | [60] |
| | $V_o$-COOH | Ar plasma | The low overpotential of 262 mV at 10 mA/$cm^2$ | [61] |
| | $Co_3O_{4-x}$ | Ar plasma | The overpotential of 330 mV and Tafel slope of 58 mV/dec | [62] |
| | $N-Co_3O_4$ | $N_2$ plasma | The required potential of 1.54 V to reach the current density of 10 mA $cm^{-2}$ | [64] |
| | $P-Co_3O_4$ | Ar plasma | The overpotential of 280 mV and Tafel slope of 51.6 mV/dec | [67] |
| | $CoO_x$-ZIF | $O_2$ plasma | The required potential of 1.548 V to reach the current density of 10 mA $cm^{-2}$ | [69] |
| | SnCoFe-Ar | Ar plasma | The overpotential of 300 mV and Tafel slope of 42.3 mV/dec | [71] |
| | $SrTi_{0.5}Fe_{0.5}O_{3-y}$ | HT-PVD using $O_2$ plasma | The onset potential of 1.40 at 100 µA | [72] |
| | SCFP-NF | High-energy argon plasma sputtering | The ultra high mass activity of 1000 mA/mg | [73] |
| | $CoFe-LDHs-H_2O$ | $H_2O$ plasma | The overpotential of only 232 mV | [74] |
| | CoFe-LDHs-Ar | Ar plasma | The overpotential of only 232 mV at a current density of 10 mA/$cm^2$ | [77] |
| | $CoFe-LDHs-N_2$ | $N_2$ plasma | The overpotential of only 233 mV at a current density of 10 mA/$cm^2$ | [78] |
| | $FeO_x$/C | pulsed arc plasma deposition | The discharge specific capacity of 500 mAh/g under 100 mA/g | [80] |
| | $Co_9S_8$/G | $NH_3$ plasma | The Tafel slope of 82.7 mV/dec | [91] |
| | $Cu_2S$/CF | active iodine DBD plasma | The overpotential of 290 mV at a current density of 10 mA/$cm^2$ | [93] |
| | $CoS/Ni_3S_2$-FeS/PNFF | Air plasma | The overpotential of 136 mV at a current density of 10 mA/$cm^2$ | [96] |

**Table 1.** *Cont.*

| Reaction Type | Samples | Methods | Electrochemical Performance | Ref. |
|---|---|---|---|---|
| | $hNi_3N$ | $N_2$ plasma | The overpotential of 325 mV at a current density of 10 mA/cm$^2$ | [101] |
| | NiCoP | $PH_3$ Plasma | The overpotential of 280 mV at a current density of 10 mA/cm$^2$ | [108] |
| | NiFePi/P | PECVD | The overpotential of 230 mV at a current density of 10 mA/cm$^2$ | [109] |
| | $O_3$-$V_{10}$-$Ni_2P$ | $O_2$ plasam | The Tafel slope of 43.5 mV/dec | [110] |
| | PA-CoPO | $H_2$ plasma | The overpotential of 240 mV and Tafel slope of 53 mV/dec | [113] |
| | Co-PBA-plasma 2 h | Air plasma | The overpotential of 330 mV at a current density of 100 mA/cm$^2$ | [117] |
| | NGF-CFP | $N_2$ plasma | The current density of 10 mA/cm$^2$ at 2.16 V | [128] |
| SO$_2$OR | $Pt_3Pd_2$ and $PtPd_4$ | Plasma sputtering | The lower onset potential of 0.587 ± 0.004 V | [18] |
| HOR | Pt/C | SPS | The ECSAs value of 266 cm$^2$/mg | [38] |
| AMXOR | $Ti_4O_7$ | Plasma deposition | A quick oxidation of 0.1 mM AMX | [79] |
| | Pt/C | CAPD | The half-wave potential of 0.87 V | [16] |
| | Pt-Ir/TiC | Ar plasma | Nyquist plots of 0.6 V | [22] |
| | Au@Pt | Ar plasma | The peak potential of 0.75 V | [23] |
| | Pt/XC72 | SPS | The onset potential of 1.04 V | [31] |
| | PdAu/KB | SPS | The reduction peak disappeared after about 700 cycles | [34] |
| | AgNW/C | SPS | The high electron density of $13.7 \times 10 - 22$ m$^{-3}$ | [37] |
| | Pt/rGO-N | SPS | The half-wave potential of 0.87 V | [39] |
| | Ag/1400-15 | spark plasma sintering | The electron transfer number of 3.9 | [41] |
| | PtNW/GDL | $N_2$ + $H_2$ plasma | The power density of 64 mW/cm$^2$ | [47] |
| | CoO–N/C | Cold plasma deposition | A 2-electron process producing $H_2O_2$ | [55] |
| | Ag@$Co_3O_4$ | SPS | The half-wave potential of 0.799 V | [56] |
| | 15$Co_3O_4$/N-AP/800 | microwave-induced plasma | The Tafel slope of 42 mV/dec | [57] |
| | Co-La-Pt | DC arc discharge plasma | The specific capacity of 3250.2 mAh/g and energy density of 8574.2 Wh/kg at 0.025 mA/cm$^2$ | [58] |
| ORR | $Co_3O_{4-x}$ | Ar plasma | The half-wave potential of 0.84 V | [62] |
| | FeO$_x$/C | Pulsed arc plasma deposition | The discharge specific capacity of 500 mAh/g under 100 mA/g | [80] |
| | MnO$_x$@C-D | Air plasma | The electron transfer number of 3.81 | [81] |
| | A-$MnO_2$ | Ar plasma | The power density of 159 mW/cm$^2$ at a current density of 157 mA/cm$^2$ | [82] |
| | $Cu_3N_{200}$/C | PEALD | The half-wave potential of 0.684 V | [103] |
| | Fe-N-CNP-CN | SPS | The onset potential of −0.10 V | [104] |
| | Fe–N/C | Air plasma | The onset potential of 0.88 V at a loading of 0.60 mg/cm$^2$ | [105] |
| | P-Graphene | Ar plasma | The onset potential of 0.912 V and half-wave potential of 0.737 V | [124] |
| | P-CNTs | Ar plasma | The onset potential of 0.83 V | [124] |
| | P-CC | Ar Plasma | The exchange current density of $2.57 \times 10^{-9}$ A/cm$^2$ | [125] |
| | hCNW-60 | PECVD | The onset potential of 830 mV in 0.1 KOH | [126] |

**Table 1.** *Cont.*

| Reaction Type | Samples | Methods | Electrochemical Performance | Ref. |
| --- | --- | --- | --- | --- |
| | N-PEGO | DBD plasma | The onset potential of 0.89 V | [127] |
| | NCNTs | MPCVD | The onset potential of 0.87 V and the electron transfer number of 4.1 | [130] |
| | VA-NCNTs | $N_2$ plasma | The ORR peak at the potentials of about $-0.3$ V | [131] |
| | NCNPs | SPS | The onset potential of $-0.17$ V | [137] |
| | NCNP-3 | SPS | The onset potential of $-0.143$ V | [138] |
| | BZ90 + DO10 | SPS | The samples were held at 0.5 V with a rotation speed of 1500 rpm for 45,000 s in an $O_2$-saturated 0.1 M KOH solution | [140] |
| | O-rGO | $O_2$ Plasma | The current density of 10 $\mu$A/cm$^2$ | [141] |
| | BCNP | SPS | The current densities of 3.15 mA/cm$^2$ at $-0.60$ V | [145] |
| | FCNP-4 | SPS | The onset potential of 0.22 V and limiting current density of 2.76 mA/cm$^2$ at 0.6 V | [147] |
| | BCN nanocarbon | SPS | 15.1% current decrease after 20000 s | [149] |
| $CO_2RR$ | Au island | $O_2$ plasma | The faradaic efficiency over 95% | [24] |
| | CNT/Cu | $O_2$ plasma | Carbon monoxide yields of 178 mmol cm$^2$ mA$^{-1}$ h$^{-1}$ | [51] |
| | Cu foil | $O_2$ plasma | The ethylene selectivity of 60% | [52] |
| | Cu nanocube | $O_2$ plasma | The ethylene selectivity of 60% | [53] |
| | ZnO | $H_2$ plasma | The current density of $-16.1$ mA/cm$^2$ and faradaic efficiency of 83% | [84] |
| | PEI-NCNT/GC | $NH_3$ plasma | The current density of 2.2 mA/cm$^2$ | [132] |
| HER | Ni–Fe–C | $CH_4$+$H_2$ Plasma carburizing | The activation potential of 57 mV at a current density of 10 mA/cm$^2$ | [49] |
| | CoNPs@ C | MPECVD | The overpotential of 153 mV at a current density of 10 mA/cm$^2$ | [50] |
| | P-$Co_3O_4$ | Ar plasma | The overpotential of 120 mV and Tafel slope of 52 mV/dec | [67] |
| | SCFP-NF | high-energy argon plasma sputtering | The onset potential of $-0.01$ V and Tafel slope of 94 mV/dec | [73] |
| | C-30s | C plasma | The Tafel slope of 44 mV/dec | [83] |
| | $TaS_2$-15 min | $O_2$ plasma | The onset potential of 310 mV | [85] |
| | $O_2$-$MoS_2$ | $O_2$ plasma | The current density of 16.3 mA/cm$^2$ at $-350$ mV | [86] |
| | $MoS_2$ | O2 plasma | The overpotential of 131 mV at a current density of 10 mA/cm$^2$ | [87] |
| | $H_3Mo_{12}O_{40}P/MoS_2$ | $O_2$ plasma | The Tafel slope of 44 mV/dec | [88] |
| | $MoS_2$-15 min | $H_2$ plasma | The overpotential of 240 mV at a current density of 10 mA/cm$^2$ | [89] |
| | $MoS_{1.7}$ | $H_2$ plasma | The overpotential of 143 mV at a current density of 10 mA/cm$^2$ | [90] |
| | $Co_3S_4$ PNSvac | Ar plasma | The mass activity of 1056.6 A/g at an overpotential of 200 mV | [92] |
| | $WS_2$ | PEALD | The overpotential of 90 mV at a current density of 100 mA/cm$^2$ | [94] |
| | $WS_2$ | $SF_6$/$C_4F_8$ plasma-etched | The overpotential of 100 mV and Tafel slope of 50 mV/dec | [95] |
| | CoS/$Ni_3S_2$-FeS/PNFF | Air plasma | The overpotential of 75 mV at a current density of 10 mA/cm$^2$ | [96] |
| | N-$MoSe_2$/VG | MPECVD | The onset potential of 45 mV and overpotential of 98 mV at 10 mA/cm$^2$ | [98] |
| | $MoSe_2$/Mo | $N_2$/$H_2$ plasma | The Tafel slope of 34.7 mV/dec | [99] |
| | $Ni_3N_{1-x}$/NF | Microwave plasma | The overpotential of 55 mV and Tafel slope of 54 mV/dec at a current density of 10 mA/cm$^2$ | [102] |

**Table 1.** *Cont.*

| Reaction Type | Samples | Methods | Electrochemical Performance | Ref. |
| --- | --- | --- | --- | --- |
| | NiCoP | $PH_3$ plasma | The overpotential of 32 mV at a current density of $-10$ mA/cm$^2$ | [108] |
| | $O_3$-$V_{10}$-$Ni_2P$ | $O_2$ plasma | The overpotential of 108 mV and Tafel slope of 72.3 mV/dec | [110] |
| | Ni-FeP/TiN/CC | Plasma-implanted method | The overpotential of 75 mV at a current density of 10 mA/cm$^2$ | [111] |
| | $CoP_x$ | PEALD | The exchange current density of $-8.9 \times 105$ A/cm$^2$ | [112] |
| | PA-CoPO | $H_2$ plasma | The overpotential of 50 mV at a current density of 10 mA/cm$^2$ | [113] |
| | WC nanowalls | DC-PACVD | The Tafel slope of 67 mV/dec | [116] |
| | Co-PBA-plasma 2 h | Air plasma | The overpotential of 77 mV at a current density of 20 mA/cm$^2$ | [117] |
| | SG-P | Ar plasma | The overpotential of 178 mV at a current density of 10 mA/cm$^2$ | [143] |
| | 3DSG-Ar | Ar plasma | The Tafel slope of 64 mV/dec | [144] |
| | P-NSG | Ar plasma | The onset potential of 58 mV | [148] |

### 4.6. Fluorine-Doped Carbon Materials

Fluorine-doped carbon nanoparticles (FCNPs) can also be prepared by the solution plasma method. Panomsuwan et al. [147] used the mixture of toluene and trifluorotoluene as the precursor to prepare FCNPs, whose fluorine doping content can range from 0.95 to 4.52 at%. The obtained FCNPs mainly exhibited disordered amorphous structure, and the incorporation of fluorine atoms resulted in more defect sites and disordered structures in carbon particles. With the increase of fluorine doping content, the ORR electrocatalytic activity of FCNPs had been significantly improved, which was mainly due to the intercalation of ionic C–F and semi-ionic C–F bonds in the carbon structure. Compared with commercial Pt-based catalysts, FCNP exhibited excellent long-term operational durability and strong tolerance to methanol oxidation.

### 4.7. Heteroatom Co-Doped Carbon Materials

Diatomic co-doping is also considered as one of the effective strategies to improve the catalytic activity of carbon-based electrocatalysts. Tian et al. [148] obtained defect-rich P–NSG samples by the etching of N and S co-doped graphene (NSG) with Ar plasma. The synergistic coupling of N and S co-doping and plasma-induced structural defects maximized the number of active sites of graphene, which significantly improved the HER catalytic activity of P–NSG in both acidic and alkaline media. Lee et al. [149] synthesized boron–carbon–nitrogen (BCN) nanocarbon materials by solution plasma method in one step. The synergistic effects of N and B at the state of uncoupled bonding changed the electronic structure of basic carbon and promoted the formation of new ORR active sites. Although the electron transfer number of BCN nanocarbon was 3.43, and the ORR activity was not as good as commercial Pt/C, the current only decreased by 15.1% after 20,000 s, which was obviously better than commercial Pt/C (under the same conditions, the current decreased by 61.5%).

## 5. Conclusions and Prospects

Electrocatalytic materials have been widely used in energy and environmental fields, such as electrocatalytic hydrogen production, the reduction of carbon dioxide, fuel cells, the $N_2$ reduction of ammonia, and so on. Noble metals have high energy utilization efficiency and excellent catalytic performance, but they have the disadvantages of poor availability and high prices. Therefore, it is important to develop highly active metal catalysts with low mass loading and high dispersibility. In recent years, researchers have focused on non-noble metal-based catalysts with low costs, high catalytic activities and long lives, such as transition metal catalysts, oxides, sulfides, nitrides, phosphides, and carbides. At the same time, carbon-based catalysts have been favored by researchers because of their superior electrocatalytic performances in both acidic and alkaline systems. The plasma device has a simple and adjustable structure, which can be used to prepare and modify the electrocatalysts.

In the preparation of electrocatalysts by plasma, different types of plasma may have different effects on the materials. The plasma can assist the precursor of gas or liquid-derived atoms, molecules, ions, or radicals to drive the synthesis of electrocatalysts. For example, the plasma deposition method can directly prepare a catalyst material with low loading and excellent dispersibility; the solution plasma can directly synthesize noble metal catalysts, noble metal alloy catalysts, and hetero atom-doped carbon-based catalysts; as a "bottom-up" synthesis technology, low-temperature plasma can be directly used to prepare specific nanostructured electrocatalysts. However, how to further improve the efficiency of plasma-assisted synthesis and expand its application in large-scale industrial production, is still challenging.

Plasma can achieve "top-down" denudation or surface treatment in plasma modification of electrocatalysts. Since the electrocatalytic process often occurs on the surface of catalysts, the structure regulation of the surfaces of catalysts by plasma can effectively improve the catalytic activities of electrocatalysts. Firstly, electrocatalysts with uniform dispersion and small particle sizes of active components can be obtained by plasma modification of the carrier or active component of electrocatalysts; secondly, plasma etching can be used to expose more interfacial defects on the surface of electrocatalysts and increase the catalytic active sites on the edge; thirdly, some heteroatoms (such as N, S, B, and P) can be doped by plasma to enhance the intrinsic defects, realizing the catalytic activity of electrocatalysts. Defects have been recognized as active sites with higher activity in electrocatalytic reactions (vacancies, marginal sites, and lattice defects). At present, how to precisely control and tailor the defects of the electrocatalysts, the surface modifications, and the atomic doping by plasma techniques, remains challenging.

**Author Contributions:** F.Y., B.D. and L.Z. investigated and designed the review. M.L., C.M., and L.D. wrote the original draft preparation. F.Y., B.D., and L.Z. edited and reviewed the manuscript.

**Funding:** This research was funded the National Natural Science Foundation of China (21663022 and 21773020), the Program for Changjiang Scholars and Innovative Research Team in University (IRT_15R46), and the Science and Technology Innovation Talents Program of Bingtuan (2019CB025).

**Conflicts of Interest:** The authors declare no competing financial interests.

## References


1. Seh, Z.W.; Kibsgaard, J.; Dickens, C.F.; Chorkendorff, I.; Nørskov, J.K.; Jaramillo, T.F. Combining theory and experiment in electrocatalysis: Insights into materials design. *Science* **2017**, *355*, eaad4998. [CrossRef] [PubMed]

2. Voiry, D.; Chhowalla, M.; Gogotsi, Y.; Kotov, N.A.; Li, Y.; Penner, R.M.; Schaak, R.E.; Weiss, P.S. Best Practices for Reporting Electrocatalytic Performance of Nanomaterials. *ACS Nano* **2018**, *12*, 9635–9638. [CrossRef] [PubMed]

3. Li, Q.; Rao, X.; Sheng, J.; Xu, J.; Yi, J.; Liu, Y.; Zhang, J. Energy storage through $CO_2$ electroreduction: A brief review of advanced Sn-based electrocatalysts and electrodes. *J. CO$_2$ Util.* **2018**, *27*, 48–59. [CrossRef]

4. Wang, Y.; Han, P.; Lv, X.; Zhang, L.; Zheng, G. Defect and Interface Engineering for Aqueous Electrocatalytic $CO_2$ Reduction. *Joule* **2018**, *2*, 2251–2587. [CrossRef]

5. Yan, D.F.; Li, Y.X.; Huo, J.; Chen, R.; Dai, L.M.; Wang, S.Y. Defect chemistry of nonprecious-metal electrocatalysts for oxygen reactions. *Adv. Mater.* **2017**, *29*, 1606459. [CrossRef] [PubMed]

6. Dou, S.; Wang, X.; Wang, S. Rational design of transition metal-based materials for highly efficient electrocatalysis. *Small Methods* **2018**, *3*, 1800211. [CrossRef]

7. Vogiatzis, K.D.; Polynski, M.V.; Kirkland, J.K.; Townsend, J.; Hashemi, A.; Liu, C.; Pidko, E.A. Computational approach to molecular catalysis by 3d transition metals: Challenges and opportunities. *Chem. Rev.* **2018**. [CrossRef] [PubMed]

8. Hong, W.T.; Risch, M.; Stoerzinger, K.A.; Grimaud, A.; Suntivich, J.; Shao-Horn, Y. Toward the rational design of non-precious transition metal oxides for oxygen electrocatalysis. *Energy Environ. Sci.* **2015**, *8*, 1404–1427. [CrossRef]

9. Liu, D.; Tao, L.; Yan, D.; Zou, Y.; Wang, S. Recent Advances on Non-precious Metal Porous Carbon-based Electrocatalysts for Oxygen Reduction Reaction. *ChemElectroChem* **2018**, *5*, 1775–1785. [CrossRef]

10. Wang, Z.; Zhang, Y.; Neyts, E.C.; Cao, X.; Zhang, X.; Jang, B.W.L.; Liu, C.-J. Catalyst Preparation with Plasmas: How Does It Work? *ACS Catal.* **2018**, *8*, 2093–2110. [CrossRef]

11. Liang, H.; Ming, F.; Alshareef, H.N. Applications of Plasma in Energy Conversion and Storage Materials. *Adv. Energy Mater.* **2018**. [CrossRef]

12. Keijser, M.D.; Opdorp, C.V. Atomic layer epitaxy of gallium arsenide with the use of atomic hydrogen. *Appl. Phys. Lett.* **1991**, *58*, 1187–1189. [CrossRef]

13. Heil, S.B.S.; Hemmen, J.L.V.; Hodson, C.J.; Singh, N.; Klootwijk, J.H.; Roozeboom, F.; Sanden, M.C.M.V.D.; Kessels, W.M.M. Deposition of TiN and $HfO_2$ in a commercial 200mm remote plasma atomic layer deposition reactor. *J. Vac. Sci. Technol. A* **2007**, *25*, 1357–1366. [CrossRef]

14. Profijt, H.B.; Potts, S.E.; Sanden, M.C.M.V.D.; Kessels, W.M.M. Plasma-Assisted Atomic Layer Deposition: Basics, Opportunities, and Challenges. *J. Vac. Sci. Technol. A* **2011**, *29*, 050801. [CrossRef]

15. Ting, C.-C.; Liu, C.-H.; Tai, C.-Y.; Hsu, S.-C.; Chao, C.-S.; Pan, F.-M. The size effect of titania-supported Pt nanoparticles on the electrocatalytic activity towards methanol oxidation reaction primarily via the bifunctional mechanism. *J. Power Sources* **2015**, *280*, 166–172. [CrossRef]

16. Yoshiaki, A.; Hiroyuki, T.; Shigemitsu, T.; Satoshi, E.; Akihiro, T.; Narishi, G.; Victor, M.; Ali, A.; Saad, M.A.; Yuichiro, K.; et al. Preparation of a platinum electrocatalyst by coaxial pulse arc plasma deposition. *Sci. Technol. Adv. Mater.* **2015**, *16*, 024804.

17. Grigoriev, S.A.; Fedotov, A.A.; Martemianov, S.A.; Fateev, V.N. Synthesis of nanostructural electrocatalytic materials on various carbon substrates by ion plasma sputtering of platinum metals. *Russ. J. Electrochem.* **2014**, *50*, 638–646. [CrossRef]

18. Falch, A.; Lates, V.; Kriek, R.J. Combinatorial Plasma Sputtering of $Pt_xPd_y$ Thin Film Electrocatalysts for Aqueous SO2 Electro-oxidation. *Electrocatalysis* **2015**, *6*, 322–330. [CrossRef]

19. Estevez, L.; Reed, D.; Nie, Z.; Schwarz, A.M.; Nandasiri, M.I.; Kizewski, J.P.; Wang, W.; Thomsen, E.; Liu, J.; Zhang, J.-G.; et al. Tunable Oxygen Functional Groups as Electrocatalysts on Graphite Felt Surfaces for All-Vanadium Flow Batteries. *ChemSusChem* **2016**, *9*, 1455–1461. [CrossRef]

20. Ma, Y.; Wang, Q.; Miao, Y.; Lin, Y.; Li, R. Plasma synthesis of Pt nanoparticles on 3D reduced graphene oxide-carbon nanotubes nanocomposites towards methanol oxidation reaction. *Appl. Surf. Sci.* **2018**, *450*, 413–421. [CrossRef]

21. Xu, J.L.; Wang, S.G.; Deng, Q.R.; Liu, Y.; Zhu, J.L.; Xu, C.B.; Wang, J.H. High-performance Pt/CNTs catalysts via hydrogen plasma for methanol electrooxidation. *Nano* **2014**, *9*, 1450018. [CrossRef]

22. Sui, S.; Ma, L.; Zhai, Y. TiC supported Pt–Ir electrocatalyst prepared by a plasma process for the oxygen electrode in unitized regenerative fuel cells. *J. Power Sources* **2011**, *196*, 5416–5422. [CrossRef]

23. Ipshita, B.; Kumaran, V.; Venugopal, S. Fabrication of electrodes with ultralow platinum loading by RF plasma processing of self-assembled arrays of Au@Pt nanoparticles. *Nanotechnology* **2016**, *27*, 305401.

24. Koh, J.H.; Jeon, H.S.; Jee, M.S.; Nursanto, E.B.; Lee, H.; Hwang, Y.J.; Min, B.K. Oxygen Plasma Induced Hierarchically Structured Gold Electrocatalyst for Selective Reduction of Carbon Dioxide to Carbon Monoxide. *J. Phys. Chem. C* **2015**, *119*, 883–889. [CrossRef]

25. Zhang, R.-C.; Sun, D.; Zhang, R.; Lin, W.-F.; Macias-Montero, M.; Patel, J.; Askari, S.; McDonald, C.; Mariotti, D.; Maguire, P. Gold nanoparticle-polymer nanocomposites synthesized by room temperature atmospheric pressure plasma and their potential for fuel cell electrocatalytic application. *Sci. Rep.* **2017**, *7*, 46682. [CrossRef]

26. Oi Lun, L.; Hoonseung, L.; Takahiro, I. Recent progress in solution plasma-synthesized-carbon-supported catalysts for energy conversion systems. *Jpn. J. Appl. Phys.* **2018**, *57*, 0102A2.

27. Kim, S.-M.; Cho, A.-R.; Lee, S.-Y. Characterization and electrocatalytic activity of Pt–M (M=Cu, Ag, and Pd) bimetallic nanoparticles synthesized by pulsed plasma discharge in water. *J. Nanopart. Res.* **2015**, *17*, 284. [CrossRef]

28. Kim, S.-M.; Lee, Y.-J.; Kim, J.-W.; Lee, S.-Y. Facile synthesis of Pt–Pd bimetallic nanoparticles by plasma discharge in liquid and their electrocatalytic activity toward methanol oxidation in alkaline media. *Thin Solid Films* **2014**, *572*, 260–265. [CrossRef]

29. Cho, A.-R.; Kim, S.-M.; Kim, S.-C.; Kim, J.-W.; Lee, S.-Y. The Facile Synthesis of Composition-Tunable Pt-Pd Bimetallic Nanocatalysts and Their Electrocatalytic Properties in Formic Acid. *J. Nanosci. Nanotechnol.* **2016**, *16*, 11443–11447. [CrossRef]

30. Zhang, J.; Hu, X.; Yang, B.; Su, N.; Huang, H.; Cheng, J.; Yang, H.; Saito, N. Novel synthesis of PtPd nanoparticles with good electrocatalytic activity and durability. *J. Alloy. Compd.* **2017**, *709*, 588–595. [CrossRef]

31. Horiguchi, G.; Chikaoka, Y.; Shiroishi, H.; Kosaka, S.; Saito, M.; Kameta, N.; Matsuda, N. Synthesis of Pt nanoparticles as catalysts of oxygen reduction with microbubble-assisted low-voltage and low-frequency solution plasma processing. *J. Power Sources* **2018**, *382*, 69–76. [CrossRef]

32. Huang, H.; Hu, X.; Zhang, J.; Su, N.; Cheng, J. Facile Fabrication of Platinum-Cobalt Alloy Nanoparticles with Enhanced Electrocatalytic Activity for a Methanol Oxidation Reaction. *Sci. Rep.* **2017**, *7*, 45555. [CrossRef] [PubMed]

33. Su, N.; Hu, X.; Zhang, J.; Huang, H.; Cheng, J.; Yu, J.; Ge, C. Plasma-induced synthesis of Pt nanoparticles supported on $TiO_2$ nanotubes for enhanced methanol electro-oxidation. *Appl. Surf. Sci.* **2017**, *399*, 403–410. [CrossRef]

34. Hu, X.; Shi, J.; Zhang, J.; Tang, W.; Zhu, H.; Shen, X.; Saito, N. One-step facile synthesis of carbon-supported PdAu nanoparticles and their electrochemical property and stability. *J. Alloy. Compd.* **2015**, *619*, 452–457. [CrossRef]

35. Kang, J.; Li, O.L.; Saito, N. A simple synthesis method for nano-metal catalyst supported on mesoporous carbon: The solution plasma process. *Nanoscale* **2013**, *5*, 6874–6882. [CrossRef] [PubMed]

36. Kang, J.; Saito, N. In situ solution plasma synthesis of mesoporous nanocarbon-supported bimetallic nanoparticles. *RSC Adv.* **2015**, *5*, 29131–29134. [CrossRef]

37. Sung-Min, K.; Sang-Yul, L. The plasma-induced formation of silver nanocrystals in aqueous solution and their catalytic activity for oxygen reduction. *Nanotechnology* **2018**, *29*, 085602.

38. Lee, Y.-J.; Kim, S.-M.; Kim, J.-W.; Lee, S.-Y. The characterization and electrocatalytic activities of carbon-supported Pt nanoparticles synthesized by the solution plasma process. *Mater. Lett.* **2014**, *123*, 184–186. [CrossRef]

39. Hussain, S.; Erikson, H.; Kongi, N.; Treshchalov, A.; Rähn, M.; Kook, M.; Merisalu, M.; Matisen, L.; Sammelselg, V.; Tammeveski, K. Oxygen Electroreduction on Pt Nanoparticles Deposited on Reduced Graphene Oxide and N-doped Reduced Graphene Oxide Prepared by Plasma-assisted Synthesis in Aqueous Solution. *ChemElectroChem* **2018**, *5*, 2902–2911. [CrossRef]

40. Cui, B.; Hu, B.; Liu, J.; Wang, M.; Song, Y.; Tian, K.; Zhang, Z.; He, L. Solution-Plasma-Assisted Bimetallic Oxide Alloy Nanoparticles of Pt and Pd Embedded within Two-Dimensional $Ti_3C_2T_x$ Nanosheets as Highly Active Electrocatalysts for Overall Water Splitting. *ACS Appl. Mater. Interfaces* **2018**, *10*, 23858–23873. [CrossRef]

41. Huang, H.; Liu, Y.; Ma, X.; Hu, J. Silver nanoparticles supported on graphitization micro-diamond as an electrocatalyst in alkaline medium. *Vacuum* **2018**, *155*, 380–386. [CrossRef]

42. Ding, G.; Jiao, W.; Wang, R.; Niu, Y.; Chen, L.; Hao, L. Ultrafast, Reversible Transition of Superwettability of Graphene Network and Controllable Underwater Oil Adhesion for Oil Microdroplet Transportation. *Adv. Funct. Mater.* **2018**, *28*, 1706686. [CrossRef]

43. Li, J.; Chen, G.; Zhu, Y.; Liang, Z.; Pei, A.; Wu, C.-L.; Wang, H.; Lee, H.R.; Liu, K.; Chu, S.; et al. Efficient electrocatalytic $CO_2$ reduction on a three-phase interface. *Nat. Catal.* **2018**, *1*, 592–600. [CrossRef]

44. Chetty, R.; Maniam, K.K.; Schuhmann, W.; Muhler, M. Oxygen-Plasma-Functionalized Carbon Nanotubes as Supports for Platinum–Ruthenium Catalysts Applied in Electrochemical Methanol Oxidation. *ChemPlusChem* **2015**, *80*, 130–135. [CrossRef]

45. Ding, D.; Song, Z.-L.; Cheng, Z.-Q.; Liu, W.-N.; Nie, X.-K.; Bian, X.; Chen, Z.; Tan, W. Plasma-assisted nitrogen doping of graphene-encapsulated Pt nanocrystals as efficient fuel cell catalysts. *J. Mater. Chem. A* **2014**, *2*, 472–477. [CrossRef]

46. Loganathan, K.; Bose, D.; Weinkauf, D. Surface modification of carbon black by nitrogen and allylamine plasma treatment for fuel cell electrocatalyst. *Int. J. Hydrogen Energy* **2014**, *39*, 15766–15771. [CrossRef]

47. Du, S.; Lin, K.; Malladi, S.K.; Lu, Y.; Sun, S.; Xu, Q.; Steinberger-Wilckens, R.; Dong, H. Plasma nitriding induced growth of Pt-nanowire arrays as high performance electrocatalysts for fuel cells. *Sci. Rep.* **2014**, *4*, 6439. [CrossRef]

48. Hu, X.; Ge, C.; Su, N.; Huang, H.; Xu, Y.; Zhang, J.; Shi, J.; Shen, X.; Saito, N. Solution plasma synthesis of Pt/ZnO/KB for photo-assisted electro-oxidation of methanol. *J. Alloy. Compd.* **2017**, *692*, 848–854. [CrossRef]

49. Flis-Kabulska, I.; Sun, Y.; Zakroczymski, T.; Flis, J. Plasma carburizing for improvement of Ni-Fe cathodes for alkaline water electrolysis. *Electrochim. Acta* **2016**, *220*, 11–19. [CrossRef]

50. Jin, Q.; Ren, B.; Li, D.; Cui, H.; Wang, C. Plasma-Assisted Synthesis of Self-Supporting Porous CoNPs@C Nanosheet as Efficient and Stable Bifunctional Electrocatalysts for Overall Water Splitting. *ACS Appl. Mater. Interfaces* **2017**, *9*, 31913–31921. [CrossRef]

51. Koo, Y.; Malik, R.; Alvarez, N.; White, L.; Shanov, V.N.; Schulz, M.; Collins, B.; Sankar, J.; Yun, Y. Aligned carbon nanotube/copper sheets: A new electrocatalyst for $CO_2$ reduction to hydrocarbons. *RSC Adv.* **2014**, *4*, 16362–16367. [CrossRef]

52. Mistry, H.; Varela, A.S.; Bonifacio, C.S.; Zegkinoglou, I.; Sinev, I.; Choi, Y.-W.; Kisslinger, K.; Stach, E.A.; Yang, J.C.; Strasser, P.; et al. Highly selective plasma-activated copper catalysts for carbon dioxide reduction to ethylene. *Nat. Commun.* **2016**, *7*, 12123. [CrossRef]

53. Gao, D.; Zegkinoglou, I.; Divins, N.J.; Scholten, F.; Sinev, I.; Grosse, P.; Roldan Cuenya, B. Plasma-Activated Copper Nanocube Catalysts for Efficient Carbon Dioxide Electroreduction to Hydrocarbons and Alcohols. *ACS Nano* **2017**, *11*, 4825–4831. [CrossRef] [PubMed]

54. Zhang, R.; Zhang, Y.-C.; Pan, L.; Shen, G.-Q.; Mahmood, N.; Ma, Y.-H.; Shi, Y.; Jia, W.; Wang, L.; Zhang, X.; et al. Engineering Cobalt Defects in Cobalt Oxide for Highly Efficient Electrocatalytic Oxygen Evolution. *ACS Catal.* **2018**, *8*, 3803–3811. [CrossRef]

55. Jozwiak, L.; Balcerzak, J.; Kubiczek, A.; Tyczkowski, J. Plasma deposited thin-film sandwich-like bifunctional electrocatalyst for oxygen reduction and evolution reactions. *Thin Solid Films* **2018**, *660*, 161–165. [CrossRef]

56. Kim, S.-M.; Jo, Y.-G.; Lee, M.-H.; Saito, N.; Kim, J.-W.; Lee, S.-Y. The plasma-assisted formation of Ag@Co$_3$O$_4$ core-shell hybrid nanocrystals for oxygen reduction reaction. *Electrochim. Acta* **2017**, *233*, 123–133. [CrossRef]

57. Taiwo, O.; Jun, M.; Yi, G.; Wei, Z.; Jian, J.; Zhao, X.S.; Zhonghua, Z. A new approach to nanoporous graphene sheets via rapid microwave-induced plasma for energy applications. *Nanotechnology* **2014**, *25*, 495604.

58. Lang, X.; Zhang, Y.; Cai, K.; Li, L.; Wang, Q.; Zhang, Q. High performance CoO nanospheres catalyst synthesized by DC arc discharge plasma method as air electrode for lithium-oxygen battery. *Ionics* **2018**, *25*, 35–40. [CrossRef]

59. Gan, Q.; He, H.; Zhao, K.; He, Z.; Liu, S.; Yang, S. Plasma-Induced Oxygen Vacancies in Urchin-Like Anatase Titania Coated by Carbon for Excellent Sodium-Ion Battery Anodes. *ACS Appl. Mater. Interfaces* **2018**, *10*, 7031–7042. [CrossRef]

60. Xu, L.; Jiang, Q.; Xiao, Z.; Li, X.; Huo, J.; Wang, S.; Dai, L. Plasma-Engraved Co$_3$O$_4$ Nanosheets with Oxygen Vacancies and High Surface Area for the Oxygen Evolution Reaction. *Angew. Chem. Int. Ed.* **2016**, *55*, 5277–5281. [CrossRef]

61. Wang, J.; Liu, J.; Zhang, B.; Wan, H.; Li, Z.; Ji, X.; Xu, K.; Chen, C.; Zha, D.; Miao, L.; et al. Synergistic effect of two actions sites on cobalt oxides towards electrochemical water-oxidation. *Nano Energy* **2017**, *42*, 98–105. [CrossRef]

62. Ma, L.; Chen, S.; Pei, Z.; Li, H.; Wang, Z.; Liu, Z.; Tang, Z.; Zapien, J.A.; Zhi, C. Flexible Waterproof Rechargeable Hybrid Zinc Batteries Initiated by Multifunctional Oxygen Vacancies-Rich Cobalt Oxide. *ACS Nano* **2018**, *12*, 8597–8605. [CrossRef] [PubMed]

63. Min, W. Plasma-induced nanoporous metal oxides with nitrogen doping for high-performance electrocatalysis. *Nanotechnology* **2017**, *28*, 242501.

64. Lei, X.; Zhimin, W.; Jialu, W.; Zhaohui, X.; Xiaobing, H.; Zhigang, L.; Shuangyin, W. N-doped nanoporous Co$_3$O$_4$ nanosheets with oxygen vacancies as oxygen evolving electrocatalysts. *Nanotechnology* **2017**, *28*, 165402.

65. Uhlig, L.M.; Sievers, G.; Brüser, V.; Dyck, A.; Wittstock, G. Characterization of different plasma-treated cobalt oxide catalysts for oxygen reduction reaction in alkaline media. *Sci. Bull.* **2016**, *61*, 612–618. [CrossRef]

66. Liang, H.; Xia, C.; Emwas, A.-H.; Anjum, D.H.; Miao, X.; Alshareef, H.N. Phosphine plasma activation of $\alpha$-Fe$_2$O$_3$ for high energy asymmetric supercapacitors. *Nano Energy* **2018**, *49*, 155–162. [CrossRef]

67. Xiao, Z.H.; Wang, Y.; Huang, Y.C.; Wei, Z.X.; Dong, C.L.; Ma, J.M.; Shen, S.H.; Li, Y.F.; Wang, S.Y. Filling the oxygen vacancies in Co$_3$O$_4$ with phosphorus: An ultra-efficient electrocatalyst for overall water splitting. *Energy Environ. Sci.* **2017**, *10*, 2563–2569. [CrossRef]

68. Wang, Z.; Liu, H.; Ge, R.; Ren, X.; Ren, J.; Yang, D.; Zhang, L.; Sun, X. Phosphorus-Doped Co$_3$O$_4$ Nanowire Array: A Highly Efficient Bifunctional Electrocatalyst for Overall Water Splitting. *ACS Catal.* **2018**, *8*, 2236–2241. [CrossRef]

69. Dou, S.; Dong, C.-L.; Hu, Z.; Huang, Y.-C.; Chen, J.-l.; Tao, L.; Yan, D.; Chen, D.; Shen, S.; Chou, S.; et al. Atomic-Scale CoO$_x$ Species in Metal–Organic Frameworks for Oxygen Evolution Reaction. *Adv. Funct. Mater.* **2017**, *27*, 1702546. [CrossRef]

70. Yin, W.-J.; Weng, B.; Ge, J.; Sun, Q.; Li, Z.; Yan, Y. Oxide perovskites, double perovskites and derivatives for electrocatalysis, photocatalysis, and photovoltaics. *Energy Environ. Sci.* **2018**, *12*, 442–462. [CrossRef]

71. Chen, D.; Qiao, M.; Lu, Y.-R.; Hao, L.; Liu, D.; Dong, C.-L.; Li, Y.; Wang, S. Preferential Cation Vacancies in Perovskite Hydroxide for the Oxygen Evolution Reaction. *Angew. Chem. Int. Ed.* **2018**, *57*, 8691–8696. [CrossRef] [PubMed]

72. Hayden, B.E.; Rogers, F.K. Oxygen reduction and oxygen evolution on SrTi$_{1-x}$Fe$_x$O$_{3-y}$ (STFO) perovskite electrocatalysts. *J. Electroanal. Chem.* **2018**, *819*, 275–282. [CrossRef]

73. Chen, G.; Hu, Z.; Zhu, Y.; Gu, B.; Zhong, Y.; Lin, H.-J.; Chen, C.-T.; Zhou, W.; Shao, Z. A Universal Strategy to Design Superior Water-Splitting Electrocatalysts Based on Fast In Situ Reconstruction of Amorphous Nanofilm Precursors. *Adv. Mater.* **2018**, *30*, 1804333. [CrossRef] [PubMed]

74. Liu, R.; Wang, Y.Y.; Liu, D.D.; Zou, Y.Q.; Wang, S.Y. Water-Plasma-Enabled Exfoliation of Ultrathin Layered Double Hydroxide Nanosheets with Multivacancies for Water Oxidation. *Adv. Mater.* **2017**, *29*, 1701546. [CrossRef] [PubMed]

75. Carrasco, J.A.; Romero, J.; Varela, M.; Hauke, F.; Abellán, G.; Hirsch, A.; Coronado, E. Alkoxide-intercalated NiFe-layered double hydroxides magnetic nanosheets as efficient water oxidation electrocatalysts. *Inorg. Chem. Front.* **2016**, *3*, 478–487. [CrossRef]

76. Browne, M.P.; Sofer, Z.; Pumera, M. Layered and two dimensional metal oxides for electrochemical energy conversion. *Energy Environ. Sci.* **2018**, *12*, 41–58. [CrossRef]

77. Wang, Y.Y.; Zhang, Y.Q.; Liu, Z.J.; Xie, C.; Feng, S.; Liu, D.D.; Shao, M.F.; Wang, S.Y. Layered double hydroxide nanosheets with multiple vacancies obtained by dry exfoliation as highly efficient oxygen evolution electrocatalysts. *Angew. Chem. Int. Ed.* **2017**, *56*, 5867–5871. [CrossRef] [PubMed]

78. Wang, Y.Y.; Xie, C.; Zhang, Z.Y.; Liu, D.D.; Chen, R.; Wang, S.Y. In Situ Exfoliated, N-Doped, and Edge-Rich Ultrathin Layered Double Hydroxides Nanosheets for Oxygen Evolution Reaction. *Adv. Funct. Mater.* **2018**, *28*, 1703363. [CrossRef]

79. Oturan, N.; Ganiyu, S.O.; Raffy, S.; Oturan, M.A. Sub-stoichiometric titanium oxide as a new anode material for electro-Fenton process: Application to electrocatalytic destruction of antibiotic amoxicillin. *Appl. Catal. B Environ.* **2017**, *217*, 214–223. [CrossRef]

80. Luo, X.; Lu, J.; Sohm, E.; Ma, L.; Wu, T.; Wen, J.; Qiu, D.; Xu, Y.; Ren, Y.; Miller, D.J.; et al. Uniformly dispersed $FeO_x$ atomic clusters by pulsed arc plasma deposition: An efficient electrocatalyst for improving the performance of $Li–O_2$ battery. *Nano Res.* **2016**, *9*, 1913–1920. [CrossRef]

81. Peng, X.; Wang, Z.; Wang, Z.; Pan, Y. Multivalent manganese oxides with high electrocatalytic activity for oxygen reduction reaction. *Front. Chem. Sci. Eng.* **2018**, *12*, 790–797. [CrossRef]

82. Jiang, M.; Fu, C.; Yang, J.; Liu, Q.; Zhang, J.; Sun, B. Defect-engineered $MnO_2$ enhancing oxygen reduction reaction for high performance Al-air batteries. *Energy Storage Mater.* **2018**, *18*, 34–42. [CrossRef]

83. Zhang, Y.; Ouyang, B.; Xu, K.; Xia, X.; Zhang, Z.; Rawat, R.S.; Fan, H.J. Prereduction of Metal Oxides via Carbon Plasma Treatment for Efficient and Stable Electrocatalytic Hydrogen Evolution. *Small* **2018**, *14*, 1800340. [CrossRef] [PubMed]

84. Geng, Z.; Kong, X.; Chen, W.; Su, H.; Liu, Y.; Cai, F.; Wang, G.; Zeng, J. Oxygen Vacancies in ZnO Nanosheets Enhance $CO_2$ Electrochemical Reduction to CO. *Angew. Chem. Int. Ed.* **2018**, *57*, 6054–6059. [CrossRef] [PubMed]

85. Li, H.; Tan, Y.; Liu, P.; Guo, C.; Luo, M.; Han, J.; Lin, T.; Huang, F.; Chen, M. Atomic-Sized Pores Enhanced Electrocatalysis of $TaS_2$ Nanosheets for Hydrogen Evolution. *Adv. Mater.* **2016**, *28*, 8945–8949. [CrossRef] [PubMed]

86. Tao, L.; Duan, X.; Wang, C.; Duan, X.; Wang, S. Plasma-engineered $MoS_2$ thin-film as an efficient electrocatalyst for hydrogen evolution reaction. *Chem. Commun.* **2015**, *51*, 7470–7473. [CrossRef]

87. Zhang, C.; Jiang, L.; Zhang, Y.; Hu, J.; Leung, M.K.H. Janus effect of $O_2$ plasma modification on the electrocatalytic hydrogen evolution reaction of $MoS_2$. *J. Catal.* **2018**, *361*, 384–392. [CrossRef]

88. Huang, J.; Deng, X.; Wan, H.; Chen, F.; Lin, Y.; Xu, X.; Ma, R.; Sasaki, T. Liquid Phase Exfoliation of $MoS_2$ Assisted by Formamide Solvothermal Treatment and Enhanced Electrocatalytic Activity Based on $(H_3Mo_{12}O_{40}P/MoS_2)$n Multilayer Structure. *ACS Sustain. Chem. Eng.* **2018**, *6*, 5227–5237. [CrossRef]

89. Cheng, C.-C.; Lu, A.-Y.; Tseng, C.-C.; Yang, X.; Hedhili, M.N.; Chen, M.-C.; Wei, K.-H.; Li, L.-J. Activating basal-plane catalytic activity of two-dimensional $MoS_2$ monolayer with remote hydrogen plasma. *Nano Energy* **2016**, *30*, 846–852. [CrossRef]

90. Lu, A.-Y.; Yang, X.; Tseng, C.-C.; Min, S.; Lin, S.-H.; Hsu, C.-L.; Li, H.; Idriss, H.; Kuo, J.-L.; Huang, K.-W.; et al. High-Sulfur-Vacancy Amorphous Molybdenum Sulfide as a High Current Electrocatalyst in Hydrogen Evolution. *Small* **2016**, *12*, 5530–5537. [CrossRef]

91. Dou, S.; Tao, L.; Huo, J.; Wang, S.; Dai, L. Etched and doped $Co_9S_8$/graphene hybrid for oxygen electrocatalysis. *Energy Environ. Sci.* **2016**, *9*, 1320–1326. [CrossRef]

92. Zhang, C.; Shi, Y.; Yu, Y.; Du, Y.; Zhang, B. Engineering Sulfur Defects, Atomic Thickness, and Porous Structures into Cobalt Sulfide Nanosheets for Efficient Electrocatalytic Alkaline Hydrogen Evolution. *ACS Catal.* **2018**, *8*, 8077–8083. [CrossRef]

93. He, L.; Zhou, D.; Lin, Y.; Ge, R.; Hou, X.; Sun, X.; Zheng, C. Ultrarapid in Situ Synthesis of $Cu_2S$ Nanosheet Arrays on Copper Foam with Room-Temperature-Active Iodine Plasma for Efficient and Cost-Effective Oxygen Evolution. *ACS Catal.* **2018**, *8*, 3859–3864. [CrossRef]

94. Yeo, S.; Nandi, D.K.; Rahul, R.; Kim, T.H.; Shong, B.; Jang, Y.; Bae, J.-S.; Han, J.W.; Kim, S.-H.; Kim, H. Low-temperature direct synthesis of high quality $WS_2$ thin films by plasma-enhanced atomic layer deposition for energy related applications. *Appl. Surf. Sci.* **2018**, *459*, 596–605. [CrossRef]

95. Escalera-López, D.; Griffin, R.; Isaacs, M.; Wilson, K.; Palmer, R.E.; Rees, N.V. $MoS_2$ and $WS_2$ nanocone arrays: Impact of surface topography on the hydrogen evolution electrocatalytic activity and mass transport. *Appl. Mater. Today* **2018**, *11*, 70–81. [CrossRef]

96. Qu, S.; Chen, W.; Yu, J.; Chen, G.; Zhang, R.; Chu, S.; Huang, J.; Wang, X.; Li, C.; Ostrikov, K. Cross-linked trimetallic nanopetals for electrocatalytic water splitting. *J. Power Sources* **2018**, *390*, 224–233. [CrossRef]

97. Sun, Y.; Xu, K.; Wei, Z.; Li, H.; Zhang, T.; Li, X.; Cai, W.; Ma, J.; Fan, H.J.; Li, Y. Strong Electronic Interaction in Dual-Cation-Incorporated $NiSe_2$ Nanosheets with Lattice Distortion for Highly Efficient Overall Water Splitting. *Adv. Mater.* **2018**, *30*, 1802121. [CrossRef]

98. Deng, S.; Zhong, Y.; Zeng, Y.; Wang, Y.; Yao, Z.; Yang, F.; Lin, S.; Wang, X.; Lu, X.; Xia, X.; et al. Directional Construction of Vertical Nitrogen-Doped 1T-2H $MoSe_2$/Graphene Shell/Core Nanoflake Arrays for Efficient Hydrogen Evolution Reaction. *Adv. Mater.* **2017**, *29*, 1700748. [CrossRef]

99. Li, Y.; Liu, J.; Yuan, Q.; Tang, H.; Yu, F.; Lv, X. A green adsorbent derived from banana peel for highly effective removal of heavy metal ions from water. *RSC Adv.* **2016**, *6*, 45041–45048. [CrossRef]

100. Zhang, Y.; Ouyang, B.; Xu, J.; Chen, S.; Rawat, R.S.; Fan, H.J. 3D Porous Hierarchical Nickel–Molybdenum Nitrides Synthesized by RF Plasma as Highly Active and Stable Hydrogen-Evolution-Reaction Electrocatalysts. *Adv. Energy Mater.* **2016**, *6*, 1600221. [CrossRef]

101. Ouyang, B.; Zhang, Y.; Zhang, Z.; Fan, H.J.; Rawat, R.S. Nitrogen-Plasma-Activated Hierarchical Nickel Nitride Nanocorals for Energy Applications. *Small* **2017**, *13*, 1604265. [CrossRef] [PubMed]

102. Liu, B.; He, B.; Peng, H.-Q.; Zhao, Y.; Cheng, J.; Xia, J.; Shen, J.; Ng, T.-W.; Meng, X.; Lee, C.-S.; et al. Unconventional Nickel Nitride Enriched with Nitrogen Vacancies as a High-Efficiency Electrocatalyst for Hydrogen Evolution. *Adv. Sci.* **2018**, *5*, 1800406. [CrossRef] [PubMed]

103. Wang, L.-C.; Liu, B.-H.; Su, C.-Y.; Liu, W.-S.; Kei, C.-C.; Wang, K.-W.; Perng, T.-P. Electronic Band Structure and Electrocatalytic Performance of $Cu_3N$ Nanocrystals. *ACS Appl. Nano Mater.* **2018**, *1*, 3673–3681. [CrossRef]

104. Panomsuwan, G.; Saito, N.; Ishizaki, T. Fe–N-doped carbon-based composite as an efficient and durable electrocatalyst for the oxygen reduction reaction. *RSC Adv.* **2016**, *6*, 114553–114559. [CrossRef]

105. Zhong, W.; Chen, J.; Zhang, P.; Deng, L.; Yao, L.; Ren, X.; Li, Y.; Mi, H.; Sun, L. Air plasma etching towards rich active sites in Fe/N-porous carbon for the oxygen reduction reaction with superior catalytic performance. *J. Mater. Chem. A* **2017**, *5*, 16605–16610. [CrossRef]

106. Wu, R.; Xiao, B.; Gao, Q.; Zheng, Y.-R.; Zheng, X.-S.; Zhu, J.-F.; Gao, M.-R.; Yu, S.-H. A Janus Nickel Cobalt Phosphide Catalyst for High-Efficiency Neutral-pH Water Splitting. *Angew. Chem. Int. Ed.* **2018**, *57*, 15445–15449. [CrossRef]

107. Pan, Y.; Sun, K.; Lin, Y.; Cao, X.; Cheng, Y.; Liu, S.; Zeng, L.; Cheong, W.-C.; Zhao, D.; Wu, K.; et al. Electronic structure and d-band center control engineering over M-doped CoP (M=Ni, Mn, Fe) hollow polyhedron frames for boosting hydrogen production. *Nano Energy* **2019**, *56*, 411–419. [CrossRef]

108. Liang, H.; Gandi, A.N.; Anjum, D.H.; Wang, X.; Schwingenschlögl, U.; Alshareef, H.N. Plasma-Assisted Synthesis of NiCoP for Efficient Overall Water Splitting. *Nano Lett.* **2016**, *16*, 7718–7725. [CrossRef]

109. Zhang, Q.; Li, T.; Liang, J.; Wang, N.; Kong, X.; Wang, J.; Qian, H.; Zhou, Y.; Liu, F.; Wei, C.; et al. Highly wettable and metallic NiFe-phosphate/phosphide catalyst synthesized by plasma for highly efficient oxygen evolution reaction. *J. Mater. Chem. A* **2018**, *6*, 7509–7516. [CrossRef]

110. Dinh, K.N.; Sun, X.; Dai, Z.; Zheng, Y.; Zheng, P.; Yang, J.; Xu, J.; Wang, Z.; Yan, Q. $O_2$ plasma and cation tuned nickel phosphide nanosheets for highly efficient overall water splitting. *Nano Energy* **2018**, *54*, 82–90. [CrossRef]

111. Peng, X.; Qasim, A.M.; Jin, W.; Wang, L.; Hu, L.; Miao, Y.; Li, W.; Li, Y.; Liu, Z.; Huo, K.; et al. Ni-doped amorphous iron phosphide nanoparticles on TiN nanowire arrays: An advanced alkaline hydrogen evolution electrocatalyst. *Nano Energy* **2018**, *53*, 66–73. [CrossRef]

112. Goryachev, A.; Gao, L.; Zhang, Y.; Rohling, R.Y.; Vervuurt, R.H.J.; Bol, A.A.; Hofmann, J.P.; Hensen, E.J.M. Stability of $CoP_x$ Electrocatalysts in Continuous and Interrupted Acidic Electrolysis of Water. *ChemElectroChem* **2018**, *5*, 1230–1239. [CrossRef] [PubMed]

113. Liu, H.; Liu, X.; Mao, Z.; Zhao, Z.; Peng, X.; Luo, J.; Sun, X. Plasma-activated $Co_3(PO_4)_2$ nanosheet arrays with $Co^{3+}$-Rich surfaces for overall water splitting. *J. Power Sources* **2018**, *400*, 190–197. [CrossRef]

114. Guo, Q.; Guo, Z.; Shi, J.; Xiong, W.; Zhang, H.; Chen, Q.; Liu, Z.; Wang, X. Atomic Layer Deposition of Nickel Carbide from a Nickel Amidinate Precursor and Hydrogen Plasma. *ACS Appl. Mater. Interfaces* **2018**, *10*, 8384–8390. [CrossRef] [PubMed]

115. Xiong, W.; Guo, Q.; Guo, Z.; Li, H.; Zhao, R.; Chen, Q.; Liu, Z.; Wang, X. Atomic layer deposition of nickel carbide for supercapacitors and electrocatalytic hydrogen evolution. *J. Mater. Chem. A* **2018**, *6*, 4297–4304. [CrossRef]

116. Ko, Y.-J.; Cho, J.-M.; Kim, I.; Jeong, D.S.; Lee, K.-S.; Park, J.-K.; Baik, Y.-J.; Choi, H.-J.; Lee, W.-S. Tungsten carbide nanowalls as electrocatalyst for hydrogen evolution reaction: New approach to durability issue. *Appl. Catal. B Environ.* **2017**, *203*, 684–691. [CrossRef]

117. Guo, Y.; Teng, W.; Chen, J.; Jie, Z.; Ostrikov, K.K. Air Plasma Activation of Catalytic Sites in a Metal-Cyanide Framework for Efficient Oxygen Evolution Reaction. *Adv. Energy Mater.* **2018**, *8*, 1800085. [CrossRef]

118. Yan, D.; Dong, C.-L.; Huang, Y.-C.; Zou, Y.; Xie, C.; Wang, Y.; Zhang, Y.; Liu, D.; Shen, S.; Wang, S. Engineering the coordination geometry of metal–organic complex electrocatalysts for highly enhanced oxygen evolution reaction. *J. Mater. Chem. A* **2018**, *6*, 805–810. [CrossRef]

119. Ghausi, M.A.; Xie, J.; Li, Q.; Wang, X.; Yang, R.; Wu, M.; Wang, Y.; Dai, L. $CO_2$ Overall Splitting by a Bifunctional Metal-Free Electrocatalyst. *Angew. Chem. Int. Ed.* **2018**, *57*, 13135–13139. [CrossRef]

120. Lai, J.; Nsabimana, A.; Luque, R.; Xu, G. 3D Porous Carbonaceous Electrodes for Electrocatalytic Applications. *Joule* **2018**, *2*, 76–93. [CrossRef]

121. Liu, M.; Yu, F.; Ma, C.; Xue, X.; Fu, H.; Yuan, H.; Yang, S.; Wang, G.; Guo, X.; Zhang, L. Effective Oxygen Reduction Reaction Performance of FeCo Alloys In Situ Anchored on Nitrogen-Doped Carbon by the Microwave-Assistant Carbon Bath Method and Subsequent Plasma Etching. *Nanomaterials* **2019**, *9*, 1284. [CrossRef] [PubMed]

122. Hong, J.; Jin, C.; Yuan, J.; Zhang, Z. Atomic Defects in Two-Dimensional Materials: From Single-Atom Spectroscopy to Functionalities in Opto-/Electronics, Nanomagnetism, and Catalysis. *Adv. Mater.* **2017**, *29*, 1606434. [CrossRef] [PubMed]

123. Jiang, H.; Gu, J.; Zheng, X.; Liu, M.; Qiu, X.; Wang, L.; Li, W.; Chen, Z.; Ji, X.; Li, J. Defect-rich and ultrathin N doped carbon nanosheets as advanced trifunctional metal-free electrocatalysts for ORR, OER and HER. *Energy Environ. Sci.* **2019**, *12*, 322–333. [CrossRef]

124. Tao, L.; Wang, Q.; Dou, S.; Ma, Z.L.; Huo, J.; Wang, S.Y.; Dai, L.M. Edge-rich and dopant-free graphene as a highly efficient metal-free electrocatalyst for the oxygen reduction reaction. *Chem. Commun.* **2016**, *52*, 2764–2767. [CrossRef] [PubMed]

125. Liu, Z.J.; Zhao, Z.H.; Wang, Y.Y.; Dou, S.; Yan, D.F.; Liu, D.D.; Xia, Z.H.; Wang, S.Y. In Situ Exfoliated, Edge-Rich, Oxygen-Functionalized Graphene from Carbon Fibers for Oxygen Electrocatalysis. *Adv. Mater.* **2017**, *29*, 1606207. [CrossRef] [PubMed]

126. Lehmann, K.; Yurchenko, O.; Melke, J.; Fischer, A.; Urban, G. High electrocatalytic activity of metal-free and non-doped hierarchical carbon nanowalls towards oxygen reduction reaction. *Electrochim. Acta* **2018**, *269*, 657–667. [CrossRef]

127. Wang, Y.; Yu, F.; Zhu, M.; Ma, C.; Zhao, D.; Wang, C.; Zhou, A.; Dai, B.; Ji, J.; Guo, X. N-Doping of plasma exfoliated graphene oxide via dielectric barrier discharge plasma treatment for the oxygen reduction reaction. *J. Mater. Chem. A* **2018**, *6*, 2011–2017. [CrossRef]

128. Zhu, Y.-P.; Ran, J.; Qiao, S.-Z. Scalable Self-Supported Graphene Foam for High-Performance Electrocatalytic Oxygen Evolution. *ACS Appl. Mater. Interfaces* **2017**, *9*, 41980–41987. [CrossRef]

129. Yu, D.; Zhang, Q.; Dai, L. Highly Efficient Metal-Free Growth of Nitrogen-Doped Single-Walled Carbon Nanotubes on Plasma-Etched Substrates for Oxygen Reduction. *J. Am. Chem. Soc.* **2010**, *132*, 15127–15129. [CrossRef]

130. Du, Z.; Wang, S.; Kong, C.; Deng, Q.; Wang, G.; Liang, C.; Tang, H. Microwave plasma synthesized nitrogen-doped carbon nanotubes for oxygen reduction. *J. Solid State Electrochem.* **2015**, *19*, 1541–1549. [CrossRef]

131. Subramanian, P.; Cohen, A.; Teblum, E.; Nessim, G.D.; Bormasheko, E.; Schechter, A. Electrocatalytic activity of nitrogen plasma treated vertically aligned carbon nanotube carpets towards oxygen reduction reaction. *Electrochem. Commun.* **2014**, *49*, 42–46. [CrossRef]

132. Zhang, S.; Kang, P.; Ubnoske, S.; Brennaman, M.K.; Song, N.; House, R.L.; Glass, J.T.; Meyer, T.J. Polyethylenimine-Enhanced Electrocatalytic Reduction of $CO_2$ to Formate at Nitrogen-Doped Carbon Nanomaterials. *J. Am. Chem. Soc.* **2014**, *136*, 7845–7848. [CrossRef] [PubMed]

133. Li, O.L.; Wada, Y.; Kaneko, A.; Lee, H.; Ishizaki, T. Oxygen Reduction Reaction Activity of Thermally Tailored Nitrogen-Doped Carbon Electrocatalysts Prepared through Plasma Synthesis. *ChemElectroChem* **2018**, *5*, 1995–2001. [CrossRef]

134. Hyun, K.; Ueno, T.; Li, O.L.; Saito, N. Synthesis of heteroatom-carbon nanosheets by solution plasma processing using N-methyl-2-pyrrolidone as precursor. *RSC Adv.* **2016**, *6*, 6990–6996. [CrossRef]

135. Li, O.L.; Chiba, S.; Wada, Y.; Panomsuwan, G.; Ishizaki, T. Synthesis of graphitic-N and amino-N in nitrogen-doped carbon via a solution plasma process and exploration of their synergic effect for advanced oxygen reduction reaction. *J. Mater. Chem. A* **2017**, *5*, 2073–2082. [CrossRef]

136. Li, O.L.; Chiba, S.; Wada, Y.; Lee, H.; Ishizaki, T. Selective nitrogen bonding states in nitrogen-doped carbon via a solution plasma process for advanced oxygen reduction reaction. *RSC Adv.* **2016**, *6*, 109354–109360. [CrossRef]

137. Panomsuwan, G.; Chiba, S.; Kaneko, Y.; Saito, N.; Ishizaki, T. In situ solution plasma synthesis of nitrogen-doped carbon nanoparticles as metal-free electrocatalysts for the oxygen reduction reaction. *J. Mater. Chem. A* **2014**, *2*, 18677–18686. [CrossRef]

138. Panomsuwan, G.; Saito, N.; Ishizaki, T. Electrocatalytic oxygen reduction on nitrogen-doped carbon nanoparticles derived from cyano-aromatic molecules via a solution plasma approach. *Carbon* **2016**, *98*, 411–420. [CrossRef]

139. Morishita, T.; Ueno, T.; Panomsuwan, G.; Hieda, J.; Yoshida, A.; Bratescu, M.A.; Saito, N. Fastest Formation Routes of Nanocarbons in Solution Plasma Processes. *Sci. Rep.* **2016**, *6*, 36880. [CrossRef]

140. Ishizaki, T.; Chiba, S.; Kaneko, Y.; Panomsuwan, G. Electrocatalytic activity for the oxygen reduction reaction of oxygen-containing nanocarbon synthesized by solution plasma. *J. Mater. Chem. A* **2014**, *2*, 10589–10598. [CrossRef]

141. Kondratowicz, I.; Nadolska, M.; Şahin, S.; Łapiński, M.; Prześniak-Welenc, M.; Sawczak, M.; Yu, E.H.; Sadowski, W.; Żelechowska, K. Tailoring properties of reduced graphene oxide by oxygen plasma treatment. *Appl. Surf. Sci.* **2018**, *440*, 651–659. [CrossRef]

142. Wong, C.H.A.; Sofer, Z.; Klímová, K.; Pumera, M. Microwave Exfoliation of Graphite Oxides in $H_2S$ Plasma for the Synthesis of Sulfur-Doped Graphenes as Oxygen Reduction Catalysts. *ACS Appl. Mater. Interfaces* **2016**, *8*, 31849–31855. [CrossRef] [PubMed]

143. Tian, Y.; Wei, Z.; Wang, X.; Peng, S.; Zhang, X.; Liu, W.-m. Plasma-etched, S-doped graphene for effective hydrogen evolution reaction. *Int. J. Hydrogen Energy* **2017**, *42*, 4184–4192. [CrossRef]

144. Zhou, J.; Qi, F.; Chen, Y.; Wang, Z.; Zheng, B.; Wang, X. CVD-grown three-dimensional sulfur-doped graphene as a binder-free electrocatalytic electrode for highly effective and stable hydrogen evolution reaction. *J. Mater. Sci.* **2018**, *53*, 7767–7777. [CrossRef]

145. Panomsuwan, G.; Saito, N.; Ishizaki, T. Electrocatalytic oxygen reduction activity of boron-doped carbon nanoparticles synthesized via solution plasma process. *Electrochem. Commun.* **2015**, *59*, 81–85. [CrossRef]

146. Li, X.; Li, H.; Li, M.; Li, C.; Sun, D.; Lei, Y.; Yang, B. Preparation of a porous boron-doped diamond/Ta electrode for the electrocatalytic degradation of organic pollutants. *Carbon* **2018**, *129*, 543–551. [CrossRef]

147. Panomsuwan, G.; Saito, N.; Ishizaki, T. Simple one-step synthesis of fluorine-doped carbon nanoparticles as potential alternative metal-free electrocatalysts for oxygen reduction reaction. *J. Mater. Chem. A* **2015**, *3*, 9972–9981. [CrossRef]

148. Tian, Y.; Mei, R.; Xue, D.-Z.; Zhang, X.; Peng, W. Enhanced electrocatalytic hydrogen evolution in graphene via defect engineering and heteroatoms co-doping. *Electrochim. Acta* **2016**, *219*, 781–789. [CrossRef]
149. Lee, S.; Heo, Y.; Bratescu, M.A.; Ueno, T.; Saito, N. Solution plasma synthesis of a boron–carbon–nitrogen catalyst with a controllable bond structure. *Phys. Chem. Chem. Phys.* **2017**, *19*, 15264–15272. [CrossRef]

 *nanomaterials*

*Review*

# A Review of Recent Advances of Dielectric Barrier Discharge Plasma in Catalysis

**Ju Li [1], Cunhua Ma [1,*], Shengjie Zhu [1], Feng Yu [1], Bin Dai [1] and Dezheng Yang [2,3]**

[1]   Key Laboratory for Green Processing of Chemical Engineering of Xinjiang Bingtuan,
     School of Chemistry and Chemical Engineering, Shihezi University, Shihezi 832003, China;
     leej222@163.com (J.L.); zsj497262724@gmail.com (S.Z.); yufeng05@mail.ipc.ac.cn (F.Y.);
     db_tea@shzu.edu.cn (B.D.)
[2]   Laboratory of Plasma Physical Chemistry, School of Physics, Dalian University of Technology,
     Dalian 116024, China; yangdz@dlut.edu.cn
[3]   Key Laboratory of Ecophysics, College of Sciences, Shihezi University, Shihezi 832003, China
*   Correspondence: mchua@shzu.edu.cn; Tel.: +86-0993-205-8775

Received: 28 August 2019; Accepted: 21 September 2019; Published: 9 October 2019

**Abstract:** Dielectric barrier discharge plasma is one of the most popular methods to generate nanthermal plasma, which is made up of a host of high-energy electrons, free radicals, chemically active ions and excited species, so it has the property of being prone to chemical reactions. Due to these unique advantages, the plasma technology has been widely used in the catalytic fields. Compared with the conventional method, the heterogeneous catalyst prepared by plasma technology has good dispersion and smaller particle size, and its catalytic activity, selectivity and stability are significantly improved. In addition, the interaction between plasma and catalyst can achieve synergistic effects, so the catalytic effect is further improved. The review mainly introduces the characteristics of dielectric barrier discharge plasma, development trend and its recent advances in catalysis; then, we sum up the advantages of using plasma technology to prepare catalysts. At the same time, the synergistic effect of plasma technology combined with catalyst on methanation, $CH_4$ reforming, $NO_x$ decomposition, $H_2O_2$ synthesis, Fischer–Tropsch synthesis, volatile organic compounds removal, catalytic sterilization, wastewater treatment and degradation of pesticide residues are discussed. Finally, the properties of plasma in catalytic reaction are summarized, and the application prospect of plasma in the future catalytic field is prospected.

**Keywords:** dielectric barrier discharge plasma; heterogeneous catalyst; element doping; defect-rich; plasma catalysis

---

## 1. Introduction

The plasma followed by solids, liquids and gases was later referred to as an ionized gas of the "fourth state of matter". In 1928, Langmuir first proposed the term "plasma" [1]. Different from the properties of the three substances above, plasma is made up of a host of high-energy electrons, free radicals, chemically active ions and excited states. The total electric charge of electrons and various negative ions is roughly equal to that of positive ions, so it is generally electrically neutral. In effect, plasma has been found in several applications like materials science and microelectronics industries, and even in many emerging environments such as medical. Plasma is divided into low and high temperature plasma on the basis of the temperature of internal electrons. Based on thermodynamic equilibrium, the low temperature plasma can be further divided into thermal plasma and non-thermal plasma (NTP). Dielectric barrier discharge (DBD) is a non-thermal plasma. After the thermal plasma reaches a local equilibrium state, the electron temperature is nearly close to the overall gas temperature [2]. This group mainly contains various plasma torches.

Compared with other common gases, plasma has the following characteristics: as a collection of charged particles, plasma has the conductivity similar to that of metal in which some ions collide, and the movement of positive and negative charges generates strong electric and magnetic fields; therefore, it has heat conduction and heat radiation. The total amount of positive and negative charges in the plasma is always equal. Any small space charge density changes will generate a huge electric field intensity, which will restore its original state and maintain its neutral state. The chemical character is lively, and it is prone to chemical reactions. The plasma processes have successfully been used in the two catalytic fields, namely catalyst preparation and direct excitation of reactants [3]. Catalysts play a pivotal role in the modern industry. Currently, nearly all chemicals in life are manufactured through catalytic processes. However, heterogeneous catalysis makes an important impact in these catalytic processes [4].

DBD research has now been going on for more than a century. Siemens et al. [5] focused on the production of ozone, and the first experimental report was made in 1857. The discharge experiment device they designed has many novel features, one of which is that the electrode does not contact with the plasma and is located outside the discharge chamber. Therefore, DBD has been considered to be "ozone discharge" for a long time. Andrew and Tait [6] called it "silent discharge" due to its quiet and silent discharge process in 1860. Buss [7] made a significant contribution to characterizing the discharge. He discovered that air breakdown between parallel electrodes covered by dielectrics is always accompanied by plenty of tiny short-lived current wires. DBD is one type of the non-thermal plasma (NTP), which can substitute the conventional catalytic chemical process that operates under high temperature conditions. Figure 1 represents two typical DBD electrode devices, namely planar and cylindrical DBD [8]. The placement of one or more dielectric layers between the metal electrodes is essential for the discharge operation. Typical dielectric materials are glass, quartz and ceramic.

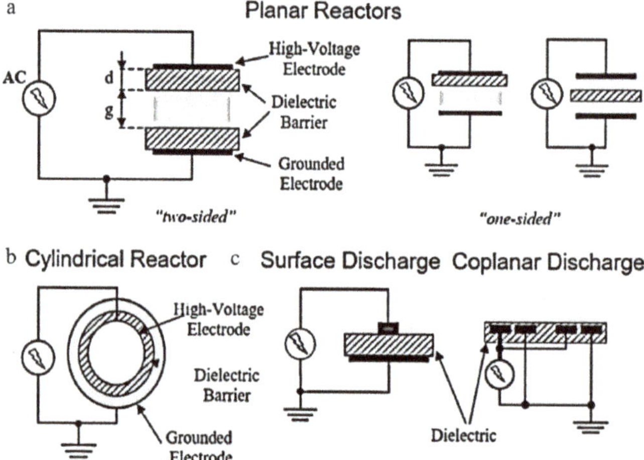


**Figure 1.** Two typical DBD electrode devices: (**a**) planar reactor; (**b**) cylindrical reactor; (**c**) two discharge types of surface discharge and coplanar discharge. Reproduced with permission from [8]; Copyright Elsevier, 2003.

In a DBD, the average temperature of high-energy electrons is very high, over the range of 10,000–100,000 K, but the actual gas temperature is close to the environment temperature. By electron impact ionization and excitation, source gases can produce active radicals, ions, excited atoms and molecular species [9]. The cold plasma characteristics of DBD, as well as the bombardment of high-energy electrons and other reactive ions generated by gas ionization, can achieve the effects that conventional means cannot achieve in the preparation and synthesis of catalysts. Compared with the control of a series of conditions such as temperature, pH and pressure and additives in liquid phase synthesis, the atmospheric pressure operation and simple structure of DBD make it widely used in the preparation of catalysts [10]. In addition, the plasma and catalyst exhibit some interdependence, which affect each other mutually [11].

Is plasma technology the solution to the problem? Despite extensive research in astrophysics and fusion plasmas has been conducted, the understanding of industrial plasmas is still limited [12]. In this work, we first introduce the properties, development trends and applications of the plasma in catalysis, and then focus on the advantages of DBD plasma preparation catalysts. The heterogeneous catalyst prepared by plasma has good dispersion and smaller particle size, and its catalytic activity, selectivity and stability are significantly improved. Secondly, we pay attention to the synergy of plasma and catalyst. Moreover, the interaction between plasma and catalyst can achieve synergistic effects, which significantly enhances the catalytic effect. Meanwhile, the application of plasma technology in the synergistic catalytic reaction such as methanation, $CH_4$ reforming, $NO_x$ decomposition, $H_2O_2$ synthesis, Fischer–Tropsch synthesis, volatile organic compounds removal, catalytic sterilization, wastewater treatment and degradation of pesticide residues is discussed. Finally, the properties of plasma in the catalytic reaction are summarized, and the application prospect of plasma in the future catalytic field is prospected.

## 2. Preparation of Catalyst by DBD Plasma

The core of catalysis mainly consists of three parts, namely chemical, chemical engineering and materials science. Due to the vigorous development of catalyst market, various types of catalyst publications are also increasing year by year. There are two types of catalysts currently in use: heterogeneous and homogeneous catalysts. Although the use of catalysts conforms to the principles of green chemistry, the preparation of the catalyst is not truly green. Innovative methods for preparing catalysts have been the focus in research. Preparation methods like microwave heating, plasma, biochemistry, etc. have been applied. Among these innovative methods, plasma has received extensive concern [13]. The plasma contains the active material prepared and processed by the catalyst. Under the action of plasma, there is a difference with conventional methods in the process of catalyst nucleation and crystal growth. Highly dispersed catalysts, defect-rich catalysts and heteroatom-doped catalysts can be prepared by plasma technology. In addition, plasma technology can also be widely used for etching, coating, and surface cleaning.

### 2.1. Highly Dispersed Catalyst

The purpose of catalyst preparation is for gaining products with high activity, selectivity and stability. The preparation of a large specific surface area catalyst requires deposition of the active metal component on the porous and thermostable material surface, which can not only enhance its thermal stability, but also improve the service life of the catalyst [14].

Researchers have applied methods such as impregnation and deposition–precipitation, etc. to prepare supported metal catalysts. These methods have significant effects on catalyst synthesis. However, they also have some disadvantages, like being time-consuming. Supported catalysts play an important role in energy and environment fields. A quick and easy way to prepare a supported metal catalyst is to combine DBD and cold plasma jets. Cold plasma preparation of supported metal catalysts mainly involves the reduction of metal ions. Depending on the types of metal element supported, it can be divided into precious metal supported catalysts and non-precious metal supported catalyst.

### 2.1.1. Precious Metals

Supported precious metal catalysts are mainly applied in commercial processes such as automobile exhaust gas catalysis [15]. Over the years, Au, Ag, Pt, Pd or metal compounds have been widely used as industrial catalysts. One particular advantage of using supported precious metal catalysts is that the carrier can disperse the metal over a larger surface area.

Kim et al. [16] used atmospheric DBD plasma to assist reduction of supported platinum catalyst and cobalt catalyst. The reduction characteristics of prepared catalysts were tested. Di et al. [17] prepared different single metal (Pt [18], Ag [19], Pd [20]) supported $TiO_2$ powders via DBD cold plasma. Moreover, they further prepared Pd/C [21], $Pd/Al_2O_3$ [22] and bimetallic $Pd-Cu/Al_2O_3$ [23,24] catalysts for CO oxidation. The photodegradation activity of methylene blue (MB) and CO oxidation activity were prominently improved. The schematic of atmospheric-pressure DBD cold plasma apparatus is demonstrated in Figure 2a. The X-ray diffraction (XRD) patterns of the activated carbon, Pd/C-PC (calcined at 300 °C in $H_2$ for 2 h) and Pd/C-PW (washed with deionized water and then dried at 120 °C for 3h) were displayed in Figure 2b. It can be observed that all samples showed the carbon structure of the activated carbon at 25.0° and 43.5° broad diffraction peaks. The histograms of the particle size distributions of Pd nanoparticles for Pd/C-PC and Pd/C-PW correspond to Figure 2c,d, respectively. Pd nanoparticles were 1.92 nm and 15.93 nm in diameter, respectively. On the activated carbon surface for Pd/C-PC, smaller Pd nanoparticles are evenly distributed. Xu et al. [25] prepared graphene with good Pd dispersion through DBD cold plasma. It has been proved that the novel catalyst has high hydrodesulfurization catalytic activity. Jang et al. [26] treated $Au/TiO_2$ catalysts by atmospheric-pressure $O_2$ DBD plasma for CO oxidation. The sample treated by plasma (S-P) has higher activity and smaller particle size distribution than the calcined (S-C). Di et al. [27] prepared high CO oxidation activity $Au/TiO_2$ catalysts by means of $O_2$ DBD plasma. Moreover, since the compound of Au ion has been entirely reduced, the catalyst shows high CO oxidation activity.

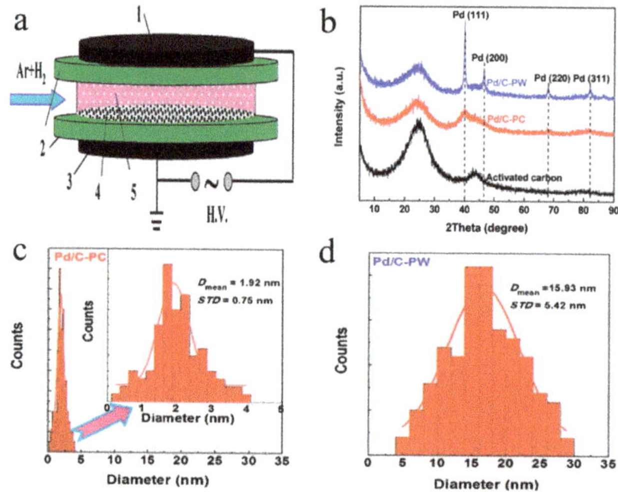


**Figure 2.** (a) the schematic of DBD cold plasma apparatus (1—high voltage electrode, 2—quartz glass, 3—low voltage electrode, 4—sample, and 5—cold plasma); (b) XRD patterns of the Pd/C-PW, Pd/C-PC and activated carbon; (c) the histograms of the particle size distributions of Pd nanoparticles corresponding to Pd/C-PC; (d) the histograms of the particle size distributions of Pd nanoparticles corresponding to Pd/C-PW. Reproduced with permission from [17,21]; Copyright Elsevier, 2013 and Royal Society of Chemistry, 2014.

### 2.1.2. Non-Precious Metals

Due to the scarcity of precious metals and high prices, people are constantly researching and developing non-precious metals or low-content precious metal catalysts. The supported metal catalysts, represented by Ni, have been widely used in the field of methanation. Hu et al. [28] used DBD plasma to treat co-precipitated $NiCO_3$-$MgCO_3$ to obtain an Ni/MgO catalyst for $CO_2$ reforming of $CH_4$. The results indicate that the catalyst prepared via plasma has a larger specific surface area and a smaller particle size, and the $CH_4$ and $CO_2$ conversion rate is increased by more than 20%. Li et al. [29] studied the preparation of $CO_2$ methanation reaction Ni/Ce/SBA-15-P catalyst through DBD plasma. From the $N_2$ adsorption–desorption, XRD and TEM characterization results, it can be concluded that the Ni/Ce/SBA-15-P catalyst has an ordered mesoporous structure. Zhao et al. [30] used DBD combined with ammonia impregnation to prepare supported Ni catalysts. The prepared catalyst exhibits excellent activity, anti-coking property and anti-sintering performances in the methanation reaction of CO. Liu et al. [31] used DBD plasma to decompose a nickel nitrate to prepare the $Ni/ZrO_2$ catalyst. The Ni particles are highly dispersed in the prepared catalyst and dramatically enhanced the methanation activity of CO. Li et al. [32] prepared an Ni/MgO catalyst for $CO_2$ reforming of $CH_4$ using DBD plasma. It was found that the plasma system can improve the catalytic property of catalysts. Liu et al. [33] prepared an $Ni/MgAl_2O_4$ catalyst by DBD decomposing Ni precursor for $CO_2$ methanation. The $CH_4$ yield of the catalyst prepared by plasma system is 71.8%, compared with 62.9% of the catalyst prepared by thermal at 300 °C. Yu et al. [34] used two-dimensional DBD plasma-treated vermiculite (2D-PVMT) to prepare an Ni-loaded catalyst system for CO methanation. The plasma treated catalyst exhibited superior performance and the CO conversion rate reached 93.5%. Liu et al. [35] decomposed an Ni precursor to prepare a $Ni/SiO_2$ catalyst through DBD plasma and used it for $CH_4$ steam reforming. The catalyst prepared this method exhibits intrinsically low activity and small size.

In addition to preparing catalysts supported Ni metal, catalysts supported other elements have been studied and applied in related fields with certain success. Ye et al. [36] found that the manganese oxide catalyst exposed to DBD plasma has lower particle size, higher dispersion and larger specific surface area. The quantity of specific sites (such as vacancies, angular atoms, edges, etc.) on the surface of the prepared catalysts increases, resulting in an increase in plasma reactivity. Li et al. [37] investigated how to use plasma precursor decomposition technology to prepare more efficient supported Co carbon nanotubes (Co/CNTs) catalyst. DBD plasma can prepare a Co/CNTs catalyst in which Co particles are evenly dispersed. Yu et al. [38] used two methods to prepare $W_2N$ catalysts for acetylene hydrochlorination. The first was prepared by depositing tungsten onto an activated carbon support, whereas the second was prepared from using the first catalyst via plasma treatment. It was found that plasma treatment increased the interaction between the support material and active components, and the catalyst activity increased by 12% compared with the untreated catalyst. El-Roz et al. [39] obtained $TiO_2$-β zeolite composite catalyst via DBD plasma under $O_2$ atmosphere. Under UV irradiation, the photooxidation rate of $TiO_2$-β in methanol is eight times that of the conventional $P25$-$TiO_2$ catalyst.

The catalysts synthesized via cold plasma typically have smaller metal nanoparticles and a high distribution, enhancing metal–carrier interactions. Hence, it usually has higher selectivity, stability and catalytic activity than catalysts synthesized by conventional means [40]. The method of preparing high dispersion catalyst by plasma system is shown in Table 1.

**Table 1.** Highly dispersed catalysts prepared by a plasma technique.

| Types of Catalyst | Types of Plasma | Characteristics | Remarks | Applications | References |
|---|---|---|---|---|---|
| Pt and Co catalyst | DBD plasma $H_2/N_2/CH_4$ | Higher selectivity | | Methane conversion | [16] |
| $Pt/TiO_2$ catalyst | DBD cold plasma Ar & $H_2$ | Smaller Pt particles homogeneously distributed | Size of particles: 1.7 nm | Mesoporous photocatalyst with enhanced activity | [17] |
| Pd/Graphene sheets | DBD plasma-assisted $H_2$ | Well-dispersed and higher catalytic efficiency | Size of particles: 2 nm | The hydrodesulfu-rization of carbonyl sulfide | [25] |
| $Au/TiO_2$ | DBD plasma $O_2$ | Higher activity small particle size and narrow size distribution | Size of particles: 2.6 nm | CO oxidation | [26] |
| Au/P25-PC | AP DBD cold plasma Ar & $H_2$ | Smaller size and higher CO oxidation activity | Size of particles: 4.6 nm | CO oxidation | [27] |
| Ni/MgO | DBD plasma $H_2$ | Smaller particle size and higher specific surface area | Size of particles: 3.4 nm | $CO_2$ reforming of methane | [28] |

**Table 1.** *Cont.*

| Types of Catalyst | Types of Plasma | Characteristics | Remarks | Applications | References |
|---|---|---|---|---|---|
| Ni/Ce/SBA-15-P | DBD plasma Ar & $H_2$ | High specific surface area and good high-temperature stability | Size of particles: 7.1 nm | Methanation of carbon oxides | [29] |
| Ni/SiO$_2$ | Ammonia impregnation combined with DBD $NH_3$ | High dispersion and high temperature stability | Size of particles: 11.2 nm | CO methanation | [30] |
| Ni/ZrO$_2$ | DBD plasma $H_2$ | Highly dispersed and long-time stability | Size of particles: 10.6 nm | CO methanation | [31] |
| Ni/MgO | DBD plasma air | Low temperature activity and good stability | Size of particles: 5.4 nm | $CO_2$ reforming of methane | [32] |
| Ni/MgAl$_2$O$_4$ | DBD plasma air | High dispersion and unique structure | Size of particles: 8.9 nm | $CO_2$ methanation | [33] |
| Ni/PVMT | DBD plasma $H_2$ | Ultralow Ni loading and excellent catalytic performance | | CO methanation | [34] |
| Ni/SiO$_2$ | DBD plasma air | Low activity with small size and coke resistance | Size of particles: 7.1 nm | Steam reforming of methane | [35] |
| MnO | Wire-plate DBD $O_2$ | Low cost, environmentally friendly and high activity | | Toluene decomposition | [36] |
| Co/CNTs | DBD plasma Ar | Uniform and dispersed Co particles | | FTs | [37] |
| W$_2$N | DBD plasma $NH_3$ | Excellent chemical stability and high performance | Size of particles: 5.99 nm | Acetylene hydrochlorina-tion | [38] |
| TiO$_2$-β zeolite | DBD plasma $O_2$ | Higher methanol photodegradation rates | The BET specific surface area: 421.9 m$^2$/g | Methanol photooxidatin | [39] |

## 2.2. Defect-Rich Catalyst

Indeed, synthesizing a perfect crystal without any defects is basically impossible. There are many defects in nanomaterials that can regulate the electronic and surface properties of materials. It is well known that point, line, plane, and volume defects are found in materials. The point defect can be further grouped vacancies, interstitial atoms, impurities and heteroatom doping according to its composition and position [41]. There have been many reports that the surface defects of electrocatalysts may have a positive effect on the electrochemical reaction [42–45]. For instance, many researchers have reported that heteroatom-doped carbon materials possess better oxygen reduction reaction (ORR) activity compared to undoped pure carbon [46–49]. Therefore, there is a great research space for surface regulation of electrocatalysts around defect chemistry. The reason why we choose plasma technology to achieve the regulation of electrocatalyst surface structure is that plasma technology has the following functions: surface cleaning, oxygen cavitation, and etching.

### 2.2.1. Carbon Material Defect

Huang et al. [50] studied the changes of surface morphology and chemical characteristics about carbon nanotubes after DBD modification. The plasma would strike the defect sites, so that the sp$^2$-hybridized carbons of nanotubes is converted into sp$^3$-hybridization, introducing COOH oxygen-containing functional groups. Meanwhile, the quantity of oxygen atom is reduced, which is due to the further oxidation of COOH group into $CO_2$ and $H_2O$ at excessive exposure time. Paredes et al. [51] experimentally and theoretically investigated the properties and characteristics of atomic-scale defects on the surface of DBD plasma oxidized graphite. Two major types of defects can be observed by STM: protrusions with a diameter of ~1–5 nm and depressions of 5–7 nm width, the latter being a novel defect on the carbon surface.

### 2.2.2. Metal and Metal Oxide Defect

Wang et al. [52] used a plasma etching technique to obtain a $Co_3O_4$-based OER electrocatalyst with high surface area and more oxygen vacancies. The Ar plasma etching can expose more surface sites. Plasma etching can adjust the electronic state by creating oxygen vacancies on $Co_3O_4$ nanosheets' surface (Figure 3a). The oxygen vacancies produced by plasma etching were verified through XRD and XPS data plots. In order to prove that the plasma etched $Co_3O_4$ can expose more surface area, the $N_2$ sorption isotherms of two nanosheets were obtained (Figure 3b). The results showed that the Brunauer–Emmett–Teller (BET) surface area of plasma etched $Co_3O_4$ nanosheets was much higher than that of the original $Co_3O_4$ nanosheets. The TEM images in Figure 3c,d show that the pristine $Co_3O_4$ nanosheets exhibit a continuous and compact surface. Figure 3e,f indicated the nanosheets have high oxygen evolution reaction (OER) activity after plasma etching. Cuenya et al. [53] used plasma treatments to prepare highly-defective nanostructured silver catalysts with selective $CO_2$ reduction to CO under low overpotential. The results illustrated that the Ag catalyst prepared by plasma-treated produced small pore-like defects, and the activity and CO selectivity of the catalyst were greatly improved. Yu et al. [54] produced lots of oxygen vacancies and $Ti^{3+}$ defects on the surface of $TiO_2$ nanoparticles by Ar DBD plasma. Compared with pristine $TiO_2$ nanoparticles, the energy band gap of Ar plasma treated $TiO_2$ nanoparticles decreased from 3.21 eV to 3.17 eV, and the photocatalytic degradation ability of organic dyes was enhanced.

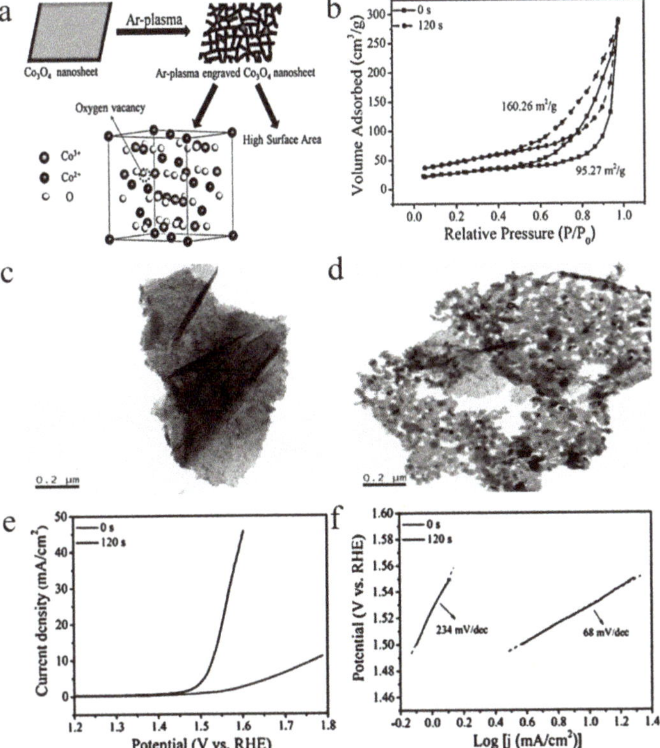


**Figure 3.** (a) graphical illustration of Ar plasma etching of $Co_3O_4$; (b) $N_2$ sorption isotherms of plasma-etched and original $Co_3O_4$; (c) TEM images of original $Co_3O_4$; (d) TEM images of Ar plasma etched $Co_3O_4$; (e) the polarization curves of OER on original $Co_3O_4$ (0 s) and the plasma etched $Co_3O_4$ (120 s); (f) Tafel plots. Reproduced with permission from [52]; Copyright Wiley, 2016.

Plasma treatment can generate more active sites and high specific surface area deficient catalysts, which increases chemical reactivity. In addition, plasma technology can be used to detect defects. Ebihara et al. [55] applied DBD to detect defects in the surface of polytetrafluoroethylene coated metal. They developed a detection system for pinholes and cracks in Teflon film and used DBD technology to reduce residual material and damage to the film. The method of preparing a defect-rich catalyst using plasma is shown in Table 2.

**Table 2.** Defect-rich catalysts prepared by a plasma technique.

| Types of Catalyst | Types of Plasma | Characteristics | Remarks | Applications | References |
|---|---|---|---|---|---|
| Multi-wall CNTs | DBD plasma $O_2$ | Generate oxygen-containing group | The BET specific surface area: 156 $m^2/g$ | Surface modification | [50] |
| Graphite | DBD plasma air | Create novel defects | Protrusions 1–5 nm in diameter and smooth depressions 5–7 nm wide | Modification of graphite surfaces | [51] |
| $Co_3O_4$ nanosheets | Plasma-engraved Ar | With high surface area and oxygen vacancies | The BET specific surface area: 160.26 $m^2/g$ | OER | [52] |
| Ag catalysts | Plasma-activated $H_2$, Ar & $O_2$ | Lower overpotential | Pore-like defects 50–100 nm in size | Carbon dioxide electroreduction | [53] |
| Ar-$TiO_2$ | DBD plasma Ar | Oxygen vacancies and $Ti^{3+}$ defects | The energy band gap reduction | Photocatalytic degradation of organic dyes | [54] |
| Teflon films | DBD $H_2O$ | Point electrode | | Prevent degradation and erosion | [55] |

### 2.3. Heteroatom-Doped Catalyst

Metals and metal oxides have a wide range of applications in many industrial processes, including synthetic materials, production and storage of clean energy. However, metal-based catalysts have many disadvantages, such as high cost, poor durability, low selectivity, easy poisoning, and environmental unfriendliness. Therefore, these shortcomings of precious metals hinder their large-scale commercial use in renewable energy [56]. Recently, heteroatom-doped catalyst materials are gradually entering the line of sight and exhibiting excellent performance in the catalytic process.

### 2.3.1. Nitrogen Atom Doping

Lee et al. [57] used DBD pulsed laser deposition technique to grow N-doped ZnO thin films. Low temperature photoluminescence spectra show that N-doped ZnO film is p-type doping status. Yu et al. [58] used fast and effective DBD plasma treatment to obtain N-doped exfoliated graphene oxide (N-PEGO). N-PEGO becomes a potential cathode ORR catalyst in fuel cells due to its high nitrogen content, large specific surface area and enough active sites. Figure 4a displays the method of synthesizing N-PEGO. In short, N-PEGO was prepared by treating the ammonium carbonate dried GO dispersion with DBD plasma. Figure 4b,c are TEM images of N-PEGO. The TEM image shows that N-PEGO flakes have a relatively smooth layer structure. In the sample's N 1s spectrum (Figure 4d), the four peaks centers of pyrrolic N, oxidic N species, pyridinic N and graphitic N are located at 400.3, 406.1, 399.0 and 401.4 eV, respectively, and the N content of N-PEGO is higher than that in GO by elemental analysis. As Figure 4e shows, linear sweep voltammetry (LSV) of GO, N-PEGO and 20 wt% Pt/C showed that the ORR onset potential of N-PEGO was 0.89 V, which was comparable to that of commercial 20 wt% Pt/C (0.93 V). Zhang et al. [59] prepared N-vacancy doped g-$C_3N_4$ with excellent photocatalytic performance through DBD plasma processing and applied it to the production of $H_2O_2$. As a result, it was found that the structure, morphology and optical properties of the catalyst were affected by the plasma treatment. Geng et al. [60] used wet impregnation to deposit CuO nanoparticles, and then processed by $N_2/Ar$ DBD plasma to obtain $Cu_2O/N\text{-}TiO_2$. Compared with $CuO/TiO_2$ and $TiO_2$, the photodegradation efficiency of methyl orange was prominently increased under visible light.

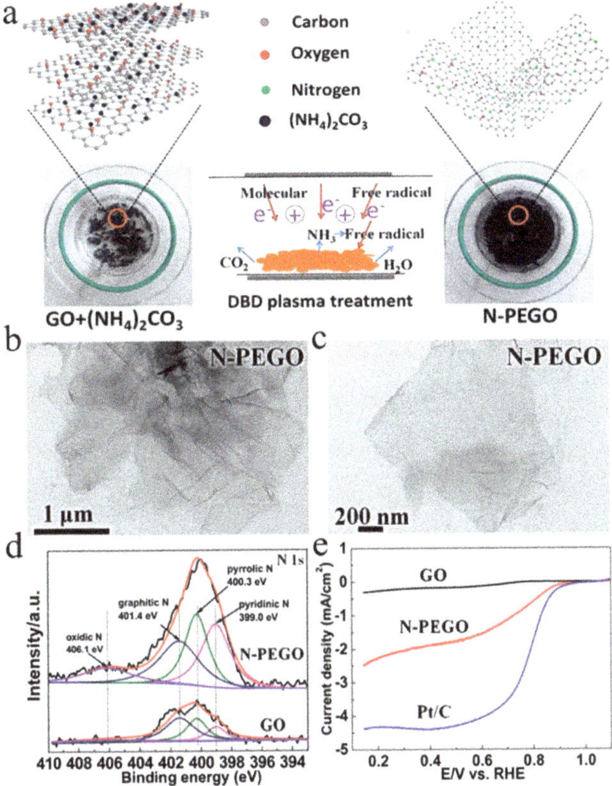


**Figure 4.** (**a**) the method of synthesizing N-PEGO; (**b,c**) TEM images of N-PEGO; (**d**) the N 1s spectrum of the samples; (**e**) LSV curves of GO, N-PEGO and 20 wt% Pt/C. Reproduced with permission from [58]; Copyright Royal Society of Chemistry, 2018.

### 2.3.2. Sulphur Atom Doping

Hu et al. [61] used DBD plasma to treat S-doped $g\text{-}C_3N_4$ catalyst with enhanced photocatalytic performance under $H_2S$ atmosphere. It was found that plasma treatment under $H_2S$ atmosphere can dope more S into the lattice of $g\text{-}C_3N_4$, which affected the morphology and electronic structure of the catalyst.

### 2.3.3. Phosphorus Atom Doping

Wang et al. [62] prepared hydrogen evolution reaction (HER) and oxygen evolution reaction (OER) electrocatalyst by processing $Co_3O_4$ containing a P precursor with Ar plasma. The P atom can significantly tune the electronic structure and adsorption characteristics of $Co_3O_4$. The method of preparing heteroatom-doped catalyst using plasma is shown in Table 3.

**Table 3.** Heteroatom-doped catalysts prepared by a plasma technique.

| Types of Catalyst | Types of Plasma | Characteristics | Remarks | Applications | References |
|---|---|---|---|---|---|
| N-doped ZnO thin films | DBD pulsed laser deposition $N_2$ | Low temperature photoluminescence spectra | The maximum hole density: $10^{-17}$–$10^{-18}$·$cm^{-3}$ | Make ZnO based electronic devices | [56] |
| N-PEGO | DBD plasma $CO_2$ & $NH_3$ | High onset potential and good electrocatalysis stability | The BET specific surface area: 380.0 $m^2/g$ | ORR | [57] |
| N-vacancy-doped g-$C_3N_4$ | DBD plasma $H_2$ | Outstanding photocatalytic | | $H_2O_2$ production | [58] |
| $Cu_2O$/N-$TiO_2$ | DBD plasma $N_2$/Ar | Dispersed uniformly and higher photodegradation efficiency | | Improved photocatalytic activity | [59] |
| S-doped g-$C_3N_4$ | DBD plasma $H_2S$ | Larger specific surface area | The BET specific surface area: 52.8 $m^2/g$ | Enhanced photocatalytic performance | [60] |
| P-$Co_3O_4$ | Ar plasma | Porous, discontinuous, and loose surface | | HER and OER | [61] |

## 3. DBD Plasma Catalysis

The synergistic effect means that the effect of combining the plasma with the catalyst is greater than the sum of their individual effects. Plasma catalysis synergy is a complex phenomenon in the interaction process between plasma and catalyst [63–67]. In general, the interaction of the plasma with the catalyst is divided into two parts. Firstly, plasma and catalyst independently affect the surface processes. Secondly, the plasma and the catalyst show some interdependence because the plasma interacts with the catalyst. This complex interdependence is graphically represented in Figure 5. It can be seen that the introduction of catalyst into the plasma discharge region may affect the discharge type and cause a change in electron distribution, which affects the generation of short-lived reactive plasma species.

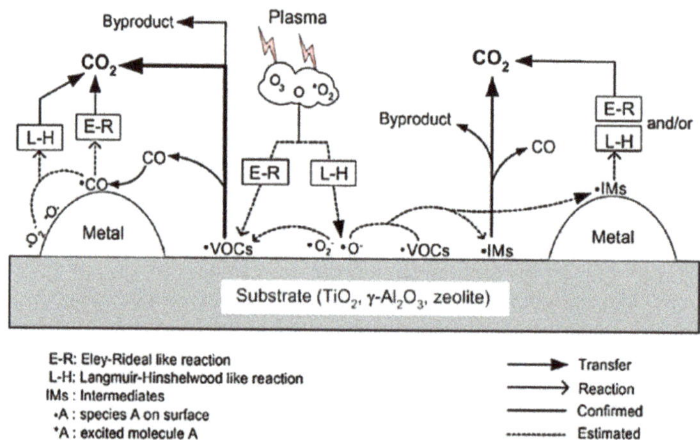


**Figure 5.** Complex interdependence of plasma–catalyst interactions. Reproduced with permission from [11]; Copyright American Chemical Society, 2015.

### 3.1. Application of C1 Chemistry

The reaction in which the reactant contains only one carbon atom during the chemical reaction is collectively referred to as carbon-one chemical (C1). The subject of C1 is a compound containing only one carbon atom in the molecule, such as CO, $CO_2$, $CH_4$, $CH_3OH$, etc. The main purpose of C1 is to save coal and oil resources, and generate more fuel with less carbon raw materials for humans.

C1 products are a good substitute for petroleum. The combination of C1 chemistry and green chemical industry is to achieve coordinated development of economy and environment. No matter whether it is from the strategic perspective of world energy development, or focusing on environmental protection and sustainable development for the benefit of future generations, the combination of green chemical industry and C1 chemistry is of great significance. With the continuous progress of science and technology, a large number of novel and environment-friendly technologies have appeared in this field, and, in some industries, even traditional industrial methods have been replaced. With the progress of the times and the leap of science, green C1 chemistry will surely exert its great potential in the future chemical industry and better benefit mankind. We have previously summarized the application of two-dimensional layered double hydroxides in methanation and methane reforming reactions of C1 chemistry [68]. This paper mainly reviews the application of DBD plasma in C1 chemistry, such as CO methanation and oxidation, $CO_2$ decomposition and $CH_4$ reforming.

### 3.1.1. Methanation

Nowadays, the gradual depletion of oil energy has become an urgent problem, which has increased the research on coal technology. Researchers have developed various technologies to convert coal into fuels such as methane and methanol [69]. The conversion of $CO_2$ into $CH_4$ is a valuable solution for carbon resource recycling in the C1 chemistry [70]. In the study of DBD plasma to convert $CO_2$, the driving catalytic method combining plasma and catalyst is widely used. The advantage is that the heat generated in the plasma can make the catalyst work in the optimal temperature range, and the effect of the catalyst also allows the plasma to operate within the proper operating range.

$CO_2$ hydromethanation is one of the important reactions in the field of energy and chemical industry. The principle of methanation reaction is that $CO_2$ reacts with $H_2$ under certain temperature and methanation catalyst to generate $CH_4$ and steam. After cooling at the back, the water vapor is condensed and separated to obtain qualified hydrogen containing only $CH_4$ impurities. Compared with other $CO_2$ conversion methods, $CO_2$ methanation has the advantages of fast reaction speed, high selectivity and few by-products. Simultaneously, the generated $CH_4$ can be directly utilized, which effectively alleviates the problem of energy shortage. Industrially, the $CO_2$ methanation reaction can also be used to prevent catalyst poisoning and crude hydrogen purification during ammonia synthesis. In addition, we have prepared different catalysts and applied them to the low-temperature CO methanation reaction, which has achieved good results [71–74]. Sabatier and Senderens first reported carbon monoxide methanation to methane in 1902 [75], which has been widely used in fuel cell industry and ammonia synthesis devices successively [76–79]. Ni catalyst is the most commonly used catalyst for methanation. Costa et al. [80] verified that DBD non-thermal plasma can improve the methanation performance of $Ni/CeO_2$ and $Ni/ZrO_2$ catalytic system, especially at lower temperatures. In the presence of plasma at 90 °C, $CO_2$ conversion rate is up to 80%, while methane selectivity is 100%. In contrast, the same conversion rate and selectivity for the same catalyst requires a temperature of about 300 °C or higher without plasma. Jwa et al. [81] applied non-thermal DBD plasma for the heterogeneously catalyzed methanation of CO and $CO_2$ over Ni-loaded alumina catalysts at 180–320 °C and atmospheric pressure. The results demonstrated that Ni-supported catalysts such as $Ni/Al_2O_3$ and $Ni-TiO_2/Al_2O_3$ were significantly affected by plasma, which markedly increased the catalytic activity and led to the enhancement of methanation rate. Costa et al. [82] measured the activity of Ni-Ce-Zr hydrotalcite-derived catalysts in a mixed plasma catalyzed process to generate $CH_4$ in low temperature. Even at very low temperatures, the methane yield is as high as 80%.

In addition, Cu, Mn and other noble metals like Ru and Rh also have excellent catalytic performance for methanation. Zeng et al. [83] reported the effect of $\gamma$-$Al_2O_3$ catalysts supported with various metals on the hydrogenation performance of $CO_2$. The $CO_2$ conversion was increased from 6.7% to 36% using both plasma and catalyst compared to plasma-catalyzed $CO_2$ hydrogenation alone. Kim et al. [84] used DBD plasma-activated $Ru/\gamma$-$Al_2O_3$ catalyst to carry out $CO_2$ methanation under an atmospheric environment. Regardless of the presence of catalyst, the $CO_2$ conversion rate increased with DBD plasma. The addition of catalysts to the plasma may affect the discharge characteristics, thus producing highly reactive plasma materials. Figure 6a,b are diagrams of the experimental set-up for $CO_2$ methanation and emission spectrum experimental device, respectively. Figure 6c illustrated the V–Q Lissajous figure of DBD plasma at 9 kV and 3 kHz on Ru/γ-$Al_2O_3$. Figure 6d presents a comparison of $CH_4$ selectivity and $CO_2$ conversion under DBD plasma conditions. In the absence of plasma, the $CH_4$ selectivity and $CO_2$ conversion are basically zero, and the Ru/γ-$Al_2O_3$ catalyst has no reactivity. In contrast, 8.21% $CO_2$ conversion and 43.84% CO selectivity are obtained with plasma alone, but $CH_4$ selectivity was reduced by 1.42%. When the DBD plasma was used together with Ru/γ-$Al_2O_3$ catalyst, CO selectivity was decreased, but the $CO_2$ conversion and $CH_4$ selectivity reach 12.80% and 73.30%, respectively. The influence of the $H_2/CO_2$ ratio on the $CH_4$ selectivity and $CO_2$ conversion is displayed in Figure 6e. The CO selectivity decreased as the $H_2/CO_2$ ratio increased; however, the $CH_4$ selectivity and $CO_2$ conversion increased. The effects of Ar on $CO_2$ conversion, $CH_4$ and CO selectivity are shown in Figure 6f. It can be seen that, when Ar is added, deoxygenation and methanation are simultaneously improved.

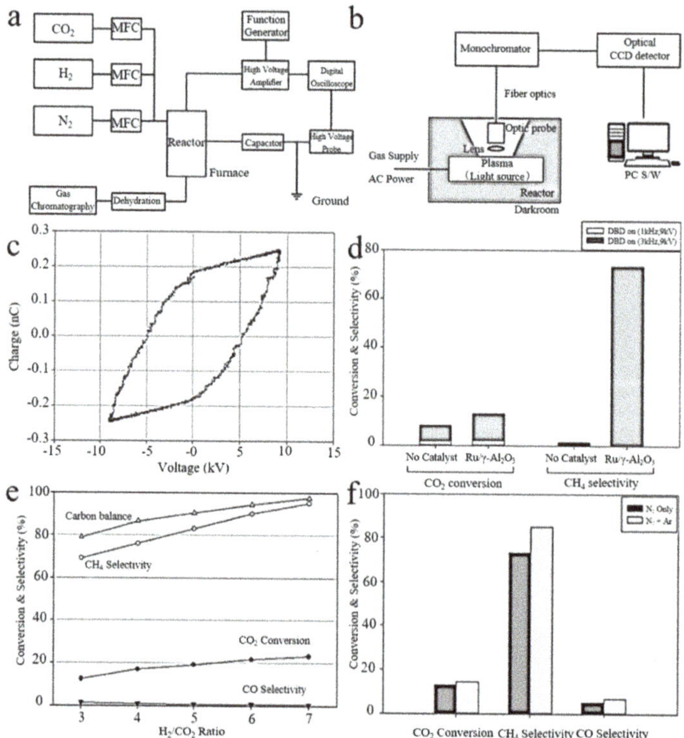

**Figure 6.** (a) the diagram of a methanation experimental device; (b) the diagram of experimental device for optical emission spectroscopy; (c) Lissajous diagram of the DBD plasma; (d) comparison of the $CH_4$ selectivity and $CO_2$ conversion; (e) effect of the $H_2/CO_2$ ratio on the $CH_4$ selectivity and $CO_2$ conversion for Ru/γ-$Al_2O_3$; (f) effect of Ar on DBD over Ru/γ-$Al_2O_3$. Reproduced with permission from [84]; Copyright Elsevier, 2017.

The results show that the introduction of catalyst into the plasma can vastly improve the conversion and selectivity of $CO_2$ methanation reaction. Compared with the thermal reaction, the plasma reaction has a higher activity and can maintain the reaction at a lower temperature, avoiding the destruction of the catalyst surface properties by high temperature.

### 3.1.2. CO Oxidation

CO oxidation has attracted much attention owing to its widespread application in automotive exhaust, indoor air and preferential CO oxidation of proton exchange membrane fuel cells [85–88]. Since the pioneering research of Haruta [89], supported gold nanometer catalysts have been considered as the most effective catalysts for CO oxidation at low temperatures. In the case of CO oxidation, it is speculated that the CO molecules may strongly adsorb on gold nanoparticles.

The supported gold nanocatalyst has an abnormally high CO oxidation activity at low temperatures, which can be widely applied to various industrial processes and instruments like environmental protection, clean energy conversion, CO sensors, air purification and gas mask. The plasma is treated under mild conditions without causing significant sintering of the calcined gold nanoparticles. Subrahmanyam et al. [90] conducted an experimental of CO monoxide oxidation by $N_2O$ in the DBD reactor. The direct oxidation of CO in the plasma reactor illustrated no apparent activity, but the presence of $N_2O$ increased the CO conversion. This is mainly attributed to the synergetic effect between plasma excitation and catalyst, which increases the performance of the reactor. Heintze et al. [91] investigated the DBD plasma-assisted partial oxidation of $CH_4$ into synthesis gas. When $Ni/Al_2O_3$ catalyst is filled into the plasma discharge region, CO is oxidized to $CO_2$ at 300 °C or higher, while the $H_2$ and $H_2O$ selectivity keep stable.

### 3.1.3. $CO_2$ Decomposition

Fossil fuels such as coal and petroleum are the main energy sources for the development of modern industrial society, but the resulting $CO_2$ gas has brought serious environmental and climate problems [92]. The use of fossil fuels produces $CO_2$, and many studies have reported the impact of greenhouse gases on the climate. There has been a great interest in DBD plasma control pollution, which is now used in ozone production. Plasma-assisted catalytic techniques have been designed in conjunction with quartz barriers to decompose $CO_2$. At the same time, some low carbon compounds such as $CH_4$ were used in these studies [93].

In general, the performance of a DBD reactor depends on the configuration of the reactor, ambient gas, packed materials, gas flow rate and input power [94–96]. Subrahmanyam et al. [97] studied the effect of packing materials in a packed DBD plasma reactor on $CO_2$ decomposition. The effects of porosity, dielectric constant and ultraviolet light on $CO_2$ decomposition were studied. The study found that DBD plasma reactors will have improved performance in the case of various fillers. Uytdenhouwen et al. [98] studied the impact of gap size reduction and filler on $CO_2$ conversion in the DBD microplasma reactor. Comparing the results with a conventional size reactor, although the energy efficiency is lower, decreasing the discharge gap can markedly enhance the $CO_2$ splitting. These two improvements have been successful, and the combined effect of the two methods may be even better. Dai et al. [99] used DBD plasma in combination with the filler $ZrO_2$ to study $CO_2$ decomposition. It was found that the filler had a significant effect on the decomposition of $CO_2$. When the catalyst is added, the CO selectivity was as high as 95% and the energy efficiency increased from 3.3% to 7%. Wang et al. [100] investigated the impact of filling materials on the $CO_2$ conversion in the DBD microplasma reactor. The results show that the $CO_2$ conversion in the dielectric filled reactor is higher than that in the without filling reactor. Tu et al. [101] performed coaxial DBD plasma photocatalytic reduction of $CO_2$ to CO and $O_2$ at low temperatures. In the case of only plasma, the maximum $CO_2$ decomposition rate reached 21.7%. Figure 7a is a diagram of a reaction experimental device. The impact of $BaTiO_3$ and $TiO_2$ photocatalysts on $CO_2$ conversion and energy efficiency is exhibited in Figure 7b. It is clear that the filling of $TiO_2$ and $BaTiO_3$ catalysts obviously increases the $CO_2$ conversion and energy efficiency. Figure 7c presents the gas temperature and catalyst surface temperature in the DBD reactor. Obviously, in the first 15 min after the plasma is ignited, the plasma gas temperature without catalyst dramatically improved from 23.3 to 123.5 °C, after which it rises slowly and remains almost stable at 25 min. In the DBD reactor filled with $BaTiO_3$ and $TiO_2$, we noticed that the plasma temperature is almost the same as the catalyst surface temperature. Figure 7d indicates the reaction mechanisms of plasma-photocatalytic

$CO_2$ decomposition. The electron-hole pairs are produced by means of high-energy electrons in the gas discharge, which moves in the opposite direction under electric field, thereby decreasing the possibility of recombination. Ma et al. [102] increased the decomposition efficiency of $CO_2$ by filling metal foam electrode in DBD plasma device. It was found that the metal foam not only functions as the energy transformation and electrode, but also consumed a part of the $O_2$ and O radicals generated in the tube, and promoted the positive equilibrium shifting of the reaction.

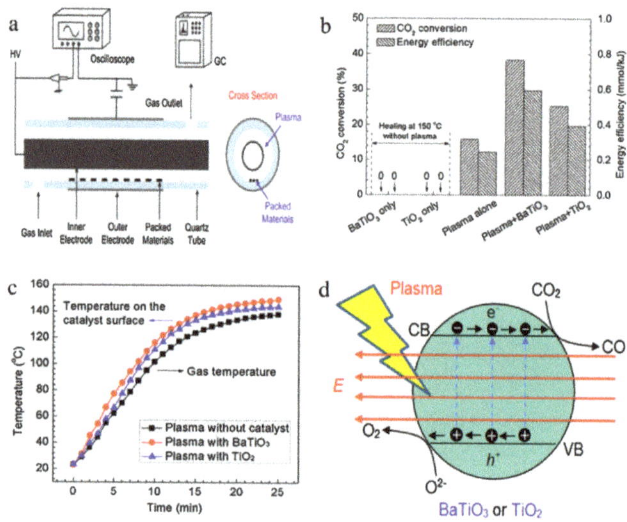


**Figure 7.** (**a**) diagram of reaction experimental device; (**b**) impact of plasma and catalyst on $CO_2$ conversion; (**c**) changes in plasma gas temperature and catalyst surface temperature; (**d**) reaction mechanisms of photocatalytic decomposition of $CO_2$ by plasma. Reproduced with permission from [101]; Copyright Elsevier, 2016.

### 3.1.4. $CH_4$ Reforming

The main component of natural gas is $CH_4$, while $CO_2$ is the final reaction product of carbon-containing compounds. Both of them are greenhouse gases and rich in carbon resources. Therefore, it is of great significance to develop the research on the basic organic chemical synthesis route of $CH_4$ and $CO_2$ as raw materials to replace coal and petroleum resources. However, due to their stable chemical properties, direct conversion requires extremely harsh reaction conditions, so plasma technology provides a new way for such conversion. The results demonstrate that the plasma reforming of $CH_4$ and $CO_2$ has the characteristics of large processing capacity, high conversion rate, high chemical energy efficiency and high calorific value yield. Simultaneously, $CH_4$ and $CO_2$ have mutual promotion effects in the reforming reaction.

Kogelschatz et al. [103] used DBD with zeolite catalyst to generate higher hydrocarbons directly from the $CH_4$ and $CO_2$ under ambient conditions, including alkanes, alkenes, oxygen-containing compounds and syngas. It can be seen that using the catalytic DBD plasma can achieve a cogeneration of syngas and higher hydrocarbons. Kraus et al. [104] conducted the combination of solid state catalysts with DBD for $CO_2$ reforming of $CH_4$, including $CH_4$ and $CO_2$ decomposition reactions. It is shown that the chemical properties of plasma can be improved by using catalytic coating in discharge. Cheng et al. [105] investigated the low temperature transformation of $CO_2$ and $CH_4$ in the coaxial DBD reactor. The experimental results show that the $CO_2$ reforming of $CH_4$ was improved by the introduction of steam through the synergistic catalysis of cold plasma and catalyst.

Furthermore, plasma technology can also be used for $CH_4$ and steam reforming to produce $H_2$. With global warming and the decline of fossil fuel, newly developed energy must satisfy both environmental and renewable requirements. $H_2$ is one of the most promising energy sources in the future society, so it is very important to develop and implement systems for hydrogen production and storage.

Steam reforming of $CH_4$ typically requires additional steam supply to control carbon formation. Catalysts with strong anti-coke properties help to decrease the amount of steam and save resources. Nozaki et al. [106] demonstrated the synergistic effect between plasma and catalyst by using $Ni/SiO_2$ catalyst for plasma $CH_4$ steam reforming. Figure 8a represents a schematic diagram of different electronic and kinetic processes. The electron collision ionization, excitation and dissociation of $CH_4$ were evaluated with electron density and field intensity. Figure 8b shows the $CH_4$ conversion curve for the DBD catalytic packed-bed reactor. The combination of Ni catalyst and barrier discharge significantly enhances product conversion. Figure 8c makes a comparison of three different reaction systems. It can observe a strong synergistic effect between the Ni catalyst and barrier discharge, and the $CH_4$ conversion rate exceeded the sum of Ni catalyst and barrier discharge used alone. Figure 8d represents enthalpy balance of the typical reforming device. Figure 8e shows a potential curve of $CH_4$. As shown in the figure, the plasma-enhanced catalytic reaction is interpreted as a two-step excitation. Kim et al. [107] realized the complete oxidation of $CH_4$ by hybridizing both catalyst and plasma in a DBD quartz tube reactor. Under the combined action of the catalyst and the plasma, $CH_4$ was oxidized at room temperature to generate $CO_2$ with lower CO selectivity, showing complete oxidation with the assistance of catalyst. Nozaki et al. [108] described the steam reforming of $CH_4$ in a plasma-catalyst hybrid reactor. It could observe a synergistic effect between Ni catalyst and the non-thermal DBD at lower reaction temperatures, and the $CH_4$ conversion far exceeded equilibrium conversion; meanwhile, the selectivity of product was prone to balance composition under given conditions.

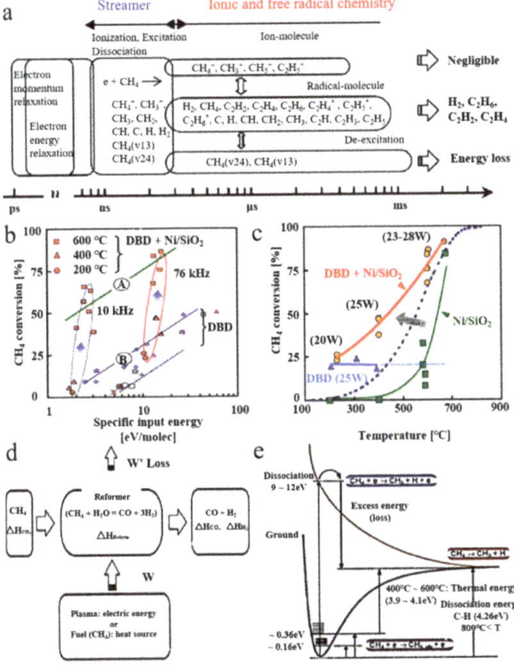

**Figure 8.** (**a**) reaction scheme; (**b**) $CH_4$ conversion: (A) barrier discharge + Ni catalyst; (B) barrier discharge; (**c**) $CH_4$ conversion; the blue dashed line represents the equilibrium conversion; (**d**) energy diagram; (**e**) methane activation. Reproduced with permission from [106]; Copyright Elsevier, 2004.

### 3.2. $NO_x$ Decomposition

Since the 21st century, with the increase in the number of automobiles and the industry development, exhaust pollutants, especially nitrogen oxides ($NO_x$), emissions have brought a great load to the environment, which causes serious harm to human health and attracts the general attention of the world's research. Plasma catalysis has made great progress in academic research and industrial applications, which is a new processing technology with very optimistic prospects.

In the past decade, many researchers concentrate on technologies that reduce emissions or control air pollution. Among these solutions, plasma is preferred because of low energy consumption and high efficiency. The non-thermal plasma low-temperature catalytic conversion of $NO_x$ to valuable chemicals or harmless gases has been effectively done [109]. Our research group also conducted experiments and reviews on selective reduction and denitrification of $MnO_x$-$Fe_2O_3$/vermiculite monolithic honeycomb catalysts and porous microspherical aggregates of Mn–Ce–Fe–Ti mixed oxide nanoparticles at low temperatures [110–113]. Yu et al. [109] used non-thermal DBD plasma and $H_2O$ as external electrode to efficiently remove $NO_x$. As illustrated in Figure 9, Yu's team studied the change in NO conversion with and without glass beads using copper foil and $H_2O$ as external electrodes, respectively. As can be seen from the picture, when the copper foil is used as the external electrode, the NO conversion rate is 14.1%; however, when $H_2O$ is the external electrode, the NO conversion rate reaches 28.8%. Meanwhile, when the glass beads are packed with $H_2O$ as the external electrode, the NO conversion rate is greatly increased to 95.9%, indicating that the glass beads could greatly improve the NO removal rate. Niu et al. [114] investigated the synergistic effect of DBD plasma and $Ag/Al_2O_3$ catalytic system on $C_2H_2$ selective catalytic reduction of $NO_x$. A significant synergistic effect was obviously observed in plasma and catalytic system. Wang et al. [115] used DBD plasma to assist with catalytic reduction of $NO_x$ on the Mn–Cu catalyst. DBD combined with Mn-Cu catalyst improved the catalytic activity of selective catalytic reduction reaction and provided a new method for selective catalytic reduction research. Niu et al. [116] reported the synergistic effects of HZSM-5 catalyst modified by indium and pulsed DC DBD plasma on the selective reduction of $NO_x$ by $C_2H_2$ at 200 °C.

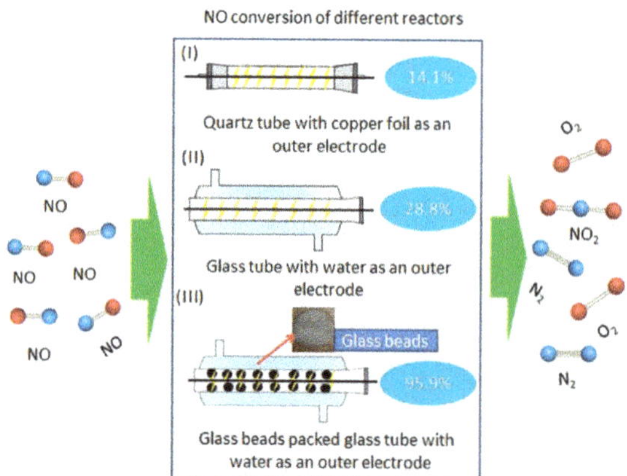


**Figure 9.** NO conversion under different reactors [109]; (Open Access).

### 3.3. $H_2O_2$ Synthesis

Hydrogen peroxide ($H_2O_2$) is an important green oxidant, which is widely used in daily life. At present, $H_2O_2$ is almost entirely produced by indirect and ungreen anthraquinone process (AQ). Some reports have investigated the direct synthesis of $H_2O_2$ through electrochemical devices and noble metal catalysts using $H_2$ and $O_2$ as raw materials. However, it is difficult to obtain high-purity $H_2O_2$ products without purification.

The semiconductor industry, including microelectronics, displays and photovoltaics, requires electronic grade $H_2O_2$. This high purity $H_2O_2$ must comply with the semiconductor equipment. Manufacturing high purity $H_2O_2$ from the commercial grade $H_2O_2$ of the AQ process requires complicated and energy-intensive purification techniques. Purification is dangerous due to the reactivity of $H_2O_2$. At the moment, reverse osmosis is considered as an ultra-purification method for $H_2O_2$, but there are no longer-life membrane materials. Therefore, it is critical to study a new technology for synthesizing high purity $H_2O_2$ [117]. Figure 10 shows the experimental device and reaction mechanism diagram of $H_2O_2$. It can be seen from the figure that, during the continuous operation, the $O_2$ conversion rate remains stable, and the volume of the $H_2O_2$ product increases linearly with time.

Studies in the 1960s showed that $H_2O_2$ can be produced by free radical reaction in the presence of any catalyst or chemical catalyst in a $H_2/O_2$ non-equilibrium plasma. However, the plasma method does not attract much attention due to the low $H_2O_2$ yield and the safety of the $H_2/O_2$ reaction caused by discharge. To prevent fire and explosion, the $O_2$ content must be strictly kept below 4 mol%.

Li et al. [118] studied the synergistic effect of $TiO_2$ photocatalysis and plasma discharge on the degradation of thiamethoxam. The results indicated that the synergy effect can obtain high concentration of $H_2O_2$. Bruggeman et al. [119] reported a detailed study of the plasma morphology and $H_2O_2$ production in DBD operating at various powers and water vapor concentrations. They evaluated the effect of discharge morphology and power on the concentration dependence of the OH and $H_2O_2$ production through a joint study. Vasko et al. [120] investigated the $H_2O_2$ production by RF glow discharge in helium–water vapor mixtures. The consistency between the experiment and model is very well, which corresponds to the uncertainty of reaction rate and experiment accuracy to some extent. Hu et al. [121] used DBD plasma to synthesize N vacancies doped g-$C_3N_4$ catalyst under $H_2$ atmosphere. The impact of doping N vacancies into g-$C_3N_4$ on the photocatalytic $H_2O_2$ production capacity was investigated. Guo et al. [122] developed a plasma reactor with multiple parallel DBD tubes to synthesize $H_2O_2$ from $H_2/O_2$. Thevenet et al. [123] synthesized $H_2O_2$ using DBD related to fibrous materials. In this work, $H_2O_2$ is synthesized by igniting a mixture of $H_2$ and $O_2$ in a DBD. This study demonstrates that non-thermal plasma can effectively synthesize $H_2O_2$, and the use of fibrous materials prominently enhances the synthesis process.

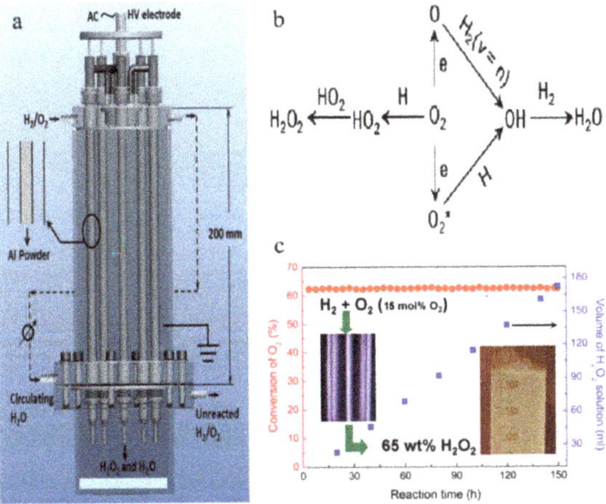


**Figure 10.** (a) diagram of double DBD reactor $H_2O_2$ synthesis device; (**b**) the main reactions network for formation of $H_2O_2$ and $H_2O$ in $H_2/O_2$ nonequilibrium plasma; (**c**) the relationship between $O_2$ conversion, $H_2O_2$ product solution volume and reaction time. Reproduced with permission from [117]; Copyright Wiley, 2014.

Research shows that the configuration of plasma reactor plays an important role in $H_2O_2$ synthesis. In a single DBD plasma reactor with a bare metal high voltage electrode and $H_2O$ ground electrode, the $O_2$ conversion rate of $H_2/O_2$ mixture containing 3% $O_2$ reached 100%, but the $H_2O_2$ selectivity was only 3.5%. However, 57.8% $O_2$ conversion and 56.3% $H_2O_2$ selectivity can be obtained by using a dual DBD plasma reactor with a pyrex-covered metal high voltage electrode. Although selectivity has been vastly enhanced, low efficiency and safety remain a major challenge due to low $O_2$ content. Even though $H_2O_2$ is less selective than the AQ process, for the direct production of high purity $H_2O_2$, simple plasma methods are attractive [124].

*3.4. Fischer–Tropsch Synthesis*

Fischer–Tropsch synthesis (FTs) is a technique that converts $CO_2$, $CH_4$ and waste biomass into value-added chemicals and fuels, which uses syngas to convert to alkenes, long-chain alkanes, oxygenates, alkenes and water [125]. With the depletion of petroleum resources and the deteriorating climate, FT synthesis will play an increasingly important role in solving energy crisis and developing low-carbon energy.

FT synthesis takes place at the Co metal site, and the total number of Co metal sites on the supported catalyst depends on the dispersion and reductibility of Co. Previous research has found that glow discharge plasma can significantly improve the Co dispersion, and the catalyst assisted by plasma jets exhibit higher Co dispersion. Plasma jet is a promising tool for controlling Co dispersion and improving the catalytic performance of cobalt FT catalyst [126].

Li et al. [127] prepared a zirconium modified $Co/SiO_2$ FT catalyst by using DBD plasma. The results showed that, compared with those treated by the calcining method, the catalyst processed by DBD plasma demonstrated higher FT activity and heavy hydrocarbon yield. Akay et al. [128] investigated CO and $H_2$ to higher hydrocarbons on Cu/Co catalyst prepared by DBD for FT synthesis. It is found that Cu/Co catalyst can improve the FT synthesis under low temperature and ambient pressure. Xu et al. [129] used DBD plasma to catalyze the conversion of $CH_4$ to advanced hydrocarbons. The experiment has confirmed that DBD plasmas produce a higher $CH_4$ conversion rate in the presence of $CO_2$. Using $CO_2$ as a coreactant obviously enhances the $CH_4$ conversion rate and inhibits the formation of carbon deposits, Saleem et al. [130] used plasma-assisted decomposition of a biomass gasification tar analogue (toluene) to decompose them into lower hydrocarbons via DBD reactor. The toluene removal rate was reached 99% at a plasma power of 40 W and a residence time of 2.82 s. Figure 11a is a diagram of the experimental device. Non-thermal plasma is generated in a cylindrical DBD, as can be seen from Figure 11b. The aromatic C–C bonds is not enough to be destroyed at low power. While at high power, the average electron energy is high, and a larger proportion of lower hydrocarbons can be observed. Figure 11c shows that the yield of $CH_4$ and the selectivity of $C_2$-$C_5$ products as a function of residence time, which increase with residence time. This may be attributed to the collisions between tar analogue and reactive substances leading to the cleavage of aromatic rings. The influence of power and temperature on toluene conversion is displayed in Figure 11d. It can be seen that, when the power is raised in the range of 20–40 W, the toluene decomposition remains unchanged. However, it ranges from 82% to 91% at 10 W. It may be reduced owing to the formation of solid deposits. Figure 11e shows the reaction mechanism diagram at high temperature. The benzene ring fragmentation increases methane production due to high energy electrons and excitations at high power.

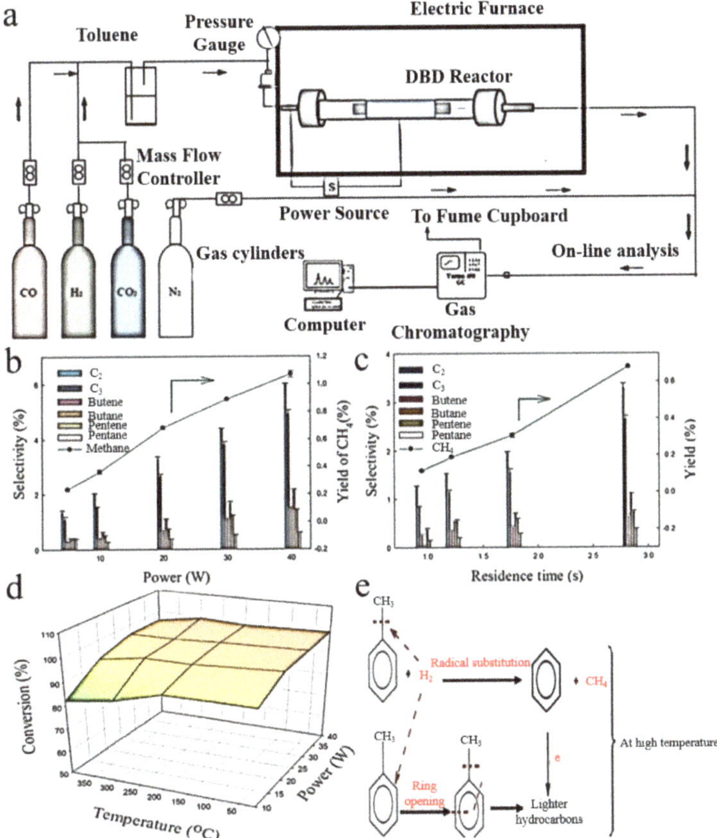

**Figure 11.** (**a**) experimental setup; (**b**) influence of plasma power on individual lower hydrocarbons; (**c**) effect of residence time on individual lower hydrocarbons; (**d**) influence of temperature on the toluene conversion; (**e**) reaction mechanism at elevated temperature. Reproduced with permission from [130]; Copyright Elsevier, 2019.

### 3.5. Volatile Organic Compounds Removal

Volatile organic compounds (VOCs) are common atmospheric pollutants, mainly those with volatile organic compounds having a boiling point of 250 °C or lower. Aromatic, alcohols, ketones and esters are typical VOCs emitted from various human activities, like printing, painting, coil coatings, wood processing, etc. In addition, the most important and common consequence of emitting VOCs into the atmosphere is that they can cause stratospheric $O_3$ consumption and tropospheric $O_3$ formation. VOCs are important precursors of atmospheric pollutants like $O_3$ and secondary organic particles. They can participate in photochemical reactions to produce secondary organic aerosols, which have an important impact on atmospheric visibility and global radiation balance, although some VOCs may in the short term will not cause serious harm to human health, but long-term exposure may cause mutagenic or carcinogenic effects [131].

Extensive research has been conducted to design low cost and effective VOC processing methods. Recently, several techniques for decomposing these organic contaminants through oxidation have been developed, especially the photocatalytic process using ultraviolet irradiation, in which HO free radicals are produced to oxidize these harmful VOCs. In recent years, an effective method for non-thermal plasma oxidation of VOCs has been reported. It has been reported that placing catalyst in the discharge area can improve the efficiency of VOCs removal and $CO_2$ formation [132].

Tatibouet et al. [132] observed synergetic effects by coupling photocatalysis with DBD plasma to remove low concentrations of isovaleraldehyde from air. The results showed that the plasma photocatalytic system obviously improved the isovaleraldehyde removal efficiency. Rtimi et al. [133] used DBD and photocatalysis in a continuous reactor to degrade butyraldehyde and dimethyl disulfide. A synergistic effect was observed with the butyraldehyde removal, but no synergistic effect was observed with dimethyl disulfide removal because of catalyst poisoning. Lu et al. [134] reported the removal efficiency of VOCs in non-thermal plasma double DBD, and also investigated plasma-catalyst synergistic effects on VOC removal. Bouzaza et al. [135] studied DBD plasma/photocatalysis combination, photocatalysis and DBD plasma to remove isovaleraldehyde from air, and the former produced a synergistic effect. Li et al. [136] investigated the synergistic effect of non-thermal DBD and catalyst for oxidation removal of toluene. The results demonstrated that the targets increasing energy efficiency and reducing $O_3$ in exhaust gas are achieved. Jo et al. [137] used a DBD reactor to decompose VOCs and found that humidity was a limiting factor in the non-thermal decomposition of VOCs. It was found that the optimum removal rate was 20% relative humidity. The graphical abstract is illustrated in Figure 12a. The impact of relative humidity on decomposition efficiency at different frequencies is displayed in Figure 12b. It can be seen from the figure that, when the frequency keeps constant, the decomposition efficiency first increases and then decreases with the relative humidity; when the relative humidity is constant, the higher the frequency, the greater the decomposition efficiency. Figure 12c presents the Lissajous figures of various discharge regions, which depends on the amount of transfer charge. Figure 12d is the impact of specific energy density on $CO_2$ selectivity and decomposition efficiency. Although higher SED results in higher decomposition efficiency, it does not exhibit any consistent linear relationship.

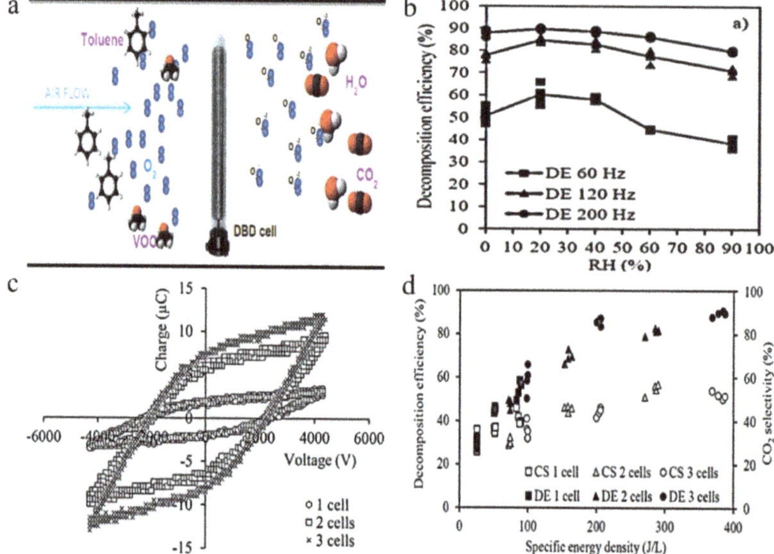


**Figure 12.** (**a**) graphical abstract; (**b**) the impact of relative humidity on decomposition efficiency at different frequencies; (**c**) Lissajous figures of different discharge regions; (**d**) effect of specific energy density on $CO_2$ selectivity and decomposition efficiency. Reproduced with permission from [137]; Copyright Elsevier, 2018.

### 3.6. Catalytic Sterilization

The indoor environment in which humans are often exposed to various types of pollution has already affected all aspects of our lives, especially organic pollution and bacterial pollution, which have severely restricted the improvement of quality of life. The existing filtration, adsorption, negative oxygen ions and ozone technology are difficult to continuously and effectively purify organic pollution, and disinfectant, ultraviolet lamp sterilization, ozone sterilization and other technologies have problems such as side effects on the human body and low sterilization efficiency. The high oxidation performance of photocatalytic technology has a strong bactericidal property, which can kill bacteria and viruses by destroying the cell wall of bacteria and coagulating proteins of viruses. Compared with photocatalyst, plasma discharge catalysis can produce more hydroxyl radicals, which can decompose harmful substances and bacteria viruses in the air more quickly and effectively.

Choi et al. [138] analyzed the sterilization effect by pulsed DBD. The experimental results showed that, when the DBD treatment time was 70 s, 99.99% *Escherichia coli* was sterilized, and the $O_3$ molecule is the main bactericidal species. Kostov et al. [139] performed bacterial sterilization by air DBD, and all bacterial cells were killed by DBD treatment for 20 min. Yi et al. [140] conducted DBD sterilization experiments on *Escherichia coli* and *Bacillus subtilis* in drinking water. The results showed that, with the increase of input voltage V and reaction time t, the sterilization rate increased significantly. The optimum sterilization effect was achieved at a pH of 7.1. Nagatsu et al. [141] studied a flexible sheet-type DBD that compared the low temperature sterilization of the wrapped material by adjusting the $N_2$ and $O_2$ ratios. Kikuchi et al. [142] studied the effects of environmental humidity and temperature on sterilization efficiency of atmospheric pressure DBD plasmas. The results showed that the inactivation of bacteria was greatly affected by humidity. Roy et al. [143] examined the role of $O_3$ in surface plasma sterilization by using DBD plasma. The results demonstrate that $O_3$ plays an important role in the plasma sterilization process, and the energy flux of electrode also plays a crucial role in plasma sterilization. Hong et al. [144] studied a novel multihole DBD (MH-DBD) plasma sterilization system for the sterilization of aqua pathogens. It was found that the MH-DBD plasma exhibited a higher sterilization performance over 6.5–7.5 log reduction. As is exhibited in the Figure 13a, the electric field is distributed symmetrically along the high voltage electrode in the coaxial DBD system. However, the MH-DBD system (Figure 13b) only shows asymmetric distribution of a strong electric field near the hole. Figure 13c displays the kill curve of Vibrio harveyi 12724. It can be observed that the presence of nitrogen in the DBD plasma system does not have a bactericidal effect on the microorganisms. In contrast, the nitrogen in the MH-DBD plasma system was continuously sterilized during the 5 min treatment period. As is demonstrated in Figure 13d, the coaxial DBD and MH-DBD plasma systems were continuously sterilized for 5 min. Figure 13e indicates the optical emission spectrum, which helps to identify the various excitations produced by the MH-DBD plasma. Figure 13f shows the kill curves of the pathogens processed by the MH-DBD plasma. Since the MH-DBD plasma can generate stable, sustainable oxidant, this remarkable sterilization effect was maintained for 4–5 days, after which the pathogen gradually proliferated due to its growth rate.

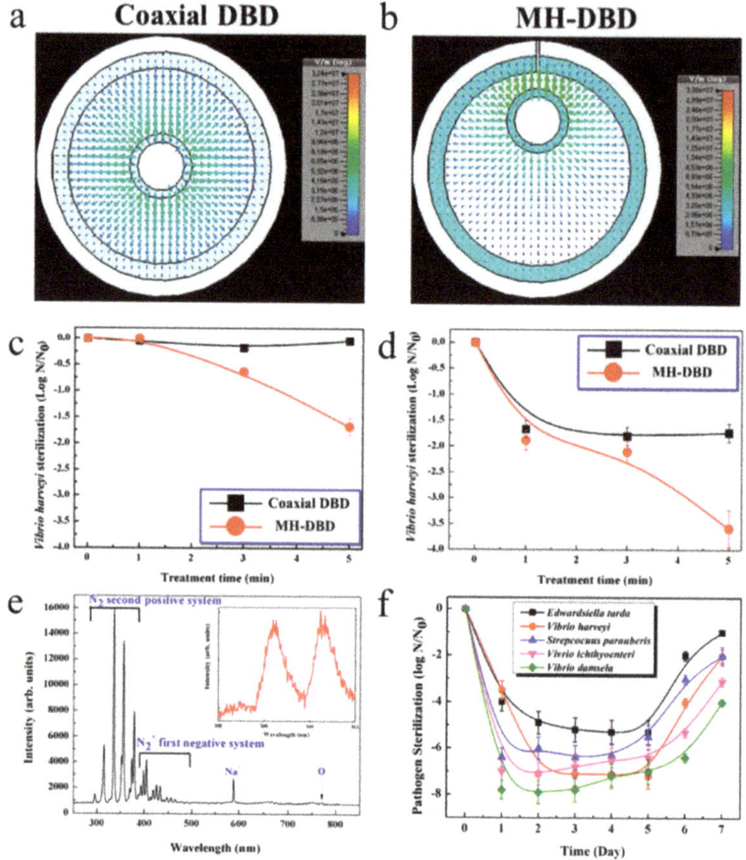



**Figure 13.** (**a**) electric filed distribution of DBD plasma; (**b**) electric filed distribution of MH-DBD plasma; (**c**) DBD and MH-DBD plasma sterilization curves comparison with nitrogen as the plasma forming gas; (**d**) DBD and MH-DBD plasma sterilization curves comparison with air as the plasma forming gas; (**e**) optical emission lines of air MH-DBD plasma; (**f**) pathogen sterilization curves. Reproduced with permission from [144]; Copyright Elsevier, 2019.

### 3.7. Wastewater Treatment

Thanks to the low removal rate of many harmful substances in wastewater, there are large amounts of man-made compounds in urban, agricultural and industrial wastewater, which can enter the natural surface and underground water bodies. Some substances owing to persistence get great attention and toxic effects after release into the receiving system [145]. The presence of hazardous chemicals in natural waters is a very serious issue. To reduce harmful effects of these harmful substances on humans and environment, it is necessary to find suitable and economical techniques to reduce their concentration.

Plasma treatment has been used to experimentally remove various substances from water, including cyanide, VOCs, phenols, organic dyes and drugs. However, there are few guidelines to remove more refractory compounds and their energy efficiency and degradation kinetics in wastewater. It is well known that plasma treatment can produce degradation by-products, so we must conduct a detailed study.

### 3.7.1. Pharmaceutical Wastewater

Fang et al. [146] reported the use of DBD to degrade aniline wastewater. The results show that the addition of a certain amount of $Na_2CO_3$ and $H_2O_2$ to the wastewater can promote the degradation of aniline.

### 3.7.2. Dyestuff Wastewater

Mok et al. [147] studied the application of $O_3$ and ultraviolet light generated by DBD reactor for the oxidative degradation of organic contaminants in an azo dye Acid Red 27 wastewater. It was found that the ultraviolet light alone can degrade quite a large number of organic pollutants. The ozonation also had a great influence on the degradation of organic pollutants, and when ozonation and photocatalytic effects were combined, the degradation effect was remarkably improved. Zhong et al. [148] used ns-pulse DBD plasma to degrade dye wastewater. Experiments indicate that the gas phase non-equilibrium plasma generated by pulse power can effectively decompose indigo carmine in atomized aqueous solution. Sun et al. [149] reported the study of ozonation and DBD plasma induced photocatalysis treatment of azo dye Acid Red 4 wastewater. The results indicate that the system has the ability of photocatalysis and ozonation to degrade organic contaminants, which is a promising wastewater processing technology. Attri et al. [150] investigated the impact of $\gamma$-ray and DBD plasma treatments on dye-polluted water treatment. The wastewater treatment efficiency at different time intervals was studied. Martuzevicius et al. [151] examined the degradation of various textile dyes by semi-continuous DBD plasma. It was found that plasma treatment can reduce the toxicity of wastewater to near zero.

### 3.7.3. Grease Wastewater

Kuraica et al. [152] studied the application of coaxial DBD for potable and oil derivative wastewater treatment, which obviously reduced the chemical oxygen demand and potassium permanganate demand in wastewater treatment.

### 3.7.4. Antibiotic Wastewater

Yuan et al. [153] added persulfate (PS) to DBD plasma device to generate and activate $SO_4^{2-}$ groups, and determined the effect of PS addition and applied voltage on tetracycline (TC) removal efficiency. The results illustrated that the addition of PS had a prominent synergistic effect, which promoted the removal rate and degradation efficiency of TC. Figure 14a shows a gas phase surface discharge device diagram. The discharge power is calculated according to the Lissajous figure (Figure 14b). Figure 14c indicates the effect of PS addition ratio on TC removal. It was concluded that, when the molar ratio increased from 5:1 to 120:1, the TC removal rate was dramatically improved due to the addition of PS. When the molar ratio was higher than 20:1, the efficiency was no longer increased and actually decreased slightly. Figure 14d,e show the influence of ethanol on the removal of TC and $H_2O_2$ production. Clearly, adding the ethanol inhibited the removal of TC in the absence or presence of PS. After 15 min of treatment, the removal efficiencies of the DBD and DBD + PS systems alone reduced from 87.5% and 83.2% to 82.6% and 70.1%, respectively. The $H_2O_2$ concentration could be used as a characterization of the oxidation capability. As can be seen from the figure, compared with no PS, the $H_2O_2$ concentration was higher when PS was added, and the $H_2O_2$ concentration in deionized water was higher than TC solution. The total ion chromatogram and HPLC/MS chromatograph of the TC degradation intermediates were depicted in Figure 14f. The TC was clearly removed after 15 min processing compared to the untreated sample TC peak intensity.

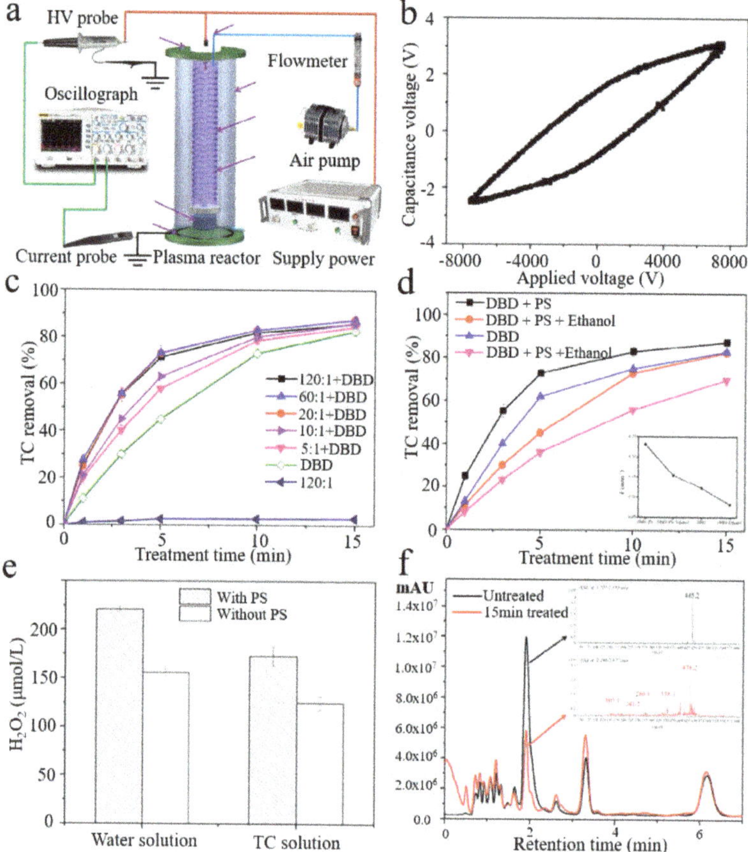



**Figure 14.** (**a**) gas phase surface discharge device diagram; (**b**) Lissajous figure; (**c**) effect of PS addition ratio on the TC removal; (**d**) effect of ethanol on TC removal; (**e**) effect of ethanol on $H_2O_2$ production; (**f**) total ion chromatograph and HPLC/MS chromatograph of TC decomposition intermediates. Reproduced with permission from [153]. Copyright Elsevier, 2018.

### 3.8. Degradation of Pesticide Residues

Pesticides are a large variety of different types of compounds that are used to delay crop spoilage and prevent pests. However, as pesticide resistance increases, their residues on food and their long-lasting effects have attracted widespread attention. Agricultural chemicals like pesticide residues on a fresh product can pose serious health risks if not washed. The possibility of pesticides' degradation in food and water by plasma discharge has recently been reported. Low-temperature plasma has become a potential biological purification technology, which is used to treat microorganisms in food like fruits and vegetables and reduce chemical risks.

### 3.8.1. Insecticide

Cullen et al. [154] selected three experimental pesticides, dichlorvos, malathion, endosulfan, and used atmospheric air high voltage DBD plasma reactor to degrade pesticides in water. After plasma treatment for 8 min at 80 kV, the degradation efficiency of pesticides were found to be 78.98% for dichlorvos, 69.62% for malathion and 57.71% for endosulfan, respectively. Shi et al. [155] processed spinach and apple samples contaminated with omethoate using low temperature plasma (LTP) generated by dielectric barrier corona discharge. It is found that omethoate residue in vegetables and fruits could be effectively degraded with proper dosage of LTP without affecting the quality of fruits and vegetables. Bai et al. [156] reported the degradation of dimethoate induced by DBD plasma in aqueous solution. The impacts of degradation pathway and parameters on dimethoate solution were studied. Li et al. [157] studied the degradation of acetamiprid in wastewater in DBD reactor. It can be concluded that acetamiprid could be effectively removed from aqueous solution and hydroxyl radicals played an important role in the degradation process.

### 3.8.2. Herbicide

Zheng et al. [158] reported the research on diuron degradation with DBD plasma and proposed the degradation mechanism. Roglic et al. [159] investigated the influence of different catalysts on mesotrione degradation in water falling film DBD reactor. Wardenier et al. [160] studied an innovative advanced oxidation processes based on a continuous-flow pulsed DBD reactor discharge.

A synthetic micropollutantsmixture containing five pesticides, two pharmaceuticals, and one plasticizer were used. The results showed that the total removal efficiency of all studied micropollutants was greater than 93.8%, and the energy efficiency varied between 2.42 and 4.25 kWh/m$^3$. Valsero et al. [161] used DBD atmospheric plasma to remove pollutants from wastewater containing atrazine. However, when the solution to be processed contains high concentrations of organic substances and mineral salts, the efficiency is decreased, which may inhibit or compete with oxidant substances generated by plasma.

### 3.8.3. Germicide

Cullen et al. [162] reported the degradation of pesticide residues on strawberries by DBD nonthermal plasma technology. After 80 kV treatment for 5 min, the contents of azoxystrobin and fludioxonil decreased by 69% and 71%, respectively. Figure 15a presents the diagram of the DBD experimental device for NTP treatments inside the package. Figure 15b shows the emission spectrums with or without strawberries. The emissions were stronger without any samples in the package. Figure 15c representative chromatogram showing retention times of the four pesticides. As can be seen from the figure, the retention time of the four pesticides, namely azoxystrobin, cyprodinil, fludioxonil and pyriproxyfen were 36.6, 20.6, 24.1 and 29.8 min, respectively. Figure 15d,e indicate the residual pesticide concentrations in strawberries before and after plasma treatment. The pesticide residue concentration of all samples was significantly reduced compared with the control.

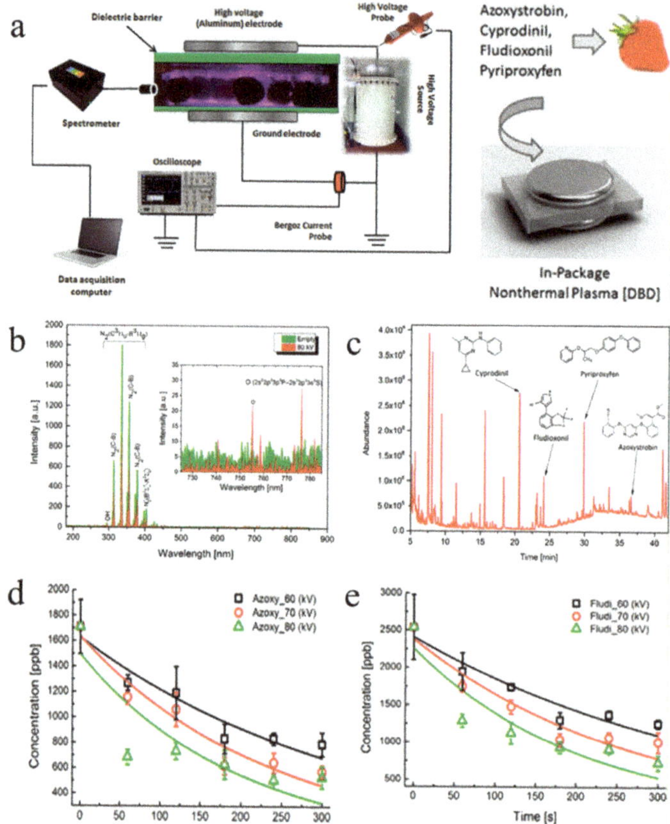

**Figure 15.** (**a**) the diagram of the DBD experimental device; (**b**) emission spectrum with or without strawberries; (**c**) retention times chromatogram of the four pesticides; (**d**) azoxystrobin concentrations before and after plasma treatment; (**e**) fludioxonil concentrations after plasma treatment. Reproduced with permission from [162]; Copyright Elsevier, 2014.

## 4. Conclusions

As an emerging frontier hotspot technology, plasma has been applied in many research fields. In this review, we first introduce the characteristics of plasma, such as conductivity, heat conduction, heat radiation, active chemical properties, easy to produce chemical reactions and its recent advances in catalysis. The advantages of preparing various catalysts by DBD plasma technology are summarized, including highly dispersed catalyst, defect-rich catalyst and heteroatom-doped catalyst. Compared with the catalyst prepared by the conventional method, the heterogeneous catalyst prepared by plasma exhibits good dispersion and smaller particle size, and its catalytic activity, selectivity and stability are significantly improved. Meanwhile, the chemical reaction activity is remarkably enhanced, attracting more and more attention from all walks of life.

Furthermore, we explored plasma catalysis, which can dramatically improve the catalytic reaction effect. For example, plasma technology can be used to generate clean fuel $CH_4$ via $CO_2$ as a raw material; under the action of an external electric field, complex macromolecular pollutants can be converted into small molecules safe and free of pollutants; plasma flue gas denitration is realized by combining electron beam technology with pulse corona discharge technology; meanwhile, plasma technology can also purify automobile exhaust and dust. Finally, the properties of plasma in the catalytic reaction are summarized, and the application prospect of plasma in the future catalytic field is prospected. At present, scientists have applied the plasma technology in the latest research fields like the "artificial sun", aerospace, stomatology, photonic crystal and spectroscopic optics, which has achieved certain results. With the development of plasma technology, low-temperature plasma will definitely be applied in a wider range and more industries, and the study on low-temperature plasma will also be more in-depth.

**Author Contributions:** Conceptualization, C.M., F.Y., B.D. and D.Y.; Methodology, C.M., F.Y. and D.Y.; Formal analysis, C.M., F.Y., D.Y. and J.L.; Investigation, C.M., F.Y., D.Y., S.Z. and J.L.; Data curation, C.M., F.Y. and D.Y.; Writing—original draft preparation, C.M., F.Y. and D.Y.; Writing—review and editing, J.L., C.M., F.Y. and D.Y.; Supervision, C.M., F.Y. and D.Y.

**Funding:** This research was funded by the National Natural Science Foundation of China (No.21663022) and Science and Technology Innovation Talents Program of Bingtuan (No.2019CB025).

**Acknowledgments:** The authors express our thanks to Cunhua Ma, Feng Yu and Dezheng Yang for providing the impetus for this article.

**Conflicts of Interest:** The authors declare no conflict of interest.

# References

1. Snoeckx, R.; Bogaerts, A. Plasma technology—a novel solution for $CO_2$ conversion? *Chem. Soc. Rev.* **2017**, *46*, 5805–5863. [CrossRef]
2. Liao, X.; Liu, D.; Xiang, Q.; Ahn, J.; Chen, S.; Ye, X.; Ding, T. Inactivation mechanisms of non-thermal plasma on microbes: a review. *Food Control* **2017**, *75*, 83–91. [CrossRef]
3. Kizling, M.B.; Järås, S.G. A review of the use of plasma techniques in catalyst preparation and catalytic reactions. *Appl. Catal. A Gen.* **1996**, *147*, 1–21. [CrossRef]
4. Liu, C.J.; Vissokov, G.P.; Jang, B.W.L. Catalyst preparation using plasma technologies. *Catal. Today* **2002**, *72*, 173–184. [CrossRef]
5. Siemens, W. Ueber die elektrostatische induction und die verzögerung des stroms in flaschendrähten. *Annalen der Physik* **1857**, *178*, 66–122. [CrossRef]
6. Andrews, T.; Tait, P.G. VII. On the volumetric relations of ozone, and the action of the electrical discharge on oxygen and other gases. *Philos. Trans. R. Soc. Lond.* **1860**, *150*, 113–131.
7. Kogelschatz, U. Dielectric-barrier discharges: their history, discharge physics, and industrial applications. *Plasma Chem. Plasma Process.* **2003**, *23*, 1–46. [CrossRef]
8. Wagner, H.E.; Brandenburg, R.; Kozlov, K.; Sonnenfeld, A.; Michel, P.; Behnke, J. The barrier discharge: basicproperties and applications to surface treatment. *Vacuum* **2003**, *71*, 417–436. [CrossRef]
9. Zhang, K.; Zhang, G.; Liu, X.; Phan, A.N.; Luo, K. A study on $CO_2$ decomposition to CO and $O_2$ by the combination of catalysis and dielectric-barrier discharges at low temperatures and ambient pressure. *Ind. Eng. Chem. Res.* **2017**, *56*, 3204–3216. [CrossRef]
10. Liu, C.J.; Ye, J.; Jiang, J.; Pan, Y. Progresses in the preparation of coke resistant Ni-based catalyst for steam and $CO_2$ reforming of methane. *Chemcatchem* **2015**, *3*, 529–541. [CrossRef]
11. Neyts, E.C.; Ostrikov, K.; Sunkara, M.K.; Bogaerts, A. Plasma catalysis: synergistic effects at the nanoscale. *Chem. Rev.* **2016**, *116*, 767. [CrossRef]
12. Kenward, M. Why plasma science means business. *Phys. World* **1995**, *8*, 31–34. [CrossRef]
13. Liu, C.; Li, M.; Wang, J.; Zhou, X.; Guo, Q.; Yan, J.; Li, Y. Plasma methods for preparing green catalysts: current status and perspective. *Chin. J. Catal.* **2016**, *37*, 340–348. [CrossRef]
14. Pinna, F. Supported metal catalysts preparation. *Catal. Today* **1998**, *41*, 129–137. [CrossRef]
15. Mojet, B.; Miller, J.; Ramaker, D.; Koningsberger, D. A new model describing the metal–support interaction in noble metal catalysts. *J. Catal.* **1999**, *186*, 373–386. [CrossRef]
16. Kim, S.S.; Lee, H.; Na, B.K.; Song, H.K. Plasma-assisted reduction of supported metal catalyst using atmospheric dielectric-barrier discharge. *Catal. Today* **2004**, *89*, 193–200. [CrossRef]
17. Di, L.; Xu, Z.; Wang, K.; Zhang, X. A facile method for preparing $Pt/TiO_2$ photocatalyst with enhanced activity using dielectric barrier discharge. *Catal. Today* **2013**, *211*, 109–113. [CrossRef]

18. Di, L.; Zhang, X.; Xu, Z.; Wang, K. Atmospheric-pressure cold plasma for preparation of high performance Pt/TiO$_2$ photocatalyst and its mechanism. *Plasma Chem. Plasma Process.* **2014**, *34*, 301–311. [CrossRef]

19. Di, L.; Xu, Z.; Zhang, X. Atmospheric-pressure cold plasma for synthesizing Ag modified Degussa P25 with visible light activity using dielectric barrier discharge. *Catal. Today* **2013**, *211*, 143–146. [CrossRef]

20. Xu, Z.; Qi, B.; Di, L.; Zhang, X. Partially crystallized Pd nanoparticles decorated TiO$_2$ prepared by atmospheric-pressure cold plasma and its enhanced photocatalytic performance. *J. Energy Chem.* **2014**, *23*, 679–683. [CrossRef]

21. Qi, B.; Di, L.; Xu, W.; Zhang, X. Dry plasma reduction to prepare a high performance Pd/C catalyst at atmospheric pressure for CO oxidation. *J. Mater. Chem. A* **2014**, *2*, 11885–11890. [CrossRef]

22. Xu, W.; Zhan, Z.; Di, L.; Zhang, X. Enhanced activity for CO oxidation over Pd/Al$_2$O$_3$ catalysts prepared by atmospheric-pressure cold plasma. *Catal. Today* **2015**, *256*, 148–152. [CrossRef]

23. Di, L.; Xu, W.; Zhan, Z.; Zhang, X. Synthesis of alumina supported Pd–Cu alloy nanoparticles for CO oxidation via a fast and facile method. *RSC Adv.* **2015**, *5*, 71854–71858. [CrossRef]

24. Di, L.; Duan, D.; Park, D.W.; Ahn, W.S.; Lee, B.J.; Zhang, X. Cold plasma for synthesizing high performance bimetallic PdCu catalysts: effect of reduction sequence and Pd/Cu atomic ratios. *Top. Catal.* **2017**, *60*, 925–933. [CrossRef]

25. Xu, W.; Wang, X.; Zhou, Q.; Meng, B.; Zhao, J.; Qiu, J.; Gogotsi, Y. Low-temperature plasma-assisted preparation of graphene supported palladium nanoparticles with high hydrodesulfurization activity. *J. Mater. Chem.* **2012**, *22*, 14363–14368. [CrossRef]

26. Zhang, S.; Li, X.S.; Zhu, B.; Liu, J.L.; Zhu, X.; Zhu, A.M.; Jang, W.L. Atmospheric-pressure O$_2$ plasma treatment of Au/TiO$_2$ catalysts for CO oxidation. *Catal. Today* **2015**, *256*, 142–147. [CrossRef]

27. Lanbo, D.; Zhibin, Z.; Xiuling, Z.; Bin, Q.; Weijie, X. Atmospheric-pressure DBD cold plasma for preparation of high active Au/P25 catalysts for low-temperature CO oxidation. *Plasma Sci. Technol.* **2016**, *18*, 544–548.

28. Hua, W.; Jin, L.; He, X.; Liu, J.; Hu, H. Preparation of Ni/MgO catalyst for CO$_2$ reforming of methane by dielectric-barrier discharge plasma. *Catal. Commun.* **2010**, *11*, 968–972. [CrossRef]

29. Bian, L.; Zhang, L.; Zhu, Z.; Li, Z. Methanation of carbon oxides on Ni/Ce/SBA-15 pretreated with dielectric barrier discharge plasma. *Mol. Catal.* **2018**, *446*, 131–139. [CrossRef]

30. Zhao, B.; Chen, Z.; Yan, X.; Ma, X.; Hao, Q. CO methanation over Ni/SiO$_2$ catalyst prepared by ammonia impregnation and plasma decomposition. *Top. Catal.* **2017**, *60*, 879–889. [CrossRef]

31. Jia, X.; Rui, N.; Zhang, X.; Hu, X.; Liu, C.J. Ni/ZrO$_2$ by dielectric barrier discharge plasma decomposition with improved activity and enhanced coke resistance for CO methanation. *Catal. Today* **2018**, *334*, 215–222. [CrossRef]

32. Li, Y.; Wei, Z.; Wang, Y. Ni/MgO catalyst prepared via dielectric-barrier discharge plasma with improved catalytic performance for carbon dioxide reforming of methane. *Front. Chem. Sci. Eng.* **2014**, *8*, 133–140. [CrossRef]

33. Fan, Z.; Sun, K.; Rui, N.; Zhao, B.; Liu, C.J. Improved activity of Ni/MgAl$_2$O$_4$ for CO$_2$ methanation by the plasma decomposition. *J. Energy Chem.* **2015**, *24*, 655–659. [CrossRef]

34. Zhang, M.; Li, P.; Zhu, M.; Tian, Z.; Dan, J.; Li, J.; Dai, B.; Yu, F. Ultralow-weight loading Ni catalyst supported on two-dimensional vermiculite for carbon monoxide methanation. *Chin. J. Chem. Eng.* **2018**, *26*, 1873–1878. [CrossRef]

35. Guo, X.; Sun, Y.; Yu, Y.; Zhu, X.; Liu, C.J. Carbon formation and steam reforming of methane on silica supported nickel catalysts. *Catal. Commun.* **2012**, *19*, 61–65. [CrossRef]

36. Guo, Y.F.; Ye, D.Q.; Chen, K.F.; He, J.C.; Chen, W.L. Toluene decomposition using a wire-plate dielectric barrier discharge reactor with manganese oxide catalyst in situ. *J. Mol. Catal. A-Chem.* **2006**, *245*, 93–100. [CrossRef]

37. Fu, T.; Huang, C.; Lv, J.; Li, Z. Fuel production through Fischer–Tropsch synthesis on carbon nanotubes supported Co catalyst prepared by plasma. *Fuel* **2014**, *121*, 225–231. [CrossRef]

38. Dai, H.; Zhu, M.; Zhao, D.; Yu, F.; Dai, B. Effective catalytic performance of plasma-enhanced W$_2$N/AC as catalysts for acetylene hydrochlorination. *Top. Catal.* **2017**, *60*, 1016–1023. [CrossRef]

39. El-Roz, M.; Lakiss, L.; El, F.J.; Lebedev, O.I.; Thibault-Starzyk, F.; Valtchev, V. Incorporation of clusters of titanium oxide in Beta zeolite structure by a new cold TiCl$_4$-plasma process: physicochemical properties and photocatalytic activity. *Phys. Chem. Chem. Phys.* **2013**, *15*, 16198–16207. [CrossRef]

40. Di, L.; Zhang, J.; Zhang, X. A review on the recent progress, challenges, and perspectives of atmospheric-pressure cold plasma for preparation of supported metal catalysts. *Plasma Process. Polym.* **2018**, *15*, 1700234. [CrossRef]

41. Yan, D.; Li, Y.; Huo, J.; Chen, R.; Dai, L.; Wang, S. Defect chemistry of nonprecious-metal electrocatalysts for oxygen reactions. *Adv. Mater.* **2017**, *29*, 1606459. [CrossRef] [PubMed]

42. Cai, L.; He, J.; Liu, Q.; Yao, T.; Chen, L.; Yan, W.; Hu, F.; Jiang, Y.; Zhao, Y.; Hu, T. Vacancy-induced ferromagnetism of $MoS_2$ nanosheets. *J. Am. Chem. Soc.* **2015**, *137*, 2622–2627. [CrossRef] [PubMed]

43. Ling, T.; Yan, D.Y.; Jiao, Y.; Wang, H.; Zheng, Y.; Zheng, X.; Mao, J.; Du, X.W.; Hu, Z.; Jaroniec, M. Engineering surface atomic structure of single-crystal cobalt (II) oxide nanorods for superior electrocatalysis. *Nat. Commun.* **2016**, *7*, 12876. [CrossRef] [PubMed]

44. Chen, J.; Han, Y.; Kong, X.; Deng, X.; Park, H.J.; Guo, Y.; Jin, S.; Qi, Z.; Lee, Z.; Qiao, Z. The origin of improved electrical double-layer capacitance by inclusion of topological defects and dopants in graphene for supercapacitors. *Angew. Chem.* **2016**, *55*, 13822–13827. [CrossRef] [PubMed]

45. Liu, X.; Meng, C.G.; Han, Y. Defective graphene supported $MPd_{12}$ (M = Fe, Co, Ni, Cu, Zn, Pd) nanoparticles as potential oxygen reduction electrocatalysts: a first-principles study. *J. Phys. Chem. C* **2013**, *117*, 1350–1357. [CrossRef]

46. Lai, L.; Potts, J.R.; Zhan, D.; Wang, L.; Poh, C.K.; Tang, C.; Gong, H.; Shen, Z.; Lin, J.; Ruoff, R.S. Exploration of the active center structure of nitrogen-doped graphene-based catalysts for oxygen reduction reaction. *Energy Environ. Sci.* **2012**, *5*, 7936–7942. [CrossRef]

47. Yang, L.; Jiang, S.; Zhao, Y.; Zhu, L.; Chen, S.; Wang, X.; Wu, Q.; Ma, J.; Ma, Y.; Hu, Z. Boron-doped carbon nanotubes as metal-free electrocatalysts for the oxygen reduction reaction. *Angew. Chem. Int. Ed.* **2011**, *50*, 7132–7135. [CrossRef]

48. Chen, S.; Bi, J.; Zhao, Y.; Yang, L.; Zhang, C.; Ma, Y.; Wu, Q.; Wang, X.; Hu, Z. Nitrogen-doped carbon nanocages as efficient metal-free electrocatalysts for oxygen reduction reaction. *Adv. Mater.* **2012**, *24*, 5593–5597. [CrossRef]

49. Silva, R.; Al-Sharab, J.; Asefa, T. Edge-plane-rich nitrogen-doped carbon nanoneedles and efficient metal-free electrocatalysts. *Angew. Chem. Int. Ed.* **2012**, *51*, 7171–7175. [CrossRef]

50. Wang, W.H.; Huang, B.C.; Wang, L.S.; Ye, D.Q. Oxidative treatment of multi-wall carbon nanotubes with oxygen dielectric barrier discharge plasma. *Surf. Coat. Technol.* **2011**, *205*, 4896–4901. [CrossRef]

51. Solís-Fernández, P.; Paredes, J.; López, M.J.; Cabria, I.; Alonso, J.A.; Martínez-Alonso, A.; Tascón, J. A combined experimental and theoretical investigation of atomic-scale defects produced on graphite surfaces by dielectric barrier discharge plasma treatment. *J. Phys. Chem. C* **2009**, *113*, 18719–18729. [CrossRef]

52. Xu, L.; Jiang, Q.; Xiao, Z.; Li, X.; Huo, J.; Wang, S.; Dai, L. Plasma-engraved $Co_3O_4$ nanosheets with oxygen vacancies and high surface area for the oxygen evolution reaction. *Angew. Chem. Int. Ed. Engl.* **2016**, *55*, 5277–5281. [CrossRef] [PubMed]

53. Mistry, H.; Choi, Y.W.; Bagger, A.; Scholten, F.; Bonifacio, C.; Sinev, I.; Divins, N.J.; Zegkinoglou, I.; Jeon, H.S.; Kisslinger, K. Enhanced carbon dioxide electroreduction to carbon monoxide over defect rich plasma-activated silver catalysts. *Angew. Chem. Int. Ed.* **2017**, *56*, 11394–11398. [CrossRef] [PubMed]

54. Li, Y.; Wang, W.; Wang, F.; Di, L.; Yang, S.; Zhu, S.; Yao, Y.; Ma, C.; Dai, B.; Yu, F. Enhanced photocatalytic degradation of organic dyes via defect-rich $TiO_2$ prepared by dielectric barrier discharge plasma. *Nanomaterials* **2019**, *9*, 720. [CrossRef] [PubMed]

55. Ebihara, K.; Tanaka, T.; Ikegami, T.; Yamagata, Y.; Matsunaga, T.; Yamashita, K.; Oyama, Y. Application of the dielectric barrier discharge to detect defects in a teflon coated metal surface. *J. Phys. D Appl. Phys.* **2003**, *36*, 2883–2886. [CrossRef]

56. Liu, X.; Dai, L. Carbon-based metal-free catalysts. *Nat. Rev. Mater.* **2016**, *1*, 16064. [CrossRef]

57. Leem, J.H.; Lee, D.H.; Sang, Y.L. Properties of N-doped ZnO grown by DBD-PLD. *Thin Solid Film.* **2009**, *518*, 1238–1240. [CrossRef]

58. Wang, Y.; Yu, F.; Zhu, M.; Ma, C.; Zhao, D.; Wang, C.; Zhou, A.; Dai, B.; Ji, J.; Guo, X. N-doping of plasma exfoliated graphene oxide via dielectric barrier discharge plasma treatment for the oxygen reduction reaction. *J. Mater. Chem. A* **2018**, *6*, 2011–2017. [CrossRef]

59. Li, X.; Zhang, J.; Zhou, F.; Zhang, H.; Bai, J.; Wang, Y.; Wang, H. Preparation of N-vacancy-doped $gC_3N_4$ with outstanding photocatalytic $H_2O_2$ production ability by dielectric barrier discharge plasma treatment. *Chin. J. Catal.* **2018**, *39*, 1090–1098. [CrossRef]

60. Luo, Z.; Jiang, H.; Li, D.; Hu, L.; Geng, W.; Wei, P.; Ouyang, P. Improved photocatalytic activity and mechanism of $Cu_2O$/N–$TiO_2$ prepared by a two-step method. *RSC Adv.* **2014**, *4*, 17797–17804. [CrossRef]

61. Lin, M.; Hu, S.; Ping, L.; Wang, Q.; Ma, H.; Wei, L. In situ synthesis of sulfur doped carbon nitride with enhanced photocatalytic performance using DBD plasma treatment under $H_2S$ atmosphere. *J. Phys. Chem. Solids* **2018**, *118*, 166–171.

62. Xiao, Z.H.; Wang, Y.; Huang, Y.C.; Wei, Z.X.; Dong, C.L.; Ma, J.M.; Shen, S.H.; Li, Y.F.; Wang, S.Y. Filling the oxygen vacancies in $Co_3O_4$ with phosphorus: an ultra-efficient electrocatalyst for overall water splitting. *Energy Environ. Sci.* **2017**, *10*, 2563–2569. [CrossRef]

63. Neyts, E.C. Plasma-surface interactions in plasma catalysis. *Plasma Chem. Plasma Process.* **2016**, *36*, 185–212. [CrossRef]

64. Chen, H.L.; Lee, H.M.; Chen, S.H.; Chao, Y.; Chang, M.B. Review of plasma catalysis on hydrocarbon reforming for hydrogen production—interaction, integration, and prospects. *Appl. Catal. B Environ.* **2008**, *85*, 1–9. [CrossRef]

65. Van Durme, J.; Dewulf, J.; Leys, C.; Van Langenhove, H. Combining non-thermal plasma with heterogeneous catalysis in waste gas treatment: a review. *Appl. Catal. B Environ.* **2008**, *78*, 324–333. [CrossRef]

66. Neyts, E.; Bogaerts, A. Understanding plasma catalysis through modelling and simulation—a review. *J. Phys. D Appl. Phys.* **2014**, *47*, 224010. [CrossRef]

67. Vandenbroucke, A.M.; Morent, R.; De Geyter, N.; Leys, C. Non-thermal plasmas for non-catalytic and catalytic VOC abatement. *J. Hazard. Mater.* **2011**, *195*, 30–54. [CrossRef] [PubMed]

68. Li, P.; Yu, F.; Altaf, N.; Zhu, M.; Li, J.; Dai, B.; Wang, Q. Two-dimensional layered double hydroxides for reactions of methanation and methane reforming in C1 chemistry. *Materials* **2018**, *11*, 221. [CrossRef] [PubMed]

69. Mok, Y.S.; Kang, H.C.; Lee, H.J.; Koh, D.J.; Shin, D.N. Effect of nonthermal plasma on the methanation of carbon monoxide over nickel catalyst. *Plasma Chem. Plasma Process.* **2010**, *30*, 437–447. [CrossRef]

70. Guo, X.; Traitangwong, A.; Hu, M.; Zuo, C.; Meeyoo, V.; Peng, Z.; Li, C. Carbon dioxide methanation over nickel-based catalysts supported on various mesoporous material. *Energy Fuels* **2018**, *32*, 3681–3689. [CrossRef]

71. Li, P.; Wen, B.; Yu, F.; Zhu, M.; Guo, X.; Han, Y.; Kang, L.; Huang, X.; Dan, J.; Ouyang, F. High efficient nickel/vermiculite catalyst prepared via microwave irradiation-assisted synthesis for carbon monoxide methanation. *Fuel* **2016**, *171*, 263–269. [CrossRef]

72. Song, Q.; Altaf, N.; Zhu, M.; Li, J.; Ren, X.; Dan, J.; Dai, B.; Louis, B.; Wang, Q.; Yu, F. Enhanced low-temperature catalytic carbon monoxide methanation performance via vermiculite-derived silicon carbide-supported nickel nanoparticles. *Sustain. Energy Fuels* **2019**, *3*, 965–974. [CrossRef]

73. Yao, Y.; Yu, F.; Li, J.; Li, J.; Li, Y.; Wang, Z.; Zhu, M.; Shi, Y.; Dai, B.; Guo, X. Two-dimensional NiAl layered double oxides as non-noble metal catalysts for enhanced CO methanation performance at low temperature. *Fuel* **2019**, *255*, 115770. [CrossRef]

74. Zhang, M.; Li, P.; Tian, Z.; Zhu, M.; Wang, F.; Li, J.; Dai, B.; Yu, F.; Qiu, H.; Gao, H. Clarification of active sites at interfaces between silica support and nickel active components for carbon monoxide methanation. *Catalysts* **2018**, *8*, 293. [CrossRef]

75. Mills, G.A.; Steffgen, F.W. Catalytic methanation. *Catal. Rev.* **1974**, *8*, 159–210. [CrossRef]

76. Panagiotopoulou, P.; Kondarides, D.I.; Verykios, X.E. Selective methanation of CO over supported noble metal catalysts: effects of the nature of the metallic phase on catalytic performance. *Appl. Catal. A Gen.* **2008**, *344*, 45–54. [CrossRef]

77. Derekaya, F.B.; Yaşar, G. The CO methanation over NaY-zeolite supported Ni/$Co_3O_4$, Ni/$ZrO_2$, $Co_3O_4$/$ZrO_2$ and Ni/$Co_3O_4$/$ZrO_2$ catalysts. *Catal. Commun.* **2011**, *13*, 73–77. [CrossRef]

78. Kok, E.; Scott, J.; Cant, N.; Trimm, D. The impact of ruthenium, lanthanum and activation conditions on the methanation activity of alumina-supported cobalt catalysts. *Catal. Today* **2011**, *164*, 297–301. [CrossRef]

79. Habazaki, H.; Yamasaki, M.; Zhang, B.P.; Kawashima, A.; Kohno, S.; Takai, T.; Hashimoto, K. Co-methanation of carbon monoxide and carbon dioxide on supported nickel and cobalt catalysts prepared from amorphous alloys. *Appl. Catal. A Gen.* **1998**, *172*, 131–140. [CrossRef]

80. Nizio, M.; Albarazi, A.; Cavadias, S.; Amouroux, J.; Galvez, M.E.; Da Costa, P. Hybrid plasma-catalytic methanation of $CO_2$ at low temperature over ceria zirconia supported Ni catalysts. *Int. J. Hydrog. Energy* **2016**, *41*, 11584–11592. [CrossRef]

81. Jwa, E.; Mok, Y.S.; Lee, S.B. Nonthermal plasma-assisted catalytic methanation of CO and $CO_2$ over nickel-loaded alumina. In Proceedings of the Energy and Sustainability III 2011, Alicante, Spain, 11–13 April 2011; Volume 143, pp. 361–368.

82. Nizio, M.; Benrabbah, R.; Krzak, M.; Debek, R.; Motak, M.; Cavadias, S.; Gálvez, M.E.; Da Costa, P. Low temperature hybrid plasma-catalytic methanation over Ni-Ce-Zr hydrotalcite-derived catalysts. *Catal. Commun.* **2016**, *83*, 14–17. [CrossRef]

83. Zeng, Y.; Tu, X. Plasma-catalytic $CO_2$ hydrogenation at low temperatures. *IEEE Trans. Plasma Sci.* **2015**, *44*, 405–411. [CrossRef]

84. Lee, C.J.; Lee, D.H.; Kim, T. Enhancement of methanation of carbon dioxide using dielectric barrier discharge on a ruthenium catalyst at atmospheric conditions. *Catal. Today* **2017**, *293*, 97–104. [CrossRef]

85. Cheng, T.; Fang, Z.; Hu, Q.; Han, K.; Yang, X.; Zhang, Y. Low-temperature CO oxidation over $CuO/Fe_2O_3$ catalysts. *Catal. Commun.* **2007**, *8*, 1167–1171. [CrossRef]

86. Chen, M.; Goodman, D. The structure of catalytically active gold on titania. *Science* **2004**, *306*, 252–255. [CrossRef] [PubMed]

87. Jia, C.J.; Liu, Y.; Bongard, H.; Schüth, F. Very low temperature CO oxidation over colloidally deposited gold nanoparticles on $Mg(OH)_2$ and MgO. *J. Am. Chem. Soc.* **2010**, *132*, 1520–1522. [CrossRef]

88. Huang, Y.; Wang, A.; Wang, X.; Zhang, T. Preferential oxidation of CO under excess $H_2$ conditions over iridium catalysts. *Int. J. Hydrog. Energy* **2007**, *32*, 3880–3886. [CrossRef]

89. Haruta, M.; Yamada, N.; Kobayashi, T.; Iijima, S. Gold catalysts prepared by coprecipitation for low-temperature oxidation of hydrogen and of carbon monoxide. *J. Catal.* **1989**, *115*, 301–309. [CrossRef]

90. Mahammadunnisa, S.; Reddy, P.M.K.; Reddy, E.L.; Subrahmanyam, C. Catalytic DBD plasma reactor for CO oxidation by in situ $N_2O$ decomposition. *Catal. Today* **2013**, *211*, 53–57. [CrossRef]

91. Pietruszka, B.; Anklam, K.; Heintze, M. Plasma-assisted partial oxidation of methane to synthesis gas in a dielectric barrier discharge. *Appl. Catal. A Gen.* **2004**, *261*, 19–24. [CrossRef]

92. Zhou, A.; Chen, D.; Dai, B.; Ma, C.; Li, P.; Yu, F. Direct decomposition of $CO_2$ using self-cooling dielectric barrier discharge plasma. *Greenh. Gases Sci. Technol.* **2017**, *7*, 721–730. [CrossRef]

93. Li, R.; Tang, Q.; Yin, S.; Sato, T. Plasma catalysis for $CO_2$ decomposition by using different dielectric materials. *Fuel Process. Technol.* **2006**, *87*, 617–622. [CrossRef]

94. Indarto, A.; Yang, D.R.; Choi, J.W.; Lee, H.; Song, H.K. Gliding arc plasma processing of $CO_2$ conversion. *J. Hazard. Mater.* **2007**, *146*, 309–315. [CrossRef] [PubMed]

95. Tu, X.; Gallon, H.J.; Twigg, M.V.; Gorry, P.A.; Whitehead, J.C. Dry reforming of methane over a $Ni/Al_2O_3$ catalyst in a coaxial dielectric barrier discharge reactor. *J. Phys. D Appl. Phys.* **2011**, *44*, 274007. [CrossRef]

96. Paulussen, S.; Verheyde, B.; Tu, X.; De Bie, C.; Martens, T.; Petrovic, D.; Bogaerts, A.; Sels, B. Conversion of carbon dioxide to value-added chemicals in atmospheric pressure dielectric barrier discharges. *Plasma Sources Sci. Technol.* **2010**, *19*, 034015. [CrossRef]

97. Ray, D.; Subrahmanyam, C. $CO_2$ decomposition in a packed DBD plasma reactor: influence of packing materials. *RSC Adv.* **2016**, *6*, 39492–39499. [CrossRef]

98. Uytdenhouwen, Y.; Van Alphen, S.; Michielsen, I.; Meynen, V.; Cool, P.; Bogaerts, A. A packed-bed DBD micro plasma reactor for $CO_2$ dissociation: does size matter? *Chem. Eng. J.* **2018**, *348*, 557–568. [CrossRef]

99. Zhou, A.; Chen, D.; Ma, C.; Yu, F.; Dai, B. DBD plasma-$ZrO_2$ catalytic decomposition of $CO_2$ at low temperatures. *Catalysts* **2018**, *8*, 256. [CrossRef]

100. Duan, X.; Hu, Z.; Li, Y.; Wang, B. Effect of dielectric packing materials on the decomposition of carbon dioxide using DBD microplasma reactor. *AIChE J.* **2015**, *61*, 898–903. [CrossRef]

101. Mei, D.; Zhu, X.; Wu, C.; Ashford, B.; Williams, P.T.; Tu, X. Plasma-photocatalytic conversion of $CO_2$ at low temperatures: understanding the synergistic effect of plasma-catalysis. *Appl. Catal. B: Environ.* **2016**, *182*, 525–532. [CrossRef]

102. Shengjie, Z.; Amin, Z.; Feng, Y.; Bin, D.; Cunhua, M. Enhanced $CO_2$ decomposition via metallic foamed electrode packed in self-cooling DBD plasma device. *Plasma Sci. Technol.* **2019**, *21*, 085504.

103. Eliasson, B.; Liu, C.J.; Kogelschatz, U. Direct conversion of methane and carbon dioxide to higher hydrocarbons using catalytic dielectric-barrier discharges with zeolites. *Ind. Eng. Chem. Res.* **2000**, *39*, 1221–1227. [CrossRef]

104. Kraus, M.; Eliasson, B.; Kogelschatz, U.; Wokaun, A. $CO_2$ reforming of methane by the combination of dielectric-barrier discharges and catalysis. *Phys. Chem. Chem. Phys.* **2001**, *3*, 294–300. [CrossRef]

105. Wang, Q.; Shi, H.; Yan, B.; Jin, Y.; Cheng, Y. Steam enhanced carbon dioxide reforming of methane in DBD plasma reactor. *Int. J. Hydrog. Energy* **2011**, *36*, 8301–8306. [CrossRef]

106. Nozaki, T.; Muto, N.; Kado, S.; Okazaki, K. Dissociation of vibrationally excited methane on Ni catalyst: Part 1. application to methane steam reforming. *Catal. Today* **2004**, *89*, 57–65. [CrossRef]

107. Lee, H.; Lee, D.H.; Song, Y.H.; Choi, W.C.; Park, Y.K.; Kim, D.H. Synergistic effect of non-thermal plasma–catalysis hybrid system on methane complete oxidation over Pd-based catalysts. *Chem. Eng. J.* **2015**, *259*, 761–770. [CrossRef]

108. Nozaki, T.; Muto, N.; Kado, S.; Okazaki, K. Minimum energy requirement for methane steam reforming in plasma-catalyst reactor. *Am. Chem. Soc. Div. Fuel Chem.* **2004**, *49*, 179.

109. Zhao, D.; Yu, F.; Zhou, A.M.; Ma, C.H.; Dai, B. High-efficiency removal of $NO_x$ using dielectric barrier discharge nonthermal plasma with water as an outer electrode. *Plasma Sci. Technol.* **2018**, *20*, 14020. [CrossRef]

110. Zhang, K.; Yu, F.; Zhu, M.; Dan, J.; Wang, X.; Zhang, J.; Dai, B. Enhanced low temperature NO reduction performance via $MnO_x$-$Fe_2O_3$/vermiculite monolithic honeycomb catalysts. *Catalysts* **2018**, *8*, 100. [CrossRef]

111. Tian, J.; Zhang, K.; Wang, W.; Wang, F.; Dan, J.; Yang, S.; Zhang, J.; Dai, B.; Yu, F. Enhanced selective catalytic reduction of NO with $NH_3$ via porous micro-spherical aggregates of Mn-Ce-Fe-Ti mixed oxide nanoparticles. *Green Energy Environ.* **2019**, *4*, 311–321. [CrossRef]

112. Tian, J.; Wang, C.; Yu, F.; Zhou, X.; Zhang, J.; Yang, S.; Dan, J.; Cao, P.; Dai, B.; Wang, Q. Mn-Ce-Fe-Al mixed oxide nanoparticles via a high shear mixer facilitated coprecipitation method for low temperature selective catalytic reduction of NO with $NH_3$. *Appl. Catal. A Gen.* **2019**, *4*, 117237. [CrossRef]

113. Liu, Z.; Yu, F.; Ma, C.; Dan, J.; Luo, J.; Dai, B. A critical review of recent progress and perspective in practical denitration application. *Catalysts* **2019**, *9*, 771. [CrossRef]

114. Niu, C.; Niu, J.; Wang, S.; Wang, Z.; Dong, S.; Fan, H.; Hong, Y.; Liu, D. Synergistic effect in one-stage dielectric barrier discharge plasma and $Ag/Al_2O_3$ catalytic systems on $C_2H_2$-SCR of $NO_x$. *Catal. Commun.* **2019**, *123*, 49–53. [CrossRef]

115. Wang, T.; Liu, H.; Sun, B. Dielectric barrier discharge plasma-assisted catalytic reduction of $NO_x$ over Mn-Cu catalyst. In Proceedings of the 2016 International Forum on Energy, Environment and Sustainable Development, Shenzhen, China, 16–17 April 2016.

116. Jinhai, N.; Zhihui, Z.; Dongping, L.; Qi, W. Low-temperature plasma-catalytic reduction of $NO_x$ by $C_2H_2$ in the presence of excess oxygen. *Plasma Sci. Technol.* **2008**, *10*, 466. [CrossRef]

117. Yi, Y.; Zhou, J.; Gao, T.; Guo, H.; Zhou, J.; Zhang, J. Continuous and scale-up synthesis of high purity $H_2O_2$ by safe gas-phase $H_2/O_2$ plasma reaction. *AIChE J.* **2014**, *60*, 415–419. [CrossRef]

118. Li, S.; Cao, X.; Liu, L.; Ma, X. Degradation of thiamethoxam in water by the synergy effect between the plasma discharge and the $TiO_2$ photocatalysis. *Desalin. Water Treat.* **2015**, *53*, 3018–3025. [CrossRef]

119. Du, Y.; Nayak, G.; Oinuma, G.; Peng, Z.; Bruggeman, P.J. Effect of water vapor on plasma morphology, OH and $H_2O_2$ production in He and Ar atmospheric pressure dielectric barrier discharges. *J. Phys. D Appl. Phys.* **2017**, *50*, 145201. [CrossRef]

120. Vasko, C.A.; Liu, D.X.; Veldhuizen, E.M.V.; Iza, F.; Bruggeman, P.J. Hydrogen peroxide production in an atmospheric pressure RF glow discharge: comparison of models and experiments. *Plasma Chem. Plasma Process.* **2014**, *34*, 1081–1099. [CrossRef]

121. Qu, X.; Hu, S.; Li, P.; Li, Z.; Wang, H.; Ma, H.; Li, W. The effect of embedding N vacancies into g-$C_3N_4$ on the photocatalytic $H_2O_2$ production ability via $H_2$ plasma treatment. *Diam. Relat. Mater.* **2018**, *86*, 159–166. [CrossRef]

122. Àç, G.H. Scale-up synthesis of hydrogen peroxide from $H_2/O_2$ with multiple parallel DBD tubes. *Plasma Sci. Technol.* **2009**, *11*, 181–186.

123. Thevenet, F.; Couble, J.; Brandhorst, M.; Dubois, J.; Puzenat, E.; Guillard, C.; Bianchi, D. Synthesis of hydrogen peroxide using dielectric barrier discharge associated with fibrous materials. *Plasma Chem. Plasma Process.* **2010**, *30*, 489–502. [CrossRef]

124. Yi, Y.; Zhou, J.; Guo, H.; Zhao, J.; Su, J.; Wang, L.; Wang, X.; Gong, W. Safe direct synthesis of high purity $H_2O_2$ through a $H_2/O_2$ plasma reaction. *Angew. Chem. Int. Ed.* **2013**, *52*, 8446–8449. [CrossRef] [PubMed]

125. Gunasooriya, G.K.K.; Van Bavel, A.P.; Kuipers, H.P.; Saeys, M. Key role of surface hydroxyl groups in C–O activation during Fischer–Tropsch synthesis. *ACS Catal.* **2016**, *6*, 3660–3664. [CrossRef]

126. Hong, J.; Chu, W.; Ying, Y.; Chernavskii, P.A.; Khodakov, A. Plasma-assisted design of supported cobalt catalysts for Fischer-Tropsch synthesis. *Stud. Surf. Sci. Catal.* **2010**, *175*, 253–257.

127. Jiang, Y.; Fu, T.; Jing, L.; Li, Z. A zirconium modified Co/SiO$_2$ Fischer-Tropsch catalyst prepared by dielectric-barrier discharge plasma. *J. Energy Chem.* **2013**, *22*, 506–511. [CrossRef]

128. Al-Harrasi, W.S.; Zhang, K.; Akay, G. Process intensification in gas-to-liquid reactions: plasma promoted Fischer-Tropsch synthesis for hydrocarbons at low temperatures and ambient pressure. *Green Process. Synth.* **2013**, *2*, 479–490. [CrossRef]

129. Liu, C.J.; Xue, B.; Eliasson, B.; He, F.; Li, Y.; Xu, G.H. Methane conversion to higher hydrocarbons in the presence of carbon dioxide using dielectric-barrier discharge plasmas. *Plasma Chem. Plasma Process.* **2001**, *21*, 301–310. [CrossRef]

130. Saleem, F.; Zhang, K.; Harvey, A. Plasma-assisted decomposition of a biomass gasification tar analogue into lower hydrocarbons in a synthetic product gas using a dielectric barrier discharge reactor. *Fuel* **2019**, *235*, 1412–1419. [CrossRef]

131. Roso, M.; Boaretti, C.; Pelizzo, M.G.; Lauria, A.; Modesti, M.; Lorenzetti, A. Nanostructured photocatalysts based on different oxidized graphenes for VOCs removal. *Ind. Eng. Chem. Res.* **2017**, *56*, 9980–9992. [CrossRef]

132. Maciuca, A.; Batiot-Dupeyrat, C.; Tatibouët, J.M. Synergetic effect by coupling photocatalysis with plasma for low VOCs concentration removal from air. *Appl. Catal. B Environ.* **2012**, *125*, 432–438. [CrossRef]

133. Saoud, W.A.; Assadi, A.A.; Guiza, M.; Bouzaza, A.; Aboussaoud, W.; Ouederni, A.; Soutrel, I.; Wolbert, D.; Rtimi, S. Study of synergetic effect, catalytic poisoning and regeneration using dielectric barrier discharge and photocatalysis in a continuous reactor: abatement of pollutants in air mixture system. *Appl. Catal. B Environ.* **2017**, *213*, 53–61. [CrossRef]

134. Mustafa, M.F.; Fu, X.; Liu, Y.; Abbas, Y.; Wang, H.; Lu, W. Volatile organic compounds (VOCs) removal in non-thermal plasma double dielectric barrier discharge reactor. *J. Hazard. Mater.* **2018**, *347*, 317–324. [CrossRef] [PubMed]

135. Assadi, A.A.; Bouzaza, A.; Vallet, C.; Wolbert, D. Use of DBD plasma, photocatalysis, and combined DBD plasma/photocatalysis in a continuous annular reactor for isovaleraldehyde elimination—synergetic effect and byproducts identification. *Chem. Eng. J.* **2014**, *254*, 124–132. [CrossRef]

136. Zhu, T.; Li, J.; Liang, W.; Jin, Y. Synergistic effect of catalyst for oxidation removal of toluene. *J. Hazard. Mater.* **2009**, *165*, 1258–1260. [CrossRef] [PubMed]

137. Nguyen, H.P.; Park, M.J.; Kim, S.B.; Kim, H.J.; Baik, L.J.; Jo, Y.M. Effective dielectric barrier discharge reactor operation for decomposition of volatile organic compounds. *J. Clean. Prod.* **2018**, *198*, 1232–1238. [CrossRef]

138. Choi, J.H.; Han, I.; Baik, H.K.; Lee, M.H.; Han, D.W.; Park, J.C.; Lee, I.S.; Song, K.M.; Lim, Y.S. Analysis of sterilization effect by pulsed dielectric barrier discharge. *J. Electrost.* **2006**, *64*, 17–22. [CrossRef]

139. Kostov, K.; Rocha, V.; Koga-Ito, C.; Matos, B.; Algatti, M.; Honda, R.Y.; Kayama, M.; Mota, R.P. Bacterial sterilization by a dielectric barrier discharge (DBD) in air. *Surf. Coat. Technol.* **2010**, *204*, 2954–2959. [CrossRef]

140. Yang, L.; Chengwu, Y.; Jingjing, L.; Rongjie, Y.; Huijuan, W. Experimental research on the sterilization of Escherichia coli and Bacillus subtilis in drinking water by dielectric barrier discharge. *Plasma Sci. Technol.* **2016**, *18*, 173–178.

141. Eto, H.; Ono, Y.; Ogino, A.; Nagatsu, M. Low-temperature sterilization of wrapped materials using flexible sheet-type dielectric barrier discharge. *Appl. Phys. Lett.* **2008**, *93*, 221502. [CrossRef]

142. Kikuchi, Y.; Miyamae, M.; Nagata, M.; Fukumoto, N. Effects of environmental humidity and temperature on sterilization efficiency of dielectric barrier discharge plasmas in atmospheric pressure air. *Jpn. J. Appl. Phys.* **2011**, *50*, 01AH03. [CrossRef]

143. Mastanaiah, N.; Banerjee, P.; Johnson, J.A.; Roy, S. Examining the role of ozone in surface plasma sterilization using dielectric barrier discharge (DBD) plasma. *Plasma Process. Polym.* **2013**, *10*, 1120–1133. [CrossRef]

144. Hong, Y.C.; Ma, S.H.; Kim, K.; Shin, Y.W. Multihole dielectric barrier discharge with asymmetric electrode arrangement in water and application to sterilization of aqua pathogens. *Chem. Eng. J.* **2019**, *374*, 133–143. [CrossRef]

145. Richardson, S.D. Water analysis: emerging contaminants and current issues. *Anal. Chem.* **2009**, *81*, 4645–4677. [CrossRef] [PubMed]

146. Haixia, W.; Zhi, F.; Yanhua, X. Degradation of aniline wastewater using dielectric barrier discharges at atmospheric pressure. *Plasma Sci. Technol.* **2015**, *17*, 228–234.

147. Mok, Y.S.; Jo, J.O.; Lee, H.J.; Ahn, H.T.; Kim, J.T. Application of dielectric barrier discharge reactor immersed in wastewater to the oxidative degradation of organic contaminant. *Plasma Chem. Plasma Process.* **2007**, *27*, 51–64. [CrossRef]

148. Jin, G.; Pingdao, G.; Li, Y.; Fangchuan, Z. Degradation of dye wastewater by ns-pulse DBD plasma. *Plasma Sci. Technol.* **2013**, *15*, 928.

149. Sun, M.Y.; Jin-Oh, J.; Heon-Ju, L. Dielectric barrier discharge plasma-induced photocatalysis and ozonation for the treatment of wastewater. *Plasma Sci. Technol.* **2008**, *10*, 100–105. [CrossRef]

150. Attri, P.; Tochikubo, F.; Park, J.H.; Choi, E.H.; Koga, K.; Shiratani, M. Impact of Gamma rays and DBD plasma treatments on wastewater treatment. *Sci. Rep.* **2018**, *8*, 2926. [CrossRef]

151. Tichonovas, M.; Krugly, E.; Racys, V.; Hippler, R.; Kauneliene, V.; Stasiulaitiene, I.; Martuzevicius, D. Degradation of various textile dyes as wastewater pollutants under dielectric barrier discharge plasma treatment. *Chem. Eng. J.* **2013**, *229*, 9–19. [CrossRef]

152. Kuraica, M.; Obradović, B.; Manojlović, D.; Ostojić, D.; Purić, J. Application of coaxial dielectric barrier discharge for potable and waste water treatment. *Ind. Eng. Chem. Res* **2006**, *45*, 882–905.

153. Tang, S.; Yuan, D.; Rao, Y.; Li, N.; Qi, J.; Cheng, T.; Sun, Z.; Gu, J.; Huang, H. Persulfate activation in gas phase surface discharge plasma for synergetic removal of antibiotic in water. *Chem. Eng. J.* **2018**, *337*, 446–454. [CrossRef]

154. Sarangapani, C.; Misra, N.; Milosavljevic, V.; Bourke, P.; O'Regan, F.; Cullen, P. Pesticide degradation in water using atmospheric air cold plasma. *J. Water Process Eng.* **2016**, *9*, 225–232. [CrossRef]

155. Xingmin, S.; Jinren, L.; Guimin, X.; Yueming, W.; Lingge, G.; Xiaoyan, L.; Yang, Y.; Zhang, G. Effect of low-temperature plasma on the degradation of omethoate residue and quality of apple and spinach. *Plasma Sci. Technol.* **2018**, *20*, 044004.

156. Hu, Y.; Bai, Y.; Li, X.; Chen, J. Application of dielectric barrier discharge plasma for degradation and pathways of dimethoate in aqueous solution. *Sep. Purif. Technol.* **2013**, *120*, 191–197. [CrossRef]

157. Li, S.; Ma, X.; Jiang, Y.; Cao, X. Acetamiprid removal in wastewater by the low-temperature plasma using dielectric barrier discharge. *Ecotoxicol. Environ. Saf.* **2014**, *106*, 146–153. [CrossRef] [PubMed]

158. Feng, J.; Zheng, Z.; Sun, Y.; Luan, J.; Wang, Z.; Wang, L.; Feng, J. Degradation of diuron in aqueous solution by dielectric barrier discharge. *J. Hazard. Mater.* **2008**, *154*, 1081–1089. [CrossRef] [PubMed]

159. Jović, M.S.; Dojčinović, B.P.; Kovačević, V.V.; Obradović, B.M.; Kuraica, M.M.; Gašić, U.M.; Roglić, G.M. Effect of different catalysts on mesotrione degradation in water falling film DBD reactor. *Chem. Eng. J.* **2014**, *248*, 63–70. [CrossRef]

160. Wardenier, N.; Vanraes, P.; Nikiforov, A.; Van Hulle, S.W.; Leys, C. Removal of micropollutants from water in a continuous-flow electrical discharge reactor. *J. Hazard. Mater.* **2019**, *362*, 238–245. [CrossRef]

161. Hijosa-Valsero, M.; Molina, R.; Schikora, H.; Müller, M.; Bayona, J.M. Removal of priority pollutants from water by means of dielectric barrier discharge atmospheric plasma. *J. Hazard. Mater.* **2013**, *262*, 664–673. [CrossRef] [PubMed]

162. Misra, N.; Pankaj, S.; Walsh, T.; O'Regan, F.; Bourke, P.; Cullen, P. In-package nonthermal plasma degradation of pesticides on fresh produce. *J. Hazard. Mater.* **2014**, *271*, 33–40. [CrossRef] [PubMed]

MDPI

St. Alban-Anlage 66

4052 Basel

Switzerland

Tel. +41 61 683 77 34

Fax +41 61 302 89 18

www.mdpi.com

*Nanomaterials* Editorial Office

E-mail: nanomaterials@mdpi.com

www.mdpi.com/journal/nanomaterials